JOHN GLENN

A Memoir

"Fascinating…Ablaze with drama…in a world full of examples of how life should not be lived, Glenn's story exemplifies the way it should be. His memoir is a joy to read."—*Book-of-the-Month Club News*

"[Glenn]'s story is chock-full of duty, honor, patriotism, hard work and bedrock love of family. They don't make 'em like they used to."—*St. Petersburg Times*

"Recently John Glenn made history again, when he went into space at the ripe old age of 77. *John Glenn: A Memoir*…relates this pioneering event and reveals just how such an American hero is made. Glenn's life embodies fulfillment of the American dream…full of fascinating details about manned spaceflight and small-town life in Glenn's hometown, New Concord, Ohio, during the Depression, as well as observations on several decades of Presidents, Glenn's book is an unqualified winner."
—*Greensboro News & Record*

"Refreshing…unembellished, upbeat…Glenn recounts his extraordinary life in frank, sometimes folksy, but never extraordinary terms…this book could well be entitled 'The Straight Stuff.' When was the last you read that kind of memoir?"—*Detroit Free Press*

JOHN GLENN

with NICK TAYLOR

JOHN GLENN

A MEMOIR

BANTAM BOOKS

New York   Toronto   London   Sydney   Auckland

This edition contains the complete text
of the original hardcover edition.
NOT ONE WORD HAS BEEN OMITTED.

JOHN GLENN: A MEMOIR

A Bantam Book

PUBLISHING HISTORY

Bantan hardcover edition published November 1999

Bantam mass market edition / October 2000

Grateful acknowledgement for reprinting lyric excerpt from THE WIND
BENEATH MY WINGS, by Larry Henley and Jeff Silbar, © 1982 Warner House
Of Music & WB Gold Music Corp. All rights reserved. Used by permission
Warner Bros. Publications U.S. Inc., Miami, FL 33014.

Photographs appearing in this book were provided by John Glenn unless otherwise noted.

Photographs appearing on page 201, courtesy U.S. Navy;
pages 255 and 485, courtesy NASA;
page 397, © P. F. Bentley.

Library of Congress Catalog Card Number: 99-042672.
No part of this book may be reproduced or transmitted in any form or by any
means, electronic or mechanical, including photocopying, recording, or by any
information storage and retrieval system, without permission in writing from

the publisher. For information address: Bantam Books.

ISBN: 0-553-58157-0
Published simultaneously in the United States and Canada

Bantam Books are published by Bantam Books, a division of Random House, Inc.
Its trademark, consisting of the words "Bantam Books" and the portrayal of a
rooster, is Registered in U.S. Patent and Trademark Office and in other countries.
Marca Registrada. Bantam Books, 1540 Broadway, New York, New York 10036.

PRINTED IN THE UNITED STATES OF AMERICA
OPM 10 9 8 7 6 5 4 3 2 1

For Annie,
always and forever "the wind beneath my wings,"

and for Zachary and Daniel,
for whom it all began

CONTENTS

PROLOGUE

It was a sin to throw anything away.

That lesson, practiced by my parents and drummed into my head during the Depression, produced over the years an accumulation of artifacts, mementos, records, files, and documents. There were the minutes of the club that friends and I formed when we were boys, the script of my senior class play with my part underlined, old ticket stubs and menus, programs and souvenir brochures, laboriously typed school and college reports, and photographs, lots and lots of photographs. My marriage to Annie brought more memorabilia, and when I went away to World War II, and later to Korea, letters, orders, and many more photographs took their places in the mix. By then we had our children, Dave and Lyn, so there were still more letters, photographs, and records, as well as souvenirs of war and remembrances from overseas. I hung up my uniforms and flight suits and they joined the array. Flying as a test pilot added a sprinkle of newspaper clippings to the collection. Project Mercury turned the sprinkle into a flood; now there were letters, paintings, plaques, medallions, voluminous training and debriefing records, and still more photographs. Years in the U.S. Senate added their share.

It was a great pile of stuff, and it grew and spread from attic to attic and garage to garage, and finally sprawled over the better part of an entire floor of our house outside Washington, D.C.

People said, "You have to get it organized." I looked at it all and thought, "Life's too short." I was more interested in doing new things than in looking back over what Annie and I had already done.

Then our grandsons began asking questions about the lives we had led, and the telling kept calling for some illustration from among the accumulated piles of stuff. The disorganized piles were unrevealing. But for the first time the thought was there. And slowly, through the efforts of a small, heroic band, a semblance of order began to emerge. Motivation followed. I made notes on a tape recorder and sat with an oral historian as he conducted interviews.

At first the purpose was simply to provide the basis for a history our family could share among ourselves. My life had been covered well enough, I thought, and I had already had all the attention anyone could expect or want in several lifetimes. I never planned to write a book.

But the surprising and sustained attention that came with the announcement in January 1998 that I would return to space at age seventy-seven encouraged me to change my mind. Old people and young seemed to be caught up in the idea of it, some thinking of their gramps in space, others of themselves, still others recalling the nationwide bonding that the early space shots had evoked nearly forty years before. People began asking about the book I never intended to write.

The *Discovery* mission made me feel that a memoir might be appropriate after all. It would encompass hard times, war, the free world's ideological struggle with Communism, the beginnings of the space age, changing political tides, the great demographic shift produced by increased human longevity. I realized that Annie and I, in our late seventies, had lived through a third of our nation's entire history. We had been privileged to participate in some of its most trying and triumphant times, and to know some of its significant figures. I had to concede, when I thought about it, that we did indeed have a story to tell.

PART ONE
NEW
CONCORD

CHAPTER 1

Patriotism filled the air of New Concord, the small eastern Ohio town where I grew up. Memorial Day, the Fourth of July, and Armistice Day were flag-waving holidays of parades and salutes to the United States and to the soldiers, living and dead, who had fought for freedom and democracy.

My father was one of those soldiers. He served in France during World War I, delivering artillery shells to the front on trucks and horse-drawn caissons, and he came home partially deaf from a cannon blast but otherwise unharmed. He also was a bugler. He blew the bugle for reveille and taps, for mail call and mess call, and when the flag was raised.

At home, on those patriotic days that I remember, Dad was again called upon to play the bugle. He marched in the parade formations when the local veterans from the Thirty-seventh Ohio Division marched down Main Street on Armistice Day, and played the colors when they raised the flag at the American Legion hall at the end of the parade. But the bugling I remember best was the taps he played on Memorial Day. It was still called Decoration Day then, and families dressed in their Sunday best would regather at the town cemetery after the parade, carrying bundles of gladioli, irises, and peonies, red, white, and blue the dominating colors. The marching soldiers also would regather. They presented arms and fired three volleys in salute as the flags flanking the Stars and Stripes were dipped. Then my father raised his war-battered brass bugle and played those drawn-out, mournful notes in memory of the soldiers killed in action, and the sound drifted across the gravestones and sent chills up my spine. As the last notes faded into silence the families of the soldiers and descendants of men who had died in other wars moved among the gravestones and placed flowers on the graves.

We had a town band in New Concord. I was nine or ten when I joined the band and learned to play the trumpet. At home, Dad taught me the military calls. And one day after I learned to play them well enough, he came to me with a request. "Bud," he said, "Decoration Day is coming up, and I want you to play taps with me."

I hardly knew what to say. Dad was my hero. He had fought in the war. The playing of taps was a special moment in the ceremony, a final, haunting valediction for the men who had made the supreme sacrifice. To play it was a great responsibility. Dad obviously had a lot of confidence in me. That meant a great deal, but it meant even more to participate as a young boy in the remembrance of men who had fought and died for our country. It was something bigger than I was, something momentous.

Dad and I practiced at home as the end of May approached. He played in the kitchen, and I stood in another room. When the day came, I was a little scared. Before, I had always watched the parade with Mother and my sister, Jean, or marched in the band. But this spring day I went alone to the cemetery, ahead of the others. I walked across the sloping grounds, and waited out of sight in the woods where the terrain fell away beyond the graves.

Soon I heard "Present arms" as the soldiers' honor guard re-formed. Peeking through the leaves, I saw them raise their guns to fire the three volleys in salute. Then my father lifted his bugle, and the first sad notes rose in the spring air. The first phrase ended, and I was ready. I put the trumpet to my lips and echoed the clear notes. We played through taps like that, my trumpet echoing his bugle phrase by phrase, until the last notes died.

That impressed me then, as a boy, and it's impressive to me to this day. Echo taps still gives me chills. It recalls the patriotic feeling of New Concord, the pride and respect everyone in the town felt for the United States of America. Love of country was a given. Defense of its ideals was an obligation. The opportunity to join in its quests and explorations was a challenge not only to fulfill a sacred duty, but to join a joyous adventure. That feeling sums up my childhood. It formed my beliefs and my sense of responsibility. Everything that came after that just seemed to follow naturally.

A boy could not have had a more idyllic early childhood than I did. Sometimes it seems to me that Norman Rockwell must have taken all his inspiration from New

Concord, Ohio. My playmates were freckle-faced boys and girls with pigtails. We played without fear in backyards and streams and endless green fields, and climbed trees to learn the limits of our daring. The adults—most of them—were concerned and reliably caring, and we respected them. Boys learned the company of men—the way they talked and held themselves, and their concerns—at the town barbershop and hunting in the woods. Saturday afternoons were for fifteen-cent sundaes at the Ohio Valley Dairy (nuts on top cost an extra nickel), Sunday mornings were for Sunday school and church, and Sunday afternoons were for family dinners and outings. These were the orderly rituals of my early years, and I never doubted even once that I was loved.

New Concord is the hometown I remember, but I was born a few miles away in Cambridge, in my parents' white frame house. The date was July 18, 1921. A doctor attended the birth. I weighed nine pounds and had my mother's red hair.

My father, John Herschel Glenn, had been home from the battlefields of France for two and a half years. He and my mother, a schoolteacher whose maiden name was Clara Sproat, had married just before he went to war. Mother was a very beautiful woman in some of the pictures made when she was young. She had a vivacious smile and lustrous hair, although its auburn tones were lost in the old photographs. They had met at the East Union Presbyterian Church near his parents' farm outside of Claysville. The farm was eighty or so acres, too small for anything but corn, garden vegetables, and a few hogs and chickens. She was from another little town not far away, Lore City. He had been seeing her for about two years, and I imagine he didn't want to let her get away. She rode a train to Montgomery, Alabama, where he was training at Camp Sheridan, and they got married on May 25, 1918. Two weeks later he left on a troopship from Newport News, Virginia.

Dad was twenty-two years old when he went away, with a sixth-grade education acquired in a one-room country school. When he came back, he had seen the world and was by all accounts a different man. His roots still were deep in the farms and small towns of eastern Ohio, in the values of a mutually supportive community. But his perspective had broadened, and he saw the need to know and understand the world beyond the cornfields.

He worked as a fireman on the Baltimore and Ohio Railroad when he came home. Mother continued to teach elementary school in Cambridge. The B&O's locomotives were coal-fired. Dad shoveled coal on the westbound trains until they reached Newark or Columbus. It was hot, dirty work, constantly swinging between a coal tender and a firebox that glowed like the mouth of Hades. He'd go to sleep exhausted in a railroad workers' barracks, and the next day do it again in a train headed home. The work was hard, but that wasn't what he minded. He was gone about half the time, and it stuck in his craw that he had to be away from home and his beautiful young wife that much.

So he quit the railroad and what was then a lifetime job with guaranteed security. Times were good, there was quite a bit of building going on, and Dad decided to take up the plumbing trade. His apprenticeship took us from Cambridge to Zanesville and back again. By August 1923 he knew his way around a pipe wrench pretty well, and

joined up with Bertel Welch, a plumber in New Concord.
I was two years old.

The move to New Concord was a homecoming of
sorts for Mother. She had attended Muskingum College
there, riding the train from Lore City through Cambridge
to New Concord and back home every day until she
earned the two-year degree required of schoolteachers at
that time. Her father had been a teacher, too. New
Concord was smaller than Cambridge. Even when the
population doubled with students during the school year,
it was barely larger than two thousand. But New Concord
was no backwater. Muskingum's concerts, art exhibits,
speeches, theatrical productions, and debates were open
to all. Townspeople could use its library. It was a United
Presbyterian Church school, and some of the housing on
campus was set aside for church missionaries home on
sabbatical from their work overseas. At any given time,
there were families who had spent time in China, Africa,
and South America, and the missionaries gave talks on
their experiences.

Dad's partnership didn't last. When it ended, he
rented a little store at the east end of Main Street and hung
out his shingle as the Glenn Plumbing Company. Mother
had stopped teaching after I was born, and once he opened
the business she watched the shop and sold plumbing supplies while he was out working on the jobs. They worked
as a team.

Mother and Dad quickly fell in with four other couples in New Concord. They called themselves the Twice
Five Club. The Twice Fives got together monthly for a dinner at one house or another. The hosts would cook the
main dish, which was usually some kind of casserole, and

everybody else would come with their kids in tow and something to put on the table.

One of the other couples was Homer and Margaret Castor. Homer——Dr. Castor——was the town dentist. The Castors had arrived in New Concord about the same time my parents did. This, too, was a homecoming. Doc Castor had grown up in Otsego, near New Concord, and had attended Muskingum and then the Ohio State University dental school, where he got his degree. He had planned to start a children's dental practice in the state capital before deciding he preferred the small-town life. They had a daughter, Anna Margaret, whom everyone called Annie. She was about a year older than I was. They put us in a playpen together, and she was part of my life from the time of my first memory.

We lived in a big house. When Mother and Dad decided to move to New Concord, they thought they could help pay for the house by renting rooms to college students. Dad had it built with four large rooms upstairs, and installed the plumbing and coal-burning furnace himself. We always had upstairs boarders, even after the college decided that all the students should live on campus.

Our house was on a gravel road called Shadyside Terrace. It sat high on an embankment overlooking the National Road, U.S. Route 40, at the western edge of town. Route 40 was the main road from Baltimore to St. Louis, and it followed the route of the old pioneer trail called the Zane Trace. For us it linked New Concord with Cambridge to the east and Zanesville and Columbus to the west. It was a brick road. It carried quite a bit of traffic, and you could sit on the porch and watch the cars and trucks go by with their tires humming on the bricks and feel like you were at the center of the world. Down to the right, or west from our south-facing house, the road made an S as it curved over a stone bridge. The bridge crossed Crooked Creek, where I caught my first fish on a bent pin. A hundred yards beyond the road, the B&O Railroad tracks carried east- and westbound passenger and freight trains. Heavy westbound trains had to strain to make the hill at neighboring Norwich. Coming east down the long grade, they barreled along with their warning whistles in full cry. My room was in the front upstairs, and I always knew when they were coming through. A south wind would blow soot right through my window screen, that's how close we were.

The downstairs of the house had a living room and dining room across the front. The kitchen and two bedrooms, including the one my parents shared, were in the back. My favorite room was the kitchen. Mother was a good cook. She made a wonderful ham loaf, delicious corn mush, and savory pies and cakes. We ate there at a big table, where it was always warm and full of the smells of cooking. It was Dad's favorite room, too. He was a stocky, muscular man, about five-nine, and he didn't have to enjoy too much of Mother's cooking for it to start to show. If I got my red hair from Mother, from Dad I inherited a tendency to put on the pounds if I wasn't careful.

Mother took me with her to the store before I was old enough to go to school. I was happy to play there. I'd lay out lengths of pipe on the floor, and people would say I was sure to be a plumber, like my father. When it came time for my nap, Mother would put some blankets in one of the porcelain bathtubs on display, and I'd crawl in and go to sleep.

Dad's father lived with us for a short time, in one of the upstairs bedrooms at the back. I was about three, and I remember him slightly, or maybe I have learned his image from old photographs. He was big, wore a mustache, and used small reading glasses. He had been alone since his wife died. Mother had had to tell Dad about his mother's death when she welcomed him home from the war. Grandpa Glenn stayed alone on the farm for a few years, but when he began to fail, Dad moved him in with us, and Dad and his brothers sold the farm.

Grandpa Glenn's ancestors were Guernsey County pioneers who had farmed outside of Cambridge for several generations. They, like my mother's family, the Sproats, were Scotch-Irish farming stock who had settled first in the area south of Pittsburgh, and then had moved west with the opening frontier. Many of the families in our area had similar backgrounds. This accounted for the austere Protestantism practiced in our churches, and for a fierce loyalty to the American ideal as they saw it: a place where individuals were free to become what they could. The Glenns were a branch of the Mackintosh clan. Grandpa Glenn's father was one of the Guernsey and Muskingum County men who had fought for the Union in the Civil War.

Grandpa Glenn died when I was almost five. His death was a new experience for me. He had been there, living with us, and then all of a sudden he wasn't. He lay quiet and still in an open casket at the end of the living room. A steady stream of people came to look at him and shake my father's hand, and all of them brought food.

I was an only child when Grandpa Glenn died, but not for long thereafter. Mother had lost two babies due to miscarriages before I was born, and another afterward. Her doctor told her not to risk another pregnancy. But she and Dad wanted another child and decided to adopt. My younger sister was named Jean. I learned later that my parents adopted her through an agency in Columbus. As far as I know, neither they nor she ever had any contact with her natural parents. She came into our family around the time I turned five.

I loved my new sister. Life in our household got more noisy and interesting when she arrived. But I was too much older than Jean to become her playmate. I already had playmates in the small world of New Concord. Some of the older kids went swimming in a pool within Crooked Creek, above our house. I didn't know how to swim yet, and the creek was the one place I was expressly forbidden to go. Regina Regier was an older neighbor girl who took me out to play, possibly out of some maternal instinct. Somehow she got me to go to the creek with her one day. When Mother started calling me and I didn't answer, she came to look for me. She found me at the swimming hole with Regina, and she was mighty unhappy. She ordered me home, then yanked a switch off a bush and followed me. Every time I slowed down, she'd swat my bare legs with that switch. The next day my legs were red with welts. Mother was horrified. She put salve on the welts while I winced, and hugged me, and told me the reason she'd gotten so angry was because she was afraid I had drowned.

That kind of punishment was rare. I didn't do much to anger or upset my parents.

I especially enjoyed being with my father, whom most

people called Herschel, while I grew up being called Bud. He could turn very serious at times, and he didn't smoke or drink, but most of the time he was lighthearted. He joked a lot and made me laugh. Part of the fun of being with him was that he was curious. He always wanted to learn about new things, and he would go out of his way to investigate them. Although the Glenn Plumbing Company grew into a successful business of which he was very proud, I think he recognized the limitations of his education. He wanted to give me the curiosity and sense of unbounded possibility that could come from learning.

The summer I turned eight, Dad took me along to Cambridge one day when he went to check on a plumbing job. It was the time of year when wildflowers bloomed on the roadsides and in the farm fields where cattle grazed. He checked on the job, and as we drove past a grass-field airport outside of town, he spotted a plane there and we stopped.

We got out of the car to look. A man had an old open-cockpit biplane—it was a WACO, but I didn't know that then—and he was taking people up. He was a Steve Canyon–type pilot, a helmet-and-goggles sort of guy right out of the comics. We were leaning against the car and watching him, and my dad said, "You want to go up, Bud?"

I almost died. Flying was a great adventure. Everybody knew about Charles Lindbergh's transatlantic flight two years before. When Lindy came home, the papers had chronicled his every move. Dad had read that he would be flying by Cambridge and New Concord on his way to Columbus, and soon after that we were on a farm outside of town when a silver plane flying west passed high overhead. I'd always imagined it was him. I probably was scared at the idea of going up, but there wasn't any doubt about it—I wanted to do it. I thought it would be the greatest thing that ever happened. "You mean it?" I said.

"I sure do," Dad replied. "In fact, if you don't want to do it, I'm going anyway. So you better come unless you want to sit down here and watch."

We walked over to the plane. It was bigger than I had thought, with two cockpits, one in front of the other. Dad handed the guy some money. He climbed into the backseat, and the pilot helped me up after him. Dad was big, but the seat was wide enough for the two of us, and one strap fit across us both. I could barely see out. The pilot got in front and revved the engine. We bounced down that grass strip and then we were in the air. The plane banked, and I could look straight down. We flew around Cambridge a couple of times. Dad kept trying to point things out to me, but I couldn't catch his words over the sound of the engine and the rushing air. We turned back and landed.

When I got out of the plane I was elated. I couldn't get the view from the air out of my mind, and the feeling of being suspended without falling. We had gone so high, and everything on the ground looked so small, like the buildings and trees in a toy train set you'd see in a store window.

As we drove home, Dad asked me if I'd liked the flight. I told him that I had. He said he had, too. He said he'd wanted to see what flying was like ever since he'd been in France in World War I and had seen biplanes dogfighting over the lines. I realized later that it wasn't simply fun for him. Flying was progressive, just the kind of thing he would have wanted to experience so he could speak with authority about what it felt like and, just as likely, what it meant. His eagerness to experiment was one of the most important lessons of my youth.

I was hooked on flying after that, on the idea of swooping and soaring. Riding in the car with Dad on a warm day, I would hold my hand out the car window and curve it into the wind. Or I would lean out with my little tin toy airplane and watch the propeller whirl around.

To some extent, I satisfied my urge to fly by building model airplanes. I learned model building inadvertently, when I was confined at home by one of the health quarantines that were common in those days. The memory of the influenza epidemic that had killed over half a million people in 1918, including Mother's father and her older brother, still frightened public health officials. Kids were quarantined to protect their own health and that of others. Scarlet fever was the biggest scare. If you had scarlet fever, the doctor put a yellow sign in a front window warning everybody else to stay away. Whooping cough, mumps, and chicken pox could get you confined to home, too. It didn't do any good to sneak out and play, because John Louis Best, the village constable, would pick you up and take you home and give you a gruff lecture on the way.

It seemed natural to build model airplanes, as opposed to boats. Every Ohio schoolboy had a sense of pride about airplanes. We all had learned about the Wright brothers, of Dayton, almost from the first day of school. The kits came with sheets of balsa wood with patterns stamped on them. I'd use one of Dad's single-edge razor blades to cut out the pieces. My workplace was the desk in my room. Once I used the kitchen table, though, and there was a nice pattern of razor marks on it when I got through. Mother wasn't too happy about that, and she put me to work with sandpaper and steel wool to get the marks out.

When the plane parts were cut, you'd glue them together and sand off the rough edges. The wings reminded me of dragonfly wings; you covered the balsa framework with tissue paper and then wet it. The paper would shrink as it dried, and you could see the frame when you held it up against the light. The wheels were rubber, stuck on wire struts. I made a WACO, like the one I'd gone up in with Dad, and a Beech Staggerwing, which had a lower wing placed forward of the upper. I built planes modeled after racers that were flown by barnstorming pilots of the day, and a replica of the dogfighting SPAD of World War I.

The models were powered by rubber bands that you wound up by twisting the propeller. They didn't fly that well. Sometimes I could launch them with a gentle toss and they actually flew and came down smoothly. About every third flight, though, they would auger in—hit the ground nose first—and I would have to pick up the pieces and go to work again with the glue and tissue paper.

I eventually had quite a collection of airplanes that I had built, flown, and rebuilt. I hung them from the ceiling in my room, and I used to imagine what it would be like to fly the real planes the models represented.

But the Depression had come by then, and it made flying seem like a far-off dream. The idea of flying took a backseat to other things. I still thought I'd like to fly a plane myself sometime, but I never thought I'd be able to because lessons were so expensive. We had to get used to the idea that there was barely enough money for necessities, let alone the luxury that flying seemed to be.

CHAPTER 2

The effects of the Depression came on slowly. At first I hardly noticed. My friends and I were too busy roaming New Concord and the surrounding countryside. Every Saturday we would finish our morning chores and then head off looking for each other and something to do.

The town was only about a mile from one end to the other. The main residential areas were at the other end of town from where we lived, and around the college, which was a little north of the center of town. Beyond the town limits were the rolling hills at the western edge of the Appalachian plateau, where it softens and falls toward the flatter country thirty miles west where the Great Plains begin. Our world was finite and infinite at the same time—small enough to get around on foot, but filled with woods, hills, farms, and streams.

Our parents saw little to fear in our wandering once we were in elementary school and old enough to tell time and find our way home. Crime amounted to the

occasional family fracas, and required policing only by a
village constable, first John Louis Best and later Mike Cox,
who was easier to get along with. Few people locked their
doors. If a stranger knocked, you'd let him in and find out
what he needed.

This last habit of small-town openness backfired on
Mother once. I was about seven when a commotion occurred at the bridge below our house. A car speeding
down the long grade from Norwich had missed the curve
and run off the road. Everybody who lived nearby rushed
down to see if they could help. The folks in the car needed
medical attention. Nobody noticed that one of them
walked away. This one found his way to our house and
knocked on the door. Mother and Jean were there. He said
he'd been in a wreck, and asked if he could come in and
wash up. Mother let him in. When he'd cleaned himself
up, she gave him some ointment for his scratches, and
some bandages. He left, but instead of going back to check
on his friends, he went the other way and hitched a ride
out of town. It turned out later that they were a gang of
bank robbers who were terrorizing the Midwest. Mother
was indignant. She said if only she'd known, she would
have locked the fellow in and held him for the sheriff. We
always heard they were members of Ma Barker's gang.

This was so obviously a fluke that nobody paid the
slightest attention, and the habit in New Concord of keeping doors open and welcoming strangers continued as before.

Some farmers still came to town in horse-drawn bug-
gies on Saturdays to do their shopping. My friend Rex
Hoon's father was the village blacksmith; he had a smithy
with a big forge across from the railroad station, and the

farmers would come to him to have their horses and mules shod. Sometimes in the winter, when we were chilled through from playing in the snow, we would watch Mr. Hoon hammer horseshoes at his blacksmith shop while we warmed ourselves by the forge. We might cross the street and watch Fred Finley, the stationmaster, tap away on the key of his telegraph machine. I was fascinated that this rapid clicking, dots and dashes barely discernible from one another, added up to words. He wore a green eyeshade when he worked the telegraph. He'd chase you away if you distracted him, but I liked to hang around and talk to him in the slack times, and we got to be friends even though a lot of kids thought he was standoffish.

One day Mr. Finley asked me if I wanted to hold up the orders for one of the freight trains coming through. The trains didn't always stop, but the engineer needed to know what to do down the line. Mr. Finley had a long bamboo pole with a hoop on the top, and attached to the hoop was a clip that held a paper with instructions for the engineer. He'd stand on the station platform when the train went by. The engineer would stick his arm out the window and spear the hoop, get his instructions, and toss the hoop out of the train a few hundred yards down the tracks. I'd re-trieved the hoop for Mr. Finley many times. To be asked to hold the pole was special. I was about ten.

I stood there on the station platform as the eastbound train came rumbling down the grade from Norwich. It was making fifty or sixty miles an hour, its whistle blowing. I'd never seen a train look so big, or heard one sound so loud. I could feel it rumbling up through my shoes all the way to my hands holding that pole. The train loomed bigger and bigger. I could see the engineer with his head and arm out

the window. Mr. Finley had told me where to stand and ordered me not to move. The train bore down, and I watched
it come, its smokestack belching soot. I kept my eyes wide
open. I was afraid if I closed them, the engineer would see
and telegraph Mr. Finley from somewhere down the line
that I was scared and had my eyes shut. When the train was
right on top of me, I felt a tug on the pole and then the train
was roaring past. I turned to watch it, and saw the hoop
float out of the cab and fall to the ground beside the tracks.
I ran to pick it up and brought it back, and handed the hoop
and the pole over to Mr. Finley.

"Good job, Bud" was all he said, but I felt as proud as if
I had learned Morse code on the telegraph or been driving
that big train myself. I had held my ground with the train
bearing down. Somehow, realizing I could do that was an
important thing to know.

That same need took me back again and again to a
sycamore tree rising from the earth at the edge of a ravine.
It was a big, old tree that had grown out over the ravine,
so that when you climbed it, you looked straight down
fifty feet or more. Every time I climbed that tree, I forced
myself to climb to the last possible safe limb and then look
down. Every time I did it, I told myself I'd never do it
again. But I kept going back because it scared me and I had
to know I could overcome that.

New Concord saw the effects of the Depression soon
enough. Hoboes camped out near the train station or beside Crooked Creek near the National Road. Threadbare
hitchhikers stood at the roadside with their thumbs stuck
out, bundles on the ground beside them. Men trudged up

the gravel road to our house and knocked at the door to say they were hungry and request a meal. Mother always fed them, but invariably asked for some work in exchange; a pile of wood grew near the kitchen door where men transferred it from the woodpile. Muskingum College had been building a new gym, and the work just stopped; the naked girders rusted in the elements like a scaly skeleton. Mother and Dad paid close attention to the radio news broadcasts, their faces showing their concern.

The plumbing business had gone down to nothing. People couldn't even pay to have their drains unclogged, let alone add improvements such as water softeners. There were no new houses being built. Mother could sit in the shop all day and not sell a plunger or a rubber stopper for a sink drain.

Dad had two struggling businesses by then. He had become an associate dealer for Hugh White, the Chevrolet dealer in Zanesville, before the start of the Depression. He had converted part of the plumbing shop on Main Street into a little showroom, with space for one new Chevy. Used cars and trade-ins occupied a lot out back. Beside the cars was a little garden plot. We also had one behind our house.

As the hard times deepened, Dad leased a two-acre tract between the road and the railroad tracks across from the house. Now we had three gardens. I had to hoe and weed and take care of all three of them, and I hated every minute of it. We had potatoes and corn, peas and beans, melons and tomatoes, bushels and bushels of stuff. Mother put everything up in mason jars—she made preserves and applesauce, and canned peaches and tomatoes and lots of other things.

I was much more fond of hunting rabbits. I had two beagles, Ike and Mike, that I had trained to be good rabbit dogs, and I'd go out in hunting season with my dad and my cousin from Cambridge, Robert Thompson, and his dad, Ralph, who was married to my mother's sister, Florence. I started out with a .22 rifle and later graduated to a 20- and then a 12-gauge shotgun. A couple of rabbits would make a meal, or Mother would put up the meat for eating after hunting season.

We grew almost everything in those gardens, but the thing that really took off was the rhubarb that grew in the plot behind Dad's store. So in the spring, I became the rhubarb salesman for New Concord. I would pick the rhubarb, wash and cut it, and tie the pink stalks into bundles that I loaded in my wooden wagon to take around and sell for a few cents a bundle. People learned to expect me and my wagon full of rhubarb. I always went to Mrs. Grimm's place first. Her son—Pooney was his nickname—loved rhubarb. I made some spending money, and it wouldn't be long before you could smell rhubarb pie baking all over town.

When I had picked and sold all the rhubarb in our garden, I turned to washing cars. I invested some of my rhubarb money in an ad in the New Concord paper, the *Enterprise,* which came out once a week: "Kars Kleaned Kwickly, Kompletely—50 cents." The Columbus papers and the *Cambridge Jeffersonian,* the closest daily, showed pictures of families from the Oklahoma dust bowl heading west in farm trucks carrying everything they owned. But there were a few people in New Concord who still could spare fifty cents to have their cars washed. I did a good job, but one of my customers liked to make a bumper-to-bumper inspection. He always found a spot or two I'd missed, and pointed out that he expected the complete cleaning that I'd advertised. I did his car over often enough to learn that taking shortcuts could be a waste of time.

I saved most of what I made. I wanted a bicycle in the worst way, and with times the way they were I knew I was going to have to come up with the money on my own.

Dad sold a few Chevys even in the depths of the Depression. One of the best things about this was picking up the new cars. He would wait until he had two or three orders, and then would arrange a trip to Norwood, north of Cincinnati, to pick up the cars at the Chevy plant there. He'd hire students from the college to go down with him and ferry the cars back. It took most of a day to get to Cincinnati then, so they would stay overnight and come back the next day, driving slowly because that was how you broke in a new car. Sometimes he'd take me along.

My savings amounted to several dollars in the New Concord National Bank, and I had been shopping for a bicycle I could afford when he took me on one of those trips. I'd been in every bike shop in Cambridge and Zanesville, and I was starting to get frustrated. Our little convoy was passing through Chillicothe, south of Columbus, on its way back, when I spied a shop with bikes in the window.

"Dad, let's stop," I said.

He pulled over, and so did the college boys in the new Chevys. The shop had a used Iver Johnson bike that was in great shape. It had a wire basket on the front, and it cost sixteen dollars, which was about what I had saved.

"That's a good deal," I told Dad.

He pulled the money from his pocket and lent it to me until we got back home. We loaded that bike in the back of the car and continued to New Concord. Night had fallen when we reached the house, and I almost died because it was too dark to ride. I rushed out at dawn the next morning, and rode up and down the National Road over the S-shaped bridge near the house.

I put a lot of miles on that bike. One time somebody held a bike Olympics in the college stadium, and I won one of the races. But the main thing was, the bike gave me a new way of making money. I got a paper route with Columbus's afternoon paper, the Citizen. Every day after school I'd go to the bus stop to meet the bus from Columbus. The driver would unload a stack of papers tied with twine. I'd fold them one by one and load them in the basket. Then I'd take off on my bike, tossing papers onto porches from one end of New Concord to the other.

Delivering the papers was the fun part. What I hated was making people pay up. Knocking on doors, asking for money, and keeping the ledger of debits and credits left an antipathy for money collecting I never would get over.

We didn't have a Boy Scout troop in New Concord, so in October of 1932 we formed a homegrown version that we called the Ohio Rangers. Our first act was to play a football game against another club of local boys who called themselves the Midgets. We won, 12–6.

The Rangers elected me chief at our first meeting. But I, like the rest of the members, had only vague ideas of what our group should be about, so we sought the help of an advisor.

Reiss Keck taught fifth grade at New Concord elementary. He was in his early twenties, closer to the age of his students than that of their parents in most cases. A Muskingum graduate who was interested in kids, Mr. Keck lived on a farm outside of town and frequently rode a bicycle to work at the school, which was just off Main Street. He liked to talk about the legendary riders of the Tour de France, and he looked like one of them with his slim frame and muscular legs. His enthusiasm for biking took him and his sister, Ferida, on a bike tour of New England sometime after this, and he bought a three-speed bike for the occasion. Nobody in New Concord had ever heard of going all that way to ride a bicycle, and a bike with gears was an exotic contraption. He liked to say he bought the bike with the money he saved by not smoking. Some boys took this as a lesson, others as finger wagging. He agreed to act as our advisor, and we entered his name in the small green book in which we kept our minutes and our ledger of nickel-a-month dues payments.

We met in a room above New Concord's chicken hatchery. Lloyd White's father owned the hatchery, which was located just off Main Street, and Lloyd and I were good friends; we used to sit on our porch at home and watch the trains chug by across the road. The room where the Rangers met was always warm due to the incubators down below. We had egg crates to sit on, and the smell wasn't as bad as you'd expect, more live-chicken than manure.

We met once a week. We talked a lot about football, basketball, and baseball, at which we contended with the

Midgets and a team in Norwich. We needed equipment. With the Depression, nobody had a lot of money to pay for balls and baseball bats and football helmets. Mr. Keck advised us to put on a play to build up our treasury. He helped us pick the cast, and presided at our rehearsals. The name of the play is lost to memory, but not our efforts to promote it. We urged one another to sell tickets to our parents and our friends, according to our minutes, by promising a play with "music, and a little bit of everything."

We also resold tickets to the local movie theater. Roy Waller, who ran the movie house downstairs from the bowling alley in the village hall, occasionally would sell us tickets at a discount. We would go around and resell them for the normal price of fifteen cents. He showed double features on Friday and Saturday nights and cowboy serials on Saturday afternoons. The movies were a way most people could afford of forgetting the misery of the Depression.

Once we got going, we played abbreviated football games during halftime of the Muskingum College games on Saturdays. The games were well attended in New Concord, which gave us quite an audience for a bunch of little guys. At the end of football season we switched to basketball, and again provided halftime entertainment at the Muskingum Muskie games.

Although we weren't Boy Scouts, we tried to emulate them. Mr. Keck had a dog-eared copy of the handbook the Boy Scouts used, and we conducted ourselves by the rules that it set out. We obeyed the Scout instruction to always be prepared. We were courteous. We tied Scout knots, awarded merit badges, and followed Scout practices of woodcraft. We rubbed sticks together trying to light fires, and searched for moss on the north side of trees.

Inevitably, we decided that we should have a camp. One of the gang, George Ray, lived a little south of town, near a stretch of Crooked Creek downstream from us. We chose a bluff above the creek just across from his family's land as our campsite, where we staked out pup tents. We hauled out some sawdust from the local lumberyard, put down a circle in the middle of the camp for a parade ground, and laid out little walks around the camp and edged them. At the circle we added a flagpole cut from a straight young pine tree.

The camp was a second home to me and the other Rangers that summer, 1933. I could hardly wait to finish hoeing and weeding and picking vegetables in our three gardens, and washing cars on the weekends, so I could head for the camp. My friends rushed there, too, when they were finished with their chores. The heat of the afternoon would find us gathering. When we were tired of playing we swam in the creek. We cooked hot dogs and hamburgers over our campfire and ate them with baked beans heated in the can, hobo-style. We saluted when we lowered the flag at sunset and I played taps before we went to sleep. We slept in our pup tents, and emerged to a new day the next morning. I usually skipped reveille, but played the colors when we raised the flag. Then we would go home to attack our chores before gathering again in the afternoon. We must have stayed there sixty or seventy days over the course of that summer, and all of us were just in elementary school.

Near the end of the summer, Annie Castor asked me if we would lend our camp to the Girl Scouts. The little girl

I had shared a playpen with when our parents got together with the Twice Five Club had become one of my regular companions. Annie had a sister, Jane. The Castors lived out on Bloomfield Road a little north of town in a new house Dr. Castor had had built a few years before the Depression started. It had a Spanish look that caused folks in town to call it "the Alamo." Annie and I were too young to be a couple at that point, but she was pretty, with dark hair and a shy, bright smile. Her smile spoke for her in many ways, because she had a severe stutter that made it difficult for her to speak.

We Rangers were proud of our camp, and I guess a little proprietary about it. We didn't want to give it up. But Reiss Keck and our parents counseled us that lending it to them was the right thing to do. Annie inveigled me the rest of the way. So we turned the camp over to the Girl Scouts and resigned ourselves to sleeping, for one night, in our beds at home.

A giant thunderstorm swept over New Concord that night. It rained so hard I had to get up and close my window, which I rarely did, even in the winter.

When I was finished with my chores the next afternoon I headed for the camp as usual. A couple of the other Rangers joined me on the way. We were greeted by a devastating sight. The girls were gone, and our tents lay in soggy clumps of rain-soaked canvas. Our precious sawdust had all been washed away. The whole place was a sea of mud. We tried to straighten things up, and then went home to find out what had happened. It turned out that the girls' parents had gotten worried and gone out in the middle of the night to take them home.

Abandon the camp? We pretended to be shocked. We were as obnoxious as little boys can be, until Mr. Keck finally assembled us for a talking-to. He said the storm was one of the biggest to hit New Concord in some time, and the girls deserved the benefit of mitigating circumstances. So we stopped ragging them. But we never got the camp back in shape after that.

Instead, we spent more time at the little log cabin we had built at Lloyd White's farm outside of town. His dad had given us the use of a patch of woods at the back end of a cornfield. We cut small logs, notched them, built the cabin pioneer-style, and filled the chinks with mud. We layered pine boughs for the roof and hung a piece of canvas for the door. We built bunks at the back where two of us at a time could sleep, and kept the dirt floor swept. It was only about a third the size of a real log cabin, but it was a snug little place to have been built by a bunch of twelve-year-olds. Watermelons grew along the edge of the cornfield, and we ate a lot of watermelon at that cabin, and roasted ears of corn. That place we never did lend out.

With all my moneymaking schemes achieving relative success, and the alternative world that the Rangers and their activities provided, it might sound as if the Depression had no effect on me. But it did, and it was profound.

The hard times had settled in by then. Poverty had a kind of permanence. The skeleton of the college gym didn't stand out anymore; it was a part of the landscape. In 1932, the year before, twelve million workers in the

United States, a quarter of the workforce, had no jobs, and the Depression was yet to bottom out. I was reading in the living room one night after dinner when I heard my parents talking in the kitchen. There was nothing unusual about that, but this time their tone was different. They spoke in urgent whispers that made me put down my book and get close enough to listen. They were sitting at the kitchen table, talking about the possibility that the bank was going to foreclose on their mortgage.

"Clara, I'm afraid we're going to lose the house," Dad said.

That conversation struck terror in my heart. I experienced what millions of other families must have gone through in the Depression, the fear of not knowing where they would be tomorrow. I didn't know where we would go, whether we had any relatives that we could move in with, what was going to happen. There were families in our area that had broken up, with children parceled out among relatives here and there. I wondered if our family would be broken up, too, and scattered to the winds. Who would I end up with, and what would I do without my parents and my sister?

Despite my fears, I knew we were better off than many in New Concord. The gardens always gave us enough to eat. What was left over, my parents shared with whoever needed it. If Mother knew somebody was having a problem, she would make a kettle of soup and send me off with it in my wagon. "Bud, take this to so-and-so," she would say. "But don't go to the front door. Go around the back. They're proud folks and they'd be embarrassed. Nobody needs to know they're going through a rough spot." Of course, if she knew about their troubles, so did everybody in New Concord.

Annie's family had a garden, too, and so did most of the families in New Concord. We all spent the winters eating fruit and vegetables put up in mason jars.

Sometimes there were plumbing problems that simply had to be dealt with, and no money to pay for the repairs. When that happened, Dad would do the work anyway and tell the customer to pay when he got back on his feet. The farmers, though, usually had something to offer. For several winters during the Depression, we always had a side of beef or a quartered hog hanging in a corner of the garage outside. We ate a lot of steak and hamburger and ham and sausage, all on the barter system.

Little did I know that my parents' fear of losing the house was being echoed all across the country. Banks that had been used to collecting only interest on home loans year after year suddenly were pressed for cash, and they were dunning homeowners. Foreclosures were widespread. Families were being turned out into the streets. President Roosevelt proposed the Home Owners' Loan Corporation in 1933 as one of the New Deal's first initiatives. It refinanced mortgages on private homes. A year later, the Federal Housing Administration came into being, backing mortgages with federal insurance that allowed banks to spread regular payments for principal and interest out over a long term.

When those programs were put into place, my parents managed to renegotiate their mortgage, and we kept the house. The FHA helped end the Depression, not only by saving home ownership for many, but by allowing more people to build and buy new homes. Later, I would take those programs of the New Deal as object lessons in the role government can play in changing people's lives for the better.

Dad loved to travel. Even in the depths of the Depression, he was a king of the road. Sometimes on Friday afternoons he'd come home and tell Mother to throw a few things in the car, and they'd drive off overnight to some state park or campground. Jean and I usually stayed home and were looked after by Dad's niece, my cousin Gladys Davis, who lived upstairs.

He liked to go to all the little county fairs within driving distance of New Concord. And more years than not, we would go to the Ohio State Fair in Columbus. He'd usually go the day before, stay over, and spend the next day at the fair before coming home that night. He loved all the stock exhibits and the animal exhibits. I was dragged through more hog barns and chicken barns than you can imagine.

I liked the school exhibits, especially those that had to do with science. High-school science classes always put together some kind of demonstration for the county fair. The best of them went to the state fair. They explained how bees could fly, or how rain happened, and those were the kinds of things that caught my attention.

Dad didn't just travel locally. He was a staunch member of the American Legion. The Legion held a huge convention every summer, and if it was in the eastern half of the United States, he'd go and take Mother with him. The Legion held its convention in Detroit in 1931, and this time they took me along. Dad introduced me to men with whom he'd fought side by side at Verdun and Saint-Mihiel. Even at ten, I could see that they had a special bond. He laughed and joked with everybody there, but what struck me, amid the hoopla and brass bands and the Legionnaires who nipped from pocket flasks, was this closeness among the men who had been together under fire. Nobody spelled it out. They didn't even talk about it much, but it was there. And I saw how all the other Legionnaires treated Dad and the other combat veterans with special attention and respect.

The Chicago World's Fair opened in 1933 and continued into 1934. It was called A Century of Progress International Exposition. Although Dad still didn't have much work, he was an optimist, and he wouldn't let the Depression keep him away from a big event like that. He got three other men together to make the trip in the fair's second summer, the year I turned thirteen, and they took me along. Dad drove, and I rode in the front seat in the middle all the way. When we got to Chicago, we got rooms in a boardinghouse on the outskirts of town.

We had to see the General Motors exhibit at the fair, since Dad was a Chevy dealer. It envisioned the transportation of the future, with streamlined cars and wide roads and spectacular modern bridges. It looked to me, as I wandered that exhibit, as if transportation was about to take off into the air.

Chicago was the city of stockyards. We all knew of Carl Sandburg's famous poem "Chicago," describing the city as "Hog butcher for the world." Dad, the former farm boy, was less the poetry enthusiast than Mother, but he decided, partly on the strength of Sandburg's poem, that he wanted to see what went on in the stockyards. Armour and Company was one of the big meat packers, and we visited the Armour slaughterhouse.

The smell is what I remember best about the stock-yards. I had been on farms around New Concord, and around livestock. Those were nothing by comparison. Powerful odors of manure and blood hung over the vast slaughterhouse and combined into a smell that I can't describe but will never forget. You could hardly hear over the din of bawling cattle.

Men drove the cattle into a big pen, and then used wooden prods to drive individual steers or cows into narrow chutes barely wide enough for them. A man stood over the chute with one foot on each side, holding what must have been a ten- or twelve-pound sledgehammer with a point on the head. Once an animal was in the chute, the man swung the hammer, hit it right between the eyes, and killed it where it stood. Then the side of the chute would drop open, the body would roll out, and another man would hook it by a back hoof and the body would be hoisted up and the butchering would start. The amount of blood was amazing; it flowed in a torrent into concrete sluices. There wasn't much need to see the hog slaughtering after that, but we did. It went pretty much the same way.

I was glad, as we drove back to New Concord, that Dad had taken me to see the slaughterhouse. It made me think that the world was a more matter-of-fact place than I had realized. One of my favorite things to do back in New Concord was to cook hamburgers and hot dogs over a campfire. To see where it all came from was pretty startling. I'd seen the sides of beef in our garage, and dressed hogs, too, but I'd never before imagined the process that started steaks and pork chops toward the dinner table as a factory with a mass-production line.

Things improved in time around our town. Roosevelt's program for rebuilding the country gradually took hold. The Works Progress Administration (WPA) brought a project to New Concord to redo the town water system. Most everybody in town was surprised to find out, when they opened up the ground, that we had wooden water mains held together with copper bindings. The WPA went to work replacing those and installing a town sewer system. Dad, because he was the town plumber, started out working on the project. Then he was made a foreman, and that kept us going while the plumbing business gradually improved.

The Rural Electrification Administration (REA), another Roosevelt program, helped to restore Dad's business indirectly. The REA started putting lines up across the country to spread the benefits of electricity into far-flung rural areas. These included some of the farmlands of eastern Ohio. Farmers who had electric service for the first time could add pumps to their water and irrigation wells. They could install indoor plumbing and pump water to their barns. The new jobs Dad got as a result were another step toward our financial recovery.

CHAPTER 3

My friends and I had big plans for the Rangers. In 1934 we launched an attempt at global expansion. We dipped into the treasury to print stationery and a glossy newsletter. *The Ranger* described the Ranger Club—no longer the Ohio Rangers—as an organization for boys ten to eighteen based on "outdoor activities, hobbies, and correspondence." It carried a story about youthful friendship, ran articles on stamp collecting and model building, and announced our intention of recruiting five hundred new correspondence members "within the next few months." Our sales pitch offered "a chance to correspond with other members all over the world."

But as we passed into junior high and high school, this started to change. Puberty stole our interest in marching around and camping out and focused it on other things. My friends and I began to gravitate to the swimming holes along Crooked Creek, and to College Lake on the Muskingum campus, where we swam in the summer and skated in the winter, because the girls tended to be there. We started to pair off into couples. When that happened, Annie was always the one I paired with. We were natural and easy together.

Dad ran for the village school board and, once elected, was named its president. He was as proud of that, with his sixth-grade education, as he was of having started a successful business.

His election had the effect of making me focus more on school. I had always been a good student. I had been offered a chance to skip the fourth grade but passed it up because I wanted to stay with my classmates. Mother and Dad enforced the importance of my role at school. They said we all bore responsibilities within the family. Dad's responsibility was to go out and work and make money to support us. Mother's was to keep the home and help him at the store. My job, and Jean's, was to learn and get good grades in school. In the evenings, after everybody had finished their day's work, we talked around the dinner table about what we had done. Reporting on our day's activities made me feel as if I had my end of the family contract to hold up. I needed to take each day's work seriously so I would have something to report.

Entering my teens, I became preoccupied with cars. Much earlier, Dad had let me sit beside him while he drove. When I was still so small that I needed pillows to see out the windshield, he would sometimes let me steer while he worked the pedals. Now I made it my job to move the used cars around on the lot behind his one-car Chevy showroom. I asked him to let me work on them, and sometimes he agreed. I loved tinkering with them, and my tinkering became an education. I learned more about engines by taking them apart and putting them back together than I would ever learn anywhere else.

Once in a great while he would get a Model T Ford as a trade-in. Model Ts were temperamental and outdated, but once I broke the code of their foot pedals and hand controls I enjoyed working on them. I'd adopt the trade-ins and make sure they looked good and ran right. Some of them still had hand cranks. Cranking a Model T to start it was a good way to break a wrist, but I managed to avoid that pitfall.

My interest in cars had an ulterior motive. I wanted to drive. You didn't have to have a driver's license in Ohio then. My aim was to drive well enough for Dad to trust me with a car out on the road. He'd let me drive when he was in the car, but I was ready to take off by myself. Driving meant incredible new freedom. As much as I enjoyed my boyhood in New Concord, I knew the world was larger than I could explore on foot or even on my Iver Johnson bike.

I started to test possible career choices as I neared high school. I had a good feel for auto engines and mechanics, but I was also interested in medicine and chemistry.

A few years earlier, one of our roomers had been a telephone operator who went out with a Muskingum student named Julian White. I got to know Julian when he was at the house, and he took me under his wing. Julian was a chemistry major. Sometimes, on Saturdays, he'd take me with him when he went up to the college to do laboratory work. I had a lot of fun heating and bending glass tubes, and heating solutions on the Bunsen burner.

For a long time, when people asked me what I wanted to be when I grew up, I'd say, "A chemist, like Julian White."

A teacher named Ellis Duitch further exposed me to the sciences. Mr. Duitch made basic physics and its practical applications interesting to me. He taught me how a radio worked and gave me some tips when I decided to build a crystal radio. It worked off a dry-cell battery. At night, when the conditions were right, I could pick up stations from all over: WLS in Chicago, WLW in Cincinnati, and KDKA in Pittsburgh.

Other influences were shaping me as well, affecting not my career goals, but rather my evolution as a citizen. The parades and flag-waving patriotism of New Concord were one level of citizenship. They were a form of civic entertainment that everyone enjoyed and felt proud of. They mirrored the events in the high-spirited musical *The Music Man* so closely that I later came to think of New Concord as resembling the musical's fictional and more comical setting of River City, Iowa. But one of my high-school teachers brought home the idea that citizenship in a democracy was more complex, more demanding, and in many ways more exciting than I had realized up till then.

Harford Steele taught civics. Mr. Steele was a barrel-chested man who prided himself on the strength of his grip. He carried around a sponge ball that he would squeeze in one hand and then the other. Later, when he became the high-school principal and he gripped your shoulder to emphasize a point of discipline, you knew you'd been gripped. His civics course covered the fundamental institutions of the country, and he had a knack for making the whole thing come alive. He made history and government and politics into something really special.

They were never remote, the way he taught them. You could see how individuals could exercise their beliefs and actually cause change and improvement. Citizenship in his terms was a dynamic practice. The idea that you really could make a difference stimulated me, partly because it reinforced what I had learned at home with Dad's participation on the school board. Mr. Steele's course ignited a fire in me that never did go out.

High school opened new worlds of activity for a young man as eager as I was. The Rangers had given me and my friends plenty to do, but high school offered such a wealth of possibilities it was hard to choose among them. Rather than choose, I did everything. My freshman year, I served on the student council, worked as a reporter for the *Maroon and White Newslite*—the New Concord High School paper—and played trumpet in the high-school orchestra, while continuing to play the trumpet with the town marching band.

Annie's musical talents were far more impressive than mine. She had added the trombone to a repertoire that already included the piano, and was also playing in the school orchestra and the town band. We began to see each other outside the context of our families. The things we did as part of a group—the ice cream socials and the cookouts, the swimming and ice skating at College Lake—continued, but we routinely paired off now. We were a couple. I had always liked Annie's smile. It revealed all the warmth and eagerness that was there, though hidden to some by the difficulty she had speaking. Now her smile struck me in a different way. It was like a pleasant trap that held me so I couldn't look away. I started to look forward to that feeling.

High school also gave me my first taste of seriously organized sports and competition. I went out for football my freshman year.

I wasn't that big, but I was slow, and I had solid, heavy legs. That meant it was my destiny to play the line, and my particular combination of shortcomings placed me at the center of the line. Center is a position I wouldn't recommend to anyone, at least not the way football was played then. The T formation, in which the quarterback takes the ball directly from the center, had not been invented yet. Offensive football used a single- or double-wing formation in which the center looked back between his legs and snapped the ball to one of the moving backs. That practically guaranteed you'd get knocked on the back of the head to the ground just as soon as you released the ball. I chewed a lot of grass playing football.

Basketball was a lot cleaner. I went out for the freshman basketball squad and learned to love the game's speed, its constant motion, and its subtle and intricate weave of strategies and plays.

All this activity stole time from the Rangers. We guiltily tried to keep it going, but outdoor activities, hobbies, and correspondence were no competition for the full school schedule and the imperatives of adolescence. My friends and I had outgrown our club at last. Our minutes recorded a meeting on December 17, 1935, and somebody penciled in January 7, 1936, as the date of the next meeting. There were no entries after that. Apparently nobody showed up.

I was sorry when basketball season ended, but had some fun playing for the New Concord High tennis team that spring. I went to work as soon as school let out for the summer. Selling rhubarb, washing cars, and delivering papers were behind me. Now I worked for Dad on the plumbing jobs that were coming with increasing frequency as the Depression eased. I worked side by side with the men he hired, digging septic tank pits and sewer line ditches. It was my first taste of hard physical labor. Dad paid me the going rate for such work. He said, "You work as hard as anybody else, and I'll pay you the same as anybody else. You don't, and I'll hire somebody who will." That was fair enough. I made seventy-five cents an hour, and that was big money to me then.

I had some muscles when I went back to school that fall. I spent another football season at center, looking forward to basketball. But my real pursuit in the fall of 1936 was a driver's license. Ohio had finally established a licensing system. It took effect in October. Eighteen-year-olds with a year's driving experience could get a license without being tested. I was only fifteen; there was no minimum age to get a license, but those under eighteen had to be tested, and to be tested, you had to apply.

This I did at the newly established Department of Motor Vehicles office at the Muskingum County seat in Zanesville. I left with a temporary permit that allowed me to drive with a licensed older driver until I was ready to come back for testing. I was all set to drive, but there was a written test, too, based on the state's new driver's manual. I took it home to study, and after a few weeks Dad drove me back to Zanesville for the big day. I passed both tests, and at the beginning of 1937, at age fifteen, I was officially licensed by the State of Ohio to drive a car.

When the school year was over, I took off on the first long trip I made without my parents.

Reiss Keck, the teacher who had been the advisor to the Rangers, and his sister, Ferida, had asked me if I wanted to join them on a driving tour. He had a little Chevy coupe that he had bought from my dad. They planned to drive up through northern Ohio and along the St. Lawrence River, then to Montreal and back down through New England. These were new areas to me, but the prospect of all that driving interested me as much as any sights I anticipated seeing.

We headed north as soon as school let out. I should have suspected from his bicycle exploits that R.M., as I called Mr. Keck by then, was a born tourer like Dad. But he proved to be even more active than Dad at pursuing his curiosity. When we got to Youngstown on our first day, R.M. decided we should stop to see steel being made. "It should be interesting," he said. "We'll probably learn something."

That was the first time I had seen the showers of sparks and the molten steel pouring like lava. But the real lesson the tour of that steel plant reinforced was to be active in your curiosity. Idle curiosity was the same as daydreaming; it didn't get you one step closer to what you wanted to know.

We got to Niagara Falls, and the lesson was repeated.

R.M. talked somebody into taking us into one of the big electric generating plants downriver from the falls. We took an elevator 150 feet down into the bowels of the plant, where huge turbine generators spun under the incredible hydraulic force of the water. The noise was deafening. That was my introduction to the making of electricity.

We drove from there into Canada, and along the north shore of Lake Ontario. The lake narrowed into the St. Lawrence River, where for some reason I wanted to go swimming. It was still early June. R.M. observed that the ice probably hadn't been melted for very long and it might be a little cold for swimming. But I insisted, so we pulled into a park where there was a deserted swimming beach, and I put on trunks in the back of the car and headed for the water. I dove in and came right back out again, wind-milling and shouting. It was ice water, but I could say I had swum in the St. Lawrence.

At Montreal we turned east and then south into the White Mountains of New Hampshire. We took an all-day hike to the top of Mt. Washington, where the wind stabbed like a knife. Driving south through New England, we stopped in Boston to visit the USS *Constitution*. I remembered the stories of British cannonballs bouncing off "Old Ironsides" in the War of 1812, and felt a surge of pride. In New York, after checking into an inexpensive hotel, we wandered through the so-called concrete canyons, craning our necks like all the other rubes, and took a ferry to the Statue of Liberty. I navigated the subways and the Staten Island ferry. We saw the Liberty Bell in Philadelphia, and stopped at Valley Forge on our way through Pennsylvania. Not long after that we passed a "Welcome to Ohio" sign, and then we were home again in New Concord. That trip opened my eyes to many new sights and experiences, but what I remember most was Reiss Keck's attitude. And I decided that when there was something I wanted to do or see or find out, I had to be willing to push really hard to do it.

Dad had a big job that summer that required me to develop new plumbing skills. Lone Pine was a large old house being converted to a nursing home. The additional plumbing meant a lot of pipe cutting and fitting, and that became my specialty. While Dad and his men worked inside, I stayed outside to supply them with pipe cut to specifications. "Half-inch pipe, twenty-two inches to center of ell," Dad or one of his men might call from inside the rambling structure, and I would go to work, cutting the pipe, reaming it, putting a thread on the end, and adding the necessary fitting. Those were all manual operations that required considerable strength, especially threading large-diameter pipe. I worked bare-chested, and developed new muscles on that job.

But while I was proud to be able to do a man's work and earn a man's pay, I decided I didn't want to be a plumber. It wasn't cutting pipe, or wrestling with the threading dies, or even doing repair jobs under houses in crawl spaces with bugs and nests of spiders, that bothered me. It was something else. I wanted broader horizons. I didn't discuss it with Dad, but I knew in my heart that when the time came, I would move in another direction just as he had left the farm behind and moved to town after he returned from World War I.

Dad signified my status as a sixteen-year-old at the

end of that summer by giving me the use of an old Chevy roadster he had taken as a trade-in. The Chevy was a '29. It had been driven to within an inch of its life, but to me it was almost like a set of wings. Annie and I dubbed it "the cruiser."

The cruiser was a dull rusty red when Dad turned it over to me. One of the first things it needed was a paint job. Dad sent repainting jobs to a man named Harvey Haney. He was a real pro who worked with a spray compressor and did a lot of sanding between coats. People came from all over to have him paint their cars. I couldn't pay for one of his paint jobs, but I watched him work and tried to pick up his technique. Without a compressor, I had to improvise. I found an old fly-spray can with a pump handle, got some maroon paint—half of our high-school color scheme—thinned it down, and put it in the spray container. I sanded the cruiser down and went to work with the pump sprayer. I sanded it again between coats, and surprisingly enough, the cruiser looked pretty good when I was finished. Dad kidded me for years about painting that old car with a fly sprayer.

The cruiser needed more than just a paint job. The radiator leaked beyond the power of any stop-leak compound to repair, so I had to stop often for water. The convertible top was so far gone I threw it away. I carried a tarp that I could throw over the seats when it rained or snowed, and drilled holes in the floor for drainage.

Not all of the problems came with the car. The football practice field down by Crooked Creek flooded when it rained and got slippery with mud—good conditions for spinning doughnuts. That's what I was doing when I ran over a log and bent the oil pan. A bearing went not long after that, and Dad was quick to figure out the reason. He grounded me for several weeks. I did most of the repair work myself, and learned a few things about mechanics and the pitfalls of joyriding.

I made the football varsity that fall. I still played center on the offense, but put in some time at linebacker on defense, and Coach Al Baisler, who had been a Little All-American when he played at Muskingum, undertook to teach me the art of placekicking. The Little Muskies, as we were called, played our home games at the college stadium. Winter brought basketball and spring tennis, and now I was playing on the varsities of those sports, too.

My schedule apart from sports was fuller than ever. I had been elected president of the junior class. I was active in Hi-Y, which was the high-school branch of the YMCA. Its YWCA equivalent was the Girl Reserves. Annie was a member of that, and the two groups did a lot of things together. Annie and I had also joined the glee club, while we continued to play in the town band.

We were getting serious by then. We never dated anybody but each other. It was just expected that if something was going on, and boys and girls were going together, Annie and I would be a couple. There was no one event that started that pattern; that was just the way it was. We were as much in love, I guess, as any high-school kids ever have been. I don't remember the first time I told Annie that I loved her, or the first time she told me. It was just something we both knew.

I always gave her a gardenia when we went to special school events. We didn't have school dances, though. Slow dancing between boys and girls was still forbidden then by the two branches of the Presbyterian church that dominated life in New Concord. Annie and her family attended the College Drive United Presbyterian Church, which because of the United Presbyterians' connection with Muskingum was New Concord's largest congregation. Westminster Presbyterian, the church my family belonged to, was part of the larger and slightly more conservative Presbyterian Church (U.S.A.). There wasn't much difference between them. Annie attended Sabbath school instead of Sunday school, which I attended, and in her church the psalms were sung, while we recited them. But the two churches agreed on the dangers of allowing young men and women to dance close together to slow music. Neither, however, was as severe as the Reformed Presbyterians. They had a church in town where no "artificial" music could be played, and the choir sang a cappella.

The only thing that marred my recollection of that year was hearing that some girls had laughed at Annie for the way she talked. Annie's stuttering wasn't something I viewed as a problem. Her father stuttered, too, though not as severely. It was just something she did, no different from some people writing left-handed and others right-handed. I thought it was cruel and thoughtless to laugh at someone for something like that—especially Annie, whom I cared for—and I told them so.

Toward the end of that year we began to talk of marriage. We knew that we wanted to spend the rest of our lives together. Annie, a year ahead of me, would graduate soon. Our parents wanted us to go to college. But this wasn't something we were going to talk over with the folks. Kentucky was well known among Ohio teenagers for its liberal marriage laws. Kentucky beckoned, even though we would have had to borrow a car or stop fifty times to put water in the cruiser's leaky radiator. Talking about an elopement was romantic. It was a way of expressing our dreams of a shared future and our commitment to each other.

Talking was all we did, though. Common sense prevailed. We still knew we were going to get married down the line, but our parents had done a good job of impressing on us the value of a college education.

New Concord High had a junior-senior banquet every year that included a program of skits and speeches. I used my role as junior-class president to promote an aviation theme. As Roosevelt's New Deal programs took hold and broke the grip of economic hardship on Americans, flying and air travel were coming into their own. It was now eleven years since Lindbergh's famous flight. The airport in Columbus, which I begged Dad to stop at when we made family shopping trips to the state capital, was getting busier. The National Air Races in Cleveland were being held annually again after intermittent staging during the Depression, and each year's news of stunts, speed, and new airplanes whetted my appetite for flight still further. We were, after all, a class of juniors about to soar into seniorhood, and the seniors were flying off to college, jobs, or marriage. My schoolmates approved the theme; we put a picture of a Ford Trimotor airplane on the program cover, and I and my senior counterpart, as toastmasters of the event, were designated as the evening's "pilots."

I decided I should get a job on my own that summer. Dad agreed. I had heard about Camp Nelson Dodd, the state YMCA camp, through Hi-Y. I had wanted to attend the camp a summer or two earlier, but I didn't have the money. Now it occurred to me that I could get a job there. Before school was out, I wrote an application letter to the state offices of the Y in Columbus. A job interview followed, and I was hired to work in the camp's mess hall. I drove back to New Concord proud that for the first time I had landed a job all on my own.

Camp Nelson Dodd was in north-central Ohio, on the Mohican River, near a little town called Brinkhaven in Knox County. I put enough sealant in the cruiser's radiator to make the distance with only a few stops. I was in charge of the kitchen's cleanup and serving crew. There were four of us, all college and older high-school kids. We lived in an old log cabin out in the woods and slept in bunk beds. Our job was to set the tables and serve the food, bus the tables, and wash the dishes, all for 125 kids eating three meals a day.

Once we got it down to a routine, we had plenty of time to work with the kids, play sports, and take part in the counseling and inspirational programs. I made a lot less money than I had working for Dad, just a few dollars a week, but I had a ball all summer long, and would return to Camp Dodd for the next two summers.

• • •

I came home at the end of the summer to find that Dad had one of his trips planned. He was going to Cleveland for the National Air Races and asked if I was interested. He knew I would be, of course. My cousin Robert Thompson and his dad went with us.

Flamboyant daredevil pilots were the order of the day. Roscoe Turner, a distinguished dogfighting pilot during World War I, had barnstormed as a stunt flyer after the war. During the 1930s he set every flying speed record he could think of, including Los Angeles to New York via Cleveland, the apparently arbitrary choice of Seattle to Los Angeles, and, in 1933, New York to Los Angeles in eleven hours and thirty minutes. Turner was as colorful as they came. He kept a lion cub as a pet, sported a rakish mustache, strode around in puttees, and wore a helmet and goggles when he flew. His exploits made the news, and crowds loved him.

Turner spawned plenty of imitators. By the late 1930s European counts and military men attended the races to compete for trophies with American pilots. The events included pylon races, aerobatic stunt flying, parachute jumping, and gliding sailplanes that were almost as light as air and could ride thermal updrafts for hours. A German woman, Hanna Reitsch, was a master of the sailplane. Flyers from the Army Air Corps, the Naval Air Forces, and the Marines—"Uncle Sam's fighting birdmen," one program called them—showed off the latest fighters. It wasn't unusual to read, in the days before the races, that a pilot practicing for some event had crashed and been injured or killed. But danger was accepted as part of aviation. It certainly didn't quell anyone's enthusiasm for the races, and Dad, Uncle Ralph, Robert, and I were no exceptions.

The races lasted three days, but we went for only one. We camped out the night before and were in the stands beside the Cleveland Airport runways early the next morning for the flying exhibitions.

The big event was the Thompson Trophy race. The famous Jimmy Doolittle had won the Thompson Trophy in 1932 in a plane that looked like a barrel with wings, called the Gee Bee Special. Roscoe Turner had won it two years later. He had been absent from the winner's circle for three years, but this year had the fastest qualifying speed, 281.25 miles per hour, in a plane that had the biggest engine in the field. The closed-course speed race was thirty laps around a ten-mile oval course. The planes took off, and the sound was like being in a tin can with a swarm of angry bees. The pilots banked tight around high pylons and thundered right past the spectators. I loved the roar of those engines and the sheer speed of the planes, and when it was over, with Turner again the winner, flying beckoned me as never before.

Beyond the news of the stunt pilots and racers and their exploits was the more serious news of aviation's evolution. People were predicting that airliners would compete with steamships in carrying people across the Atlantic. Giant passenger planes were forecast. The flying boats of Pan American were already making trans-Pacific flights seem routine. I began thinking, as I approached my senior year in high school, that the airlines might offer a fine career for someone who knew how to fly. But I was no closer to having the money for flying lessons than I had ever been.

●   ●   ●

My senior year was another blur of activity. I lettered in football, basketball, and tennis. The football team went 6–1–1 and won the championship of the Muskingum Valley Football League. In basketball, we won the Muskingum County tournament and the sectional championship before losing to the eventual state champions. At the same time, as sports editor of the *Maroon and White Newslite,* I tried to be objective about the Little Muskies. I played Vernon Wetherell, aka Lord Bantock, in the senior-class play, a justly forgotten comedy called *Fanny and the Servant Problem* in which love overcomes the fact that Fanny, Vernon's wife, turns out to be the butler's niece. And I took, along with most of the other football players, a class called Boys' Home-Making, taught by Winifred Conley. Mrs. Conley's course aimed to teach us cooking and sewing, to save us from being helpless when we were on our own. I made a pretty tasty fudge cake in that course. In a perfect world, chocolate fudge cake would be my idea of a perfect diet.

I wasn't planning on being on my own, however. Although Annie and I had put aside our plans for a teenage marriage in Kentucky, we both were certain that we would be married. It wasn't a question of if, only when. But at my high-school graduation in 1939, I was looking forward to college.

CHAPTER 4

I never considered applying anywhere but Muskingum. Growing up in its shadow, I had met enough of its students, graduates, and professors to know that character, while it wasn't listed in the catalogue, was among its offerings. Its severe restrictions against smoking and drinking didn't bother me, since I didn't do either.

The school had served my mother well. She had brought poetry and a level of erudition into our home that raised my expectations for myself. The sad abandoned toys of "Little Boy Blue" were etched into my memory from her recitations in the kitchen. Tom Sawyer, Huckleberry Finn, and the other boy heroes of children's literature bounded through their adventures in the books she gave me to read when I was young. Annie's father and mother were also alumni of our local college. Annie had considered other small, similar Ohio colleges, including Denison and Oberlin; the latter had a fine music conservatory. But they were expensive, so she had chosen Muskingum. Since

she was a year ahead of me, I had the benefit of her freshman year to help confirm my choice. I thought, too, that
no matter where you got your education, it was what you
made of it that counted.

Finances were another factor. My family still didn't
have a lot of money in the aftermath of the Depression.
Dad said he was willing to do whatever it took to see
that I went to college. But I knew that I could attend
Muskingum and live at home to save the cost of room and
board, and I applied for a work scholarship to help with
tuition.

I approached my first year of college through one of
the most wonderful summers I can imagine a young man
having. Three friends and I took a graduation trip that mirrored the trip I had taken with Reiss Keck and his sister
two years earlier. Dane Handschy, my closest neighbor,
and Lloyd White were good friends and classmates. My
cousin from Cambridge, Robert Thompson, joined us.
Dad lent me the family car, a Chevy sedan, for the trip. We
drove up along the St. Lawrence River and back down
through New England.

It was minimal-cost traveling. We never slept under a
roof. We'd find a place that looked good and pull over, and
we'd spread our blankets in a corner of a field or at the
edge of a park somewhere. We stopped at little grocery
stores and bought bread and bologna and mustard and
milk. That was our diet. We had a great time.

The 1939 World's Fair had opened in New York. We
stopped to take it in. Its theme was "The World of
Tomorrow." A great big globe dominated everything, and
we got a taste of the future that had been updated from
what I remembered of the Chicago World's Fair. We found

camping space out on Long Island and spent two or three days at the fair.

The trip lasted about two weeks. When we got back home I took off again for Camp Nelson Dodd. I had graduated from mess hall duty to part-time work driving the camp truck. I traveled three miles to Brinkhaven every day to pick up the mail. I also drove to Mt. Vernon, almost thirty miles away, to pick up supplies from a wholesale grocer there. I knew how much 125 kids at camp could eat, since I'd served the meals and cleaned up after them the year before. Still, it amazed me that the supplies filled up a big truck three times a week.

When I wasn't ferrying mail and groceries to the camp, I worked as a counselor. This was a lot of fun, and great experience. I helped organize baseball and football games, swimming, crafts, and other activities, which included skits and plays in the lodge after dinner.

I entered Muskingum planning to major in chemistry. I was still under the influence of Julian White and the Saturdays I'd spent with him in the laboratory filling retorts and lighting Bunsen burners to help him with his experiments. I had the vague notion that I wanted to do chemical research, or more likely follow some kind of a career in medicine, perhaps in the sort of medical research that would place me in a lab.

As for the opportunities that were going to come with the expansion of air travel, they still were reserved for the lucky few who could afford the price of flying lessons. Flying kept its grip on my imagination, but the Glenns were a practical family, and I wanted to pursue realistic goals.

My enthusiasm for playing football waned that fall. I went out for the freshman team and made it. I played solidly, but without brilliance. I went out for the freshman basketball squad and made that, too, but I noticed that while I had not gotten any faster or grown any taller, the other players had. Nevertheless, I joined a campus social club, the Stags, that had a number of athletes as members.

While my so-so abilities had something to do with the lessening intensity I felt for sports, they weren't the only problem. The war in Europe that had started with Hitler's invasion of Poland in September was widening and intensifying. Hitler was grabbing everything he could. British ships and planes were under attack, and American college men could not escape the thought of what America's close ties with England might mean for them.

To fulfill the terms of my scholarship, I worked with the head of college maintenance. Ernie Wylie was an easy-going fellow with a sense of humor. I had known him in the way that you know people in a small town. Now that we were working together, we hit it off well. I started off mowing lawns around the campus. When the weather turned cold and the grass died, I worked on the steam and plumbing lines with a fellow named Jim Hurley. And I shoveled snow, an awful lot of it. There were no classes on weekends, and that was the time I spent polishing floors in the classroom buildings. I used a heavy floor polisher with a round buffing head; if you tilted it the wrong way, it would run away from you and pull you behind it down the hall. You could do serious damage.

I made twenty-five cents an hour working at the college—far from the relatively good money I had made working for Dad, or the money some students got working on state highway crews during the summer. I could look forward to making thirty cents an hour as a sophomore. The one advantage of working at the college was that, living in New Concord, I could set my own schedule. In addition to weekends, I worked during vacations, when most other students left the campus. I didn't actually take any of my meager earnings home. The college applied it against my tuition aid of $250.

Annie also had a partial working scholarship. She worked in the alumni office, where she did all the bookkeeping and typing for Dwight Balentine, the alumni director.

Annie was majoring in music. Her talents embraced not only the trombone, which she continued to play in the college band and college symphony, and the piano, but now the pipe organ as well. She grew to love its versatility and range of expression, and it became her major instrument.

She also carried double minors. Her family, like mine, had a practical side. That was one permanent mark the Depression had left on those who went through it. Parents urged their children to take up solid, respectable jobs that would give them a lifetime's employment. Dad was only half joking when he said that if he'd only stayed with the railroad, he could have retired in his forties instead of having to go out on plumbing jobs. Music wasn't exactly a frivolous career, but most college music majors could expect to teach music when they graduated. Annie couldn't teach because of her stutter, and pursuing a performing career was too risky in those insecure times. She had worked in her father's dental office in the summers, developing X rays and keeping his records. Doc Castor had encouraged this, and urged her to take up a secretarial curriculum in addition to music. She also was on the college swim team, with times in the freestyle and breast stroke that made her entertain Olympic hopes. Between swimming and a spot on the volleyball team, she expected to have enough credits for a second minor, in physical education.

I made the football varsity as a sophomore. The varsity coach was Stu Holcombe, who went on to become athletic director at Purdue. I wasn't good enough to start, but Coach Holcombe brought me in to placekick, and I did later manage a crucial extra point in a close game. My basketball career ended, however. I simply wasn't good enough to make the team, and I didn't want to do something I saw no chance at getting really good at. My involvement with sports outside of football didn't end, though. My work scholarship placed me in the locker room cage passing out towels and taking care of the equipment.

I was elected to the student council as a sophomore and was doing pretty well academically, but I was beginning to have second thoughts about my major. Organic chemistry required a level of mathematics that I had not yet taken. I had either jumped in over my head, or the college had let me take a course it shouldn't have. I was passing, but I was struggling, and I was going to have to refocus my major very soon.

The news that reached Muskingum from Europe now centered on the battle in the skies over England. British pilots in agile fighter planes called Spitfires had repelled overwhelming numbers of German planes in the late summer and early fall of 1940. Winston Churchill's accolade

"Never in the field of human conflict was so much owed by so many to so few" gave a ring of heroism to the Royal Air Force pilots we read and heard about in news reports.

Then, when classes reconvened at the start of 1941 after the Christmas and New Year's holidays, I saw a notice on the physics department bulletin board that seemed as if it had been sent from heaven.

The notice described something called the Civilian Pilot Training Program. I stood there and read it. I had to read it a second time, because I couldn't believe my eyes. The U.S. Department of Commerce, in the interest of training pilots, would pay for the cost of ground and air instruction for qualifying students. Successful trainees would get their private pilot's license. They would also get college credit in physics, since aerodynamics, combustion, and heat transfer would be among the subjects covered. This was too good to be true. Applicants were to sign up with the head of the physics department, Dr. Paul Martin. Doc Martin and my dad were good friends. I walked into the physics office and said I wanted to apply.

I was bursting with the news when I got home that evening. It was still our practice to discuss around the dinner table what we had done that day. When it was my turn, I told Mother and Dad and Jean about the opportunity to learn to fly.

It was clear as I talked that Mother and Dad didn't share my enthusiasm. Jean, who was struggling in high school, didn't offer an opinion. My parents exchanged looks and listened quietly. Then Dad dumped cold water on my plans.

"I'm not in favor of it, Bud," he said.

"Why not?" I asked. I was surprised and disappointed.

"It's too dangerous," he said.

"They're always crashing, aren't they?" Mother chimed in.

"It's not dangerous if you know what you're doing," I said. I didn't know anything about flying at that point, but it seemed to me that it had to be like driving or operating machinery—you could do it right, or not. The danger came when you cut corners, let your attention stray, or didn't prepare fully. "If I do it," I said, "I'll do it right."

They kept up their objections anyway. Dad said later he and Mother were sick at the thought of my flying, that it would be like taking me out and burying me. I knew he was thinking of a newsreel that had run over and over again in the early 1930s, in which the dangers of flying were vividly displayed. A pilot was making a low-altitude speed run in a Gee Bee Special when the plane lost a wing and hit the ground, disintegrating in a huge fireball that left nothing but wisps of smoke. Finally Doc Martin came to the house and had a long talk with them. He explained that aviation would be a major industry in a few years, and that the course would be a good way for me to get in on the ground floor. That was the kind of talk they liked to hear, the practical side, and they relented. Nobody talked about the escalating war in Europe.

I passed the physical examination, and typed answers to the questions on the application form. One question asked my preference for a flying career. I put the airlines at the top, with the Navy second and the Army third. Signing the application conveyed my agreement to apply for military flight training "when needed."

My acceptance came in a matter of weeks. There were four other Muskingum men in the program, and we started our classroom instruction right away. As I worked on the particulars of aerodynamics, the atmosphere, airplane controls, and the forces that affect planes in the air, I looked forward to our first trip to the airfield at New Philadelphia, sixty miles away, where we would actually take to the sky.

We drove north from New Concord along two-lane brick roads, the Ohio countryside beginning to turn green around us. It was April 1941. I was at the wheel of a wood-sided Ford station wagon the college owned; Doc Martin had turned it over to me for the trips the five of us in the Civilian Pilot Training Program would be making to New Philadelphia. My end of the bargain was to keep the wagon serviced. In the car with me were my friend and neighbor Dane Handschy, Willis Gaston, and a couple of kids I didn't know as well, whose names were Johnson and Wilson. We were talking nervously about what we could expect from the flight instructors we were about to meet.

The New Philadelphia Airport was a grass strip south of the town. Tuscarawas County Aviation, the outfit that had contracted with the government to train civilian pilots in that neck of Ohio, occupied a light-colored brick hangar. The building was well kept and neat, with a pilot's ready room and an office in one corner. The manager, Harry Clever, emerged from the office and introduced himself. He was in his forties, brisk and businesslike.

Clever in turn introduced us to the man who was going to teach us to fly.

If any of us had expected a grizzled ex-barnstormer in a leather flying jacket, we were in for a surprise. The flying instructor was tall and slender, with wavy brown hair and a farm boy's face that was no more weather-beaten than ours were. He wore regular street clothes—khaki pants and a shirt with the sleeves rolled up on his tanned arms. His name was Wallace Spotts. He would tell us over the next few weeks that he was twenty-three and had dropped out of college to dedicate himself to flying. We were his first class of students under the Civilian Pilot Training Program, and he was as nervous as we were.

We were all eager to see the planes we were going to fly. Spotts wasted no time leading us through a once-over of a Taylorcraft, a small, light plane with a high wing and fixed landing gear, with the third wheel at the tail. Its four-cylinder Lycoming engine put out sixty-five horsepower. Its cockpit was enclosed, and it had two seats side by side behind the windshield. It would fly at eighty or ninety miles an hour.

The little Taylorcraft proved to be a great learning plane. There was nothing complicated about it. Soon after my first flight in the pilot's seat next to Wally Spotts on April 14, I felt as if I knew its rudimentary controls.

At first we only taxied and controlled the plane in level flight and through medium turns. Before long we were taking off, climbing, gliding, and landing. These maneuvers were followed by S turns and figure eights. The course was intensive. We were on the road to New Philadelphia two and three afternoons a week, and sometimes on weekends.

The faster Tuscarawas County Aviation put us through the stages of the course, the faster they could bill the government.

I loved it. I applied myself and paid attention. The plane soon felt natural in my hands, and the increasingly complicated maneuvers Spotts put us through also seemed to come naturally to me. The total attention that flying demanded was exhilarating. I felt when I was in the air that all my senses were operating at their fullest. I enjoyed the challenge of learning just what I could do, and the more I did, the more I realized I could do. I thought that flying was a skill at which I might, in time, learn to excel.

You had to be aware of everything. The instructors brought this home to us time and time again. You'd be flying all over the countryside when Spotts would say, "Okay, let's go to the airport," and you had to be oriented over the terrain in order to know which way to turn. Sometimes—he never gave a warning—he would reach over, pull the throttle back, and say, "Okay, the engine's failed. Where are you going to go?" We had learned always to keep looking for a field within gliding distance that would make a likely runway, and the job was to set up a landing pattern as if we were going into our home airport, and do everything but land. We flew figure eights in crosswinds around pylons half a mile apart, adjusting constantly to keep them at an equal distance from the plane. Sometimes I'd find myself sitting beside Harry Clever or John Anderson, the instructor from the noncollege training group. These check rides allowed the instructors to look for each other's errors; they were tough, and it was a matter of great pride when you passed the scrutiny of an instructor other than your own.

We finished dual instruction at the beginning of May. My first solo came unexpectedly, and I'll never forget it. Coming in with Wally, I taxied onto the line and was starting to shut the engine down when he said, "You just stay in and I'll get out. Take it around the field." I took off, all alone, and made a good landing with the instructors and other students watching from in front of the hangar. It was easier than it would have been if I'd had advance notice.

After soloing for just two weeks, we moved on to precision maneuvers. This was more intense: three weeks of touchdowns, forced landings, stalls, spins, 720-degree power turns, spiral approaches, tight figure eights, and power approaches and landings. At the end of the first week of June, we made our cross-country solo flights. I flew from New Philadelphia to Cambridge to Zanesville, a triangle too tight for a modern jet to make at normal speed. We crammed on the *Civil Pilot Training Manual* for our examination. I made a 96, getting seventy-two out of seventy-five questions. Wally Spotts signed his recommendation for my private pilot's license on June 26, calling me "apt and conscientious." In July, at not quite twenty-one, I had my license.

My career as a pilot started with a near disaster. I wanted to convince Dad that flying was safer than he thought, and at the same time I wanted to thank him somehow for that first flight in the WACO all those years earlier. I had saved enough to buy an hour or two of flight time. One day I told Dad I wanted to take him up.

We drove to the airfield in Zanesville on a day toward the end of the summer, and I rented a small plane for an hour. Dad climbed in while I did the preflight run-through, checking fuel and oil levels and the condition of the plane. When he got in, the plane sagged. Dad was heavy then. He probably weighed 230 pounds, not loose fat but a thickness of muscle from hard work and a healthy appetite for Mother's cooking. I weighed 175, and his arms were almost as big as my legs. I got in, taxied to a corner of the field, and turned for takeoff.

It was a hot day, indicating less dense air, which provides less engine power and less wing lift. But things seemed all right as I started down the field. We went through a dip in the grass runway, and a row of trees at the far end started getting closer. I had the throttle wide open. The engine strained, but we weren't developing the expected speed for takeoff. The trees kept getting closer. I was afraid to look at Dad, and didn't have time to do it, anyway. I saw those trees coming toward me, and I began to have some doubt about whether we were going to get off the ground. We just barely had takeoff speed when we came to the far end of the field with the trees still ahead of us. I pulled back on the stick and the plane struggled up over the trees and dipped down again, following the land behind them that sloped down to the Muskingum River. With the engine running wide open we gradually got enough speed to finally start climbing.

When I had the nerve to look at Dad, he was unclenching his grip on the edge of the seat. But his eyes were twinkling behind his horn-rimmed glasses and he said, "Like those exciting takeoffs, do you?"

•   •   •

The war news from Europe dampened, to some extent, the enthusiasm with which I began my junior year at Muskingum. The Germans were in France, northern Africa, and the eastern Mediterranean and were driving deep into the Soviet Union after invading in June. Hitler's Nazi war machine was insatiable. England remained threatened, and it seemed inevitable that the United States would enter the war. Nevertheless, I put my limited football skills on the line again for old Muskingum. In the swirl of football, dating Annie, trying to conquer the mathematics required if I was going to continue to study chemistry, and working to save money to buy flying time, the worries of the war receded. We rarely talked about it on campus.

Annie was a senior. She had become so accomplished on the organ that the Juilliard School in New York, the nation's premier music school, had offered her a scholarship for graduate study. This was great news. Annie could make an organ talk. But it was hard to imagine her in New York, which I had seen as a fast-paced city without a lot of patience. Annie's stutter had forced her to adopt strategies for certain situations. Shopping in an unfamiliar store was torture. She had to write a description of what she wanted, or take a sample to show a clerk, because she couldn't ask for it. Without those aids, she had to hunt all over the store, and sometimes she simply went home in frustration. People often laughed at her as she tried to get words out. This happened in Cambridge and Zanesville, and I could only imagine what would happen in New York. Also, I didn't know what the effect would be on our relationship. We planned to get married as soon as we both were out of college.

Senior music majors all had to perform in recital.

Annie's was scheduled for the first Sunday in December. I borrowed Dad's car to drive up to the campus. Unlike the cruiser, it had a radio, and I turned it on for the brief spin to Brown Chapel, where she was to play.

Turning off Main Street onto College Drive, which led into the campus, a bulletin interrupted the music program I was listening to. The astonishing news that the Japanese had bombed Pearl Harbor crackled over the car radio.

I kept the news to myself while Annie was playing. I tried to keep my mind on her recital. She was performing music from *Finlandia,* by Jean Sibelius. I loved the piece. The music rises in tumult, and then recedes to the quiet of a pastoral section that is the music for the stately hymn "Be Still My Soul." As a message, it could not have been more appropriate. In the beauty of her playing, I was agitated as I thought ahead to what I had to do.

I waited for her afterward. She came down from the stage beaming at what she knew had been a great performance. But my own smile was troubled, and she saw it. "What's wrong?" she asked, and I told her.

We talked deep into that night as we listened to the continuing radio reports of damage, sailors killed, ships and planes destroyed. It was almost impossible to believe that Japan had had the audacity to attack American territory and American forces. Not only had they done it, they had inflicted enormous damage in a sneak attack. No longer was war a possibility in Europe; it was a reality in the Pacific.

"I have to go," I said.

She held my hand and nodded with tears in her eyes.

Over the next few days, all conversations revolved around the subject of war. Mother and Dad, predictably, wanted me to finish college and enter the military once I graduated. But the influence of New Concord's parades, of our pride in the flag, and of my playing echo taps with Dad was strong. I knew my responsibility, and I thought it was important to get going.

Dane Handschy and I went over to Zanesville and signed up for the Army Air Corps. We raised our hands and took the oath and then went back home to wait for orders. While we waited, I learned more about Navy flight training, and it sounded more appealing. But I was committed.

Each day that passed I expected a letter from the Army containing my orders. No orders arrived for me or Dane. The winter of 1942 eased into spring. Dane and I both had left Muskingum at the end of the fall semester, assuming that with our pilot's licenses we'd be sent for training right away. With nothing to do and no orders, we hooked up with a man who was doing custom plowing on the farms around New Concord. He wanted us to drive tractors for him. So that spring, instead of learning to fly Army Air Corps fighters, we plowed fields. We plowed hundreds of acres. From my seat on the tractor I took in the Ohio countryside in which I'd grown up and which had shaped me. The land, its contours ordered by the plow, separated by tree lines and fences, rich and bountiful, was unimaginable in the hands of some invading force bent on destroying America.

Still the orders didn't come. By March, tanned brown and with thousands of furrowed rows behind me, I returned to Zanesville and went to the Navy recruiting station at the post office. I told the recruiter I had sixty hours of flight time, and I was ready to fly with the Navy.

I was sworn in again, and orders came within two weeks instructing me to report to the U.S. Navy's pre-flight school at the University of Iowa in Iowa City. I wasn't sure that I might not be AWOL from the Army, but I made plans to go.

Annie and I wanted to get married before I left, but you couldn't be in Navy flight training and be married. Getting engaged was the next best thing. I had made good money plowing, and I had almost $130 saved. I withdrew it from the bank and went to a jewelry store in Zanesville to shop for a ring. I had an idea of the fit, because Annie wore a friendship ring that I had managed under some pretext to slip on my little finger. I bought a nice small diamond for $125, and I took it home and put it on her finger and asked her to marry me. I already knew the answer, and in all the years since, she's never let me replace that diamond with a bigger one.

We said goodbye at the train station in New Concord. My family was there, and Annie, and her family. My mother had been through this before, seeing Dad leave for the "war to end all wars." But another war had arisen, and a new generation was now called upon to fight.

Leaving was hard. Annie and I had seen each other almost every day for nearly twenty years. For the last six or seven of those years we had been learning that love is a combination of passion, friendship, and respect. I hated to leave her, not really knowing what lay ahead except the opportunity and obligation to serve my country in a time of need. We kissed for a long time before the conductor called, "All aboard." I was still touching her hand when the train started moving.

Then I was leaving New Concord, and my boyhood.

PART TWO
WAR

CHAPTER 5

I rode the train west without a clear idea of what was before me. I knew combat pilots trained for eighteen months or more. The war would dictate what happened beyond that. I was just proud that I was going off to do my duty. The train reached Chicago early in the morning after an all-night ride, and before an hour passed I was on another train that left Chicago behind and rolled through endless flat farm fields. The sun was getting low when it pulled into Iowa City.

I stepped off the train into a military organization on war footing. Guides waited on the platform to assemble cadets who were arriving for the Naval Aviation Pre-Flight School. There had been several of us on the train. We boarded a bus and rode to the campus. It was a brand-new training site, and I was the twelfth person to check in. Before long I was wearing a Naval Aviation cadet's uniform as part of First Battalion, Company A, Platoon One. Our rooms were over the main entrance to the university quadrangle.

The three months that followed were a combination of Navy boot camp and officer candidate school with a substantial academic load. The physical training was tough. We marched in platoon-sized groups of thirty-six, each of us taking turns as drill instructor. We made ten-mile fast marches through the Iowa countryside in the heat of the plains summer. Some of the cadets weren't in shape for this, and an ambulance followed to pick up the men who fell out.

The war effort had drawn people from all walks of life. One of them was Bernie Bierman, a retired Marine lieutenant colonel whose University of Minnesota Gopher football teams were perennial Big Ten and national champions. The Navy had called him back into service to head a comprehensive program of athletic training at its preflight school. His fame as a football coach allowed him to recruit all-American athletes from around the country.

They put us through the paces, a week at a time, on about every sport you have ever heard of: football, basketball, trampoline, boxing, wrestling, swimming, gymnastics, running, and something called pushball.

Pushball used a large rubber ball, seven or eight feet across, placed on the fifty-yard line, with a platoon of cadets at each end of the field. The objective was to push the ball over the opponent's goal line by any means or strategy. Bodies flew. It was mayhem.

But football was where Bierman concentrated his training talents. He assembled a team he called the Iowa Seahawks. My platoon football squad met them in a scrimmage. I played guard across from an all-American center, and I never got so beat up in my life. Bierman's Gophers had won the national and Big Ten championships the year before with an undefeated season, but the Seahawks were to beat them in a game that fall and depose his old team as national champions.

It wasn't clear what all this had to do with flying until you thought about the competition. Combat was the ultimate competition, and the Navy aimed to produce warworthy pilots. At some point we would have to fly into enemy fire or face other tests of endurance, confidence, and skill. We would have to trust ourselves and the men around us as part of a team. Most of all, we would have to compete, in some cases for our lives. The athletic training was designed to get us in shape and hone the competitive edge that we would need later on, and to weed out the men who didn't like competing.

We all paid close attention to war news, especially developments in the Pacific. The Navy was fighting its air war from aircraft carriers and island outposts against Japanese ships that were trying to finish what they had started at Pearl Harbor. When we heard that Navy pilots had destroyed four Japanese carriers that had launched a strike against the Midway Islands, we felt enormous pride. We were angry, too, because many Navy planes and pilots had been lost in the early stages of the battle.

Marching and athletic training were half of our routine. We spent the rest in classrooms studying navigation, aerodynamics, engines, recognition of enemy aircraft, and other subjects that we'd need for military flying. We underwent indoctrination into Navy regulations, customs, traditions, and the Articles of War.

The services were pushing men, pilots included, through training as fast as they could. First Battalion, Company A, Platoon One finished preflight school in

August and received orders to Olathe, Kansas, for primary flight training.

I took the few days of leave I had and rushed to Ohio on a train to visit Annie. She had graduated from Muskingum in May. I hadn't gotten to see her graduate because I had already left for Iowa City. Now she was living in Dayton. She had turned down the scholarship at Juilliard. She hadn't been afraid of living in New York, despite my fears, and those of her parents, of what life would be like for her there with the trouble she had speaking. She had simply made up her mind to work in the war effort and had gotten a job as the personal secretary to a colonel in the Army Air Corps. She took dictation and typed letters, which the colonel would read and sign without making her read them back beforehand. She had roomed briefly with another woman from New Concord before moving into a room of her own. I thought this was another example of her enormous courage. Annie's parents had not even wanted her to move to Dayton, but she insisted not only on moving but on living independently once she got there. She wouldn't let her parents or anyone else treat her as if she were handicapped.

It was not enough to say I missed her. The three months I had been gone were the longest stretch I had ever been without her. I wrote her every day, but I was just beginning to realize what a large part of me she had become.

In Dayton I stayed at the YMCA, and we went to movies and walked along the levee beside the Miami River, and as quickly as I had arrived I had to leave for Kansas.

•   •   •

The training base at Olathe was still a work in progress when I reported for primary flight training. The Navy put two hundred of us into a Naval Reserve base along the Missouri River. When we moved twenty miles south to Olathe a week later, the paint was still drying in the barracks. We walked around on duckboards while the construction crews finished pouring concrete for the sidewalks.

The planes were there and ready. They were Stearmans, open-cockpit biplanes that were bigger and more powerful than the Taylorcraft in which I'd gone through the Civilian Pilot Training Program. They had fore and aft cockpits, like the old WACO I'd gone up in with Dad. I had a jump on most of the other cadets, since I already had my pilot's license. But this was a whole different kind of flying. The Stearman had 220 horsepower compared to the Taylorcraft's 65, and would fly at 135 miles per hour compared to barely 90. This meant aerial acrobatics—loops, rolls, and other moves that mimicked combat flying—could be introduced later on as part of our training.

One of my instructors liked to roll the plane over upside-down. We would both be hanging by our seat belts, and then the engine would cut out because the fuel wouldn't feed in that position. We were flying in a glider, upside-down. When we rolled the plane back over, the engine would cut in again.

When we started, instructors took the forward seat. They moved to the one behind just before we soloed. Once we soloed, we started to learn flying in formation and flying at night. The night flying was aimed at getting us accustomed to bringing the plane down in darkness by visual cues. The Stearmans had no landing lights, and we landed by runway lights or flare pots set up along the edges of the runway. All of this was new, but I continued to be comfortable each step of the way. We flew five and six days a week, unless there was bad weather, and I looked forward to every flight. Not everybody felt that way. Some of the guys were lukewarm on the acrobatics, but I loved it. I began to develop the sense that Wally Spotts back in the Civilian Pilot Training Program had called the kinesthetics of flying—a feeling that the plane was an extension of myself. My hands moved on the controls without my having to think about it, and the plane responded to my thought.

The planes at Olathe were scattered between the main base, where our barracks and the classrooms were, and an outlying field about six miles away, dubbed the OLF in the military habit of reducing things to acronyms. A trip to the OLF required a bus ride, which always seemed slow and aggravating because I was eager to get up in the air.

The barracks we slept in were long rooms with two-level bunk beds lined up along opposite walls, and back-to-back rows of metal lockers down the middle. There were maybe forty of us to a room. We kept our personal gear in footlockers at the ends of our bunks. Olathe was so new that it didn't have a system for playing recordings of the military calls by which our lives were regulated: morning reveille, nighttime taps, and all the rest. When word got out that there was a cadet who knew the calls, the cadet training officer handed me a trumpet and made playing the calls one of my duties.

Our classroom instruction proceeded at the same hectic pace as the flying. The academics were more of what we had been doing in preflight school—studying bigger engines and aircraft and aerodynamics. We also, for the first time, studied ordnance. Learning about .30-caliber and .50-caliber machine guns brought home the fact that we were going to fly and fight at the same time. This heightened the challenge and my desire to pick up all the information I could.

There was little time to relax in this welter of activity. Our days were structured and full. War news reached us in regular briefings. These were the days when the Marines had gained a foothold on Guadalcanal and other islands in the Solomons despite fanatical Japanese resistance, and our Soviet allies were holding off the Germans at Stalingrad. As much as each day held, I always managed to find a few minutes to write Annie, to report on my activities, and to tell her that I loved her.

Toward the end of my three months at Olathe, I received orders to report to the Naval Air Training Center at Corpus Christi, Texas, for basic and advanced flight training. I wouldn't have enough leave time to go home, so Annie took a leave from her job and got on a train to Kansas City, where I met her for a weekend.

Two weeks later, at the beginning of November, I was on a train to Corpus Christi.

The Naval Air Training Center was actually a complex of airfields. No one field could have accommodated the twenty-five thousand trainees and support staff—the flight and ground instructors, administrators, mechanics, and all the rest it took to get such a crowd of pilots combat-ready. There were four auxiliary fields—Cabaniss,

Cuddihy, Rodd, and Waldron—arrayed around Naval Air Station Corpus Christi, which we called Mainside. Each had a set of barracks and hangars painted the standard Navy battleship gray. I was assigned to Cabaniss.

The flying moved to an entirely new level. At first we flew OS2U Kingfishers, a long-range single-engine plane that could be equipped with floats or wheels as landing gear. Then we got a taste of the SNV, the Vultee Valiant. This plane was nicknamed the "Vibrator" because the sound of its radial engine rattled everything within earshot, including the pilot. It was a low-wing two-seater with a greenhouse canopy over a fore-and-aft cockpit, and with a 450-horsepower engine it could fly at up to 180 miles an hour. This was another exciting leap in the capabilities of the aircraft we were learning to fly.

We were beginning to learn instrument flying now. One of the exercises in the Vultee was particularly unnerving. Cadets rode in the backseat with the instructor in the front, and flew the plane entirely by instruments. We operated under a hood that cut off all outside vision. There was no way to cheat. We were flying blind except for what the instruments told us.

As in Kansas, we split our days between flying and classroom instruction designed to increase our familiarity with the machines we were flying: tactics, weapons use, and navigation. I spent hours poring over navigation problems. There were sample problems you could work, and I tried every one of them. I wanted to know everything there was to know.

When weekend liberty came around, most of the men headed into town. I hardly ever did. You had everything you needed on the base. I went to a movie once in a while, there was a PX, and if you didn't want to eat in the mess hall, you could get something in one of the snack bars. I think I went into Corpus Christi once in the almost six months I was there.

I had a girl at home, of course. A lot of the guys didn't, and that was the attraction of civilian Corpus Christi for them. I continued writing Annie every day. We expected to marry as soon as I finished training and had my wings.

One of the other fellows who stayed around the base was Tom Miller. He was from just up the road in George West, Texas, a town that got its name from a cattleman who had a big ranch there. I had met Tom through one of my platoon mates at Iowa City, Henry Knauth, who was Tom's roommate at Corpus, and we got to be friends. Tom was a lanky six-two with a South Texas drawl and I spoke uninflected Ohioan, but there were more similarities than differences between us. Both of us were from small towns, we both had left college to enlist after Pearl Harbor, and both of us were our parents' only son. (Jean's adoption meant I was not an only child, as Tom was, but Jean had continued to struggle in school and, as she grew older, with other aspects of her life. My parents spent a lot of time and effort dealing with Jean's problems, and I was just sorry I couldn't help them more.) In addition, Tom and I were both gung-ho pilots who worked hard to improve our flying and wanted to spend as much time in the air as possible.

Between flying and increasingly demanding classroom work covering areas of physics, mathematics, and navigation by the stars, we had few breaks. When we could, we talked about flying. We were proud that we had chosen Naval aviation, because we thought it demanded higher standards of its pilots, since flying from aircraft carriers required more skill and training. There was the celestial navigation component, and we believed the math requirements were higher. We also wanted to fly from aircraft carriers, because that, too, was more demanding.

But the Marine Corps appealed to me as well. Their performance in combat was certainly impressive, but more than that was an attraction that I couldn't quite put my finger on. The Marines were different somehow. They carried themselves differently. That winter of 1942 leading into 1943, the Marines had been in the forefront of the continued fighting on Guadalcanal, steadily rooting out the Japanese, who almost always fought to the death. Some of the Marines were coming back from the fighting there, telling stories of combat and bravery that made me want to join them. One day Tom and I saw a notice that a Marine captain just back from Guadalcanal was going to talk about the fighting there, and we went to hear him.

The captain was an impressive-looking man. His hair was gray although he couldn't have been over thirty, and he looked sharp in his green winter uniform. He gave a stirring talk. He told how the Marines had fought their way to a toehold on the island and taken the airfield, and how Marine pilots had played a key role in stopping the Japanese advance in that part of the Pacific, and helped save the day by defending our outnumbered ground troops against Japanese efforts to retake the all-important airfield. The thought that pilots could play a role in the inch-by-inch fighting across Guadalcanal and the rest of the Pacific islands, that we could help the men who were fighting there advance and stay alive, excited me. It offered a new dimension to the duties of a pilot.

I felt a strong pull to request that my officer commission be granted in the Marine Corps, which Navy pilots were eligible to do at that point in our training, but I wasn't absolutely sure. Another inducement took me to the brink.

Corpus Christi, like most military bases, was alive with rumors of all kinds. Tom and I heard the Marines were going to be flying the hottest new fighter around, the Lockheed P-38 Lightning. The P-38 had twin in-line 1,425-horsepower engines, with booms that reached back from the engines to the tail, replacing the standard single fuselage. It cruised at 290 miles per hour, could reach a top speed of 414 miles per hour, had a ceiling of 44,000 feet, and could achieve a range of 2,100 miles. It had a 20-millimeter cannon and four .50-caliber machine guns, and could carry 3,200 pounds of bombs. The P-38, we decided, was the bird to fly.

Finally cadet class rankings were posted. Tom and I read our names among the top 10 percent of our class in both ground and air instruction. This pushed us over the edge. We were ready to join the elite. We applied for the Marines, and were accepted.

Right away we wanted to get the multi-engine experience we'd need if we were going to fly P-38s. The only twin around was the Consolidated PBY, the old Catalina flying boat.

There was a risk in asking for the PBYs. If we got into multi-engine planes and didn't get assigned to P-38s, or if the rumor we had heard was wrong, then we might end up in bombers or transports. I wanted to fly fighters, not the lumbering larger planes, and Tom felt the same way. Still, we thought the P-38 was worth the risk. And there was life after the war to think about. We both had left college without our degrees. We thought that if we survived the war, we could make a good living flying for the airlines even if we never finished college, and it would take multi-engine experience to do that. So we applied for advanced training in the PBYs, and once again were accepted. We hoped we could wheedle our way back into single-engine fighters if we had to.

The PBY was a big change from the planes we had flown so far in our training. It had, at 104 feet, two and a half times the wingspan of the Vultee, and at 64 feet was more than twice as long. It was heavy, and even with two 1,200-horsepower engines, getting it to lift off from Corpus Christi Bay made you feel like one of those geese that flaps its wings forever before its feet clear the water. Once in the air, it was slow, with a cruising speed of about 120 miles per hour and a maximum of about 180. It took forever to land, too, especially at night, when there were absolutely no visual cues to follow. You adopted a slow rate of descent, only 125 to 150 feet per minute, and put it in a landing attitude until you heard the water hit the hull, and only then could you pull the throttles back.

Bouncing in and out of Corpus Christi Bay was only part of the excitement. The PBY had .30- and .50-caliber machine gun turrets and a bombing station. Cadets rotated position among piloting, gunnery, and bombing on every flight. The Norden bombsight had just been developed and was still top secret. We practiced with it on a scaffold in the hangar, and used it to drop miniature practice bombs along the Gulf of Mexico. I learned to be a bombardier relaying instructions to the pilot to adjust for wind drift while I sighted on the target. We had to return the bombsight to a locked vault every night.

German U-boats had preyed on Allied shipping along the Atlantic coast a year earlier, aiming particularly at tankers carrying oil to refineries in the Northeast. The subs operated so close to the coast that the papers carried stories about Florida tourists seeing ships explode and burn. A system of convoys had halted the U-boats' East Coast attacks. Now, intermittently, one of the German subs would slip into the Gulf of Mexico, where a couple of tankers had been hit by torpedoes. Shrimpers and commercial fishermen sometimes reported U-boat sightings in the Gulf. Once in a while, the base would roust out a crew in the middle of the night that included us cadets, and put us up in the PBYs for submarine patrol. We would man the turrets and the bomb station while the instructors flew the planes. We would chug around the Gulf of Mexico at a thousand feet, looking for a German sub to bomb or depth-charge. We never did find one. But we thought it was special, because nobody else in Texas was being called into combat.

I completed my advanced training at the end of March 1943. There was only one thing I wanted to do with the fifteen-day leave I had coming. Now that I had my wings and was no longer a cadet but a Marine second lieutenant, I could get married. Annie and I had already set the date. I headed north on a train from Corpus Christi that was as slow as any train I've ever been on.

The train north was full of soldiers and sailors. There was a whole battalion going somewhere. Their laughter and the click of dice and the snap of cards being shuffled and dealt sounded deep into the night. Even with the windows open, all the cars were thick with cigarette smoke and the smell of booze. There was a lot of talk about the "Japs" and "Krauts" America was fighting, a lot of talk of girls at home. I smiled inside to hear this, because I was going home to marry my girl. Just thinking about it made me feel good, but I didn't want to brag, so I kept it to myself.

The ride home took two or three days. When I got to New Concord the preparations for the wedding were well under way. I didn't have to do anything but show up.

We were married on April 6 at three-thirty in the afternoon. It was a Tuesday, but it felt more like Easter Sunday when people started arriving at the College Drive United Presbyterian Church. The day was sunny and beautiful, and everybody in New Concord must have been there. I wore my Marine dress blues, and Annie a white street-length dress and a hat with a feather. The church was decorated with ferns and a basket of white gladioli. Annie said we didn't need to have a formal wedding with the war going on.

Lloyd White was my best man. Lloyd was still in college at Muskingum, and was planning to go into the ministry. My parents and Annie's mother sat in the front row on either side of the aisle. Annie's sister, Jane, was her attendant. I stood at the front of the church and watched Annie walk down the aisle on her father's arm. She was carrying a Bible that her mother's mother had given her when she was a little girl, and a rose orchid with it. She wore such a smile, and I guess I did, too. She was beautiful. Doc Castor handed her over to me and said under his breath, "You take good care of her now, Johnny." Dr.

Henry Evans, the pastor, read the vows, and when we said our "I dos" and exchanged rings and he pronounced us man and wife, I felt like the luckiest man alive.

Our honeymoon wasn't exotic. We drove to Columbus and spent three nights at the Deshler-Wallick Hotel at Broad and High Streets downtown. It was romantic, though; Annie wanted to eat by candlelight, and that's how we dined each night in the hotel dining room. After that we went back to New Concord and stayed at Annie's parents' house as newlyweds. It seemed a little strange to be sleeping with Annie in her room under her parents' roof. I couldn't help but feel a little self-conscious at the breakfast table in the morning. We weren't there long, in any case. The war was waiting.

I had orders to report to the Marine Corps Air Station at Cherry Point, North Carolina. Dad's wedding present to us was a black 1934 Chevy coupe he'd chosen from his lot. It had a luggage platform in place of a backseat. Annie and I packed all the stuff of our new household into it, and the gear barely came up to the windows. The car was in pretty good shape except for the tires. All the new tires were going onto military vehicles. On a cross-country trip like that, I always carried a Shaler Hot Patch Kit and a couple of flat steel tire irons so I could take the inner tube out and do my own patch job if we couldn't reach a town.

We headed south, and it seemed as if we had a flat about every ten miles. Gas was rationed, and to conserve it we joined the other speedsters on the road, clipping along at forty miles per hour.

At Cherry Point my friend Tom Miller and I were

reunited. We cast our eyes around the runways thinking we were going to see our P-38s out there on the line. When we asked, we got blank looks. The Marine Corps had never heard of P-38s. One rumor replaced another; we had also heard the Marines might be flying A-20 Havoc bombers. The films we had seen in Corpus Christi of the A-20s flying over Germany blowing up trains and ammunition dumps made it look like great flying. There were no A-20s, either. We were assigned to Operational Training Squadron Eight for training in B-25s.

The North American B-25 Mitchell was a twin-engine, medium-range bomber with a crew of five to seven. The Navy and Marine versions were called PBJs. Ready for combat, they bristled with machine gun turrets in the nose and tail, and on top and on each side of the fuselage, and carried three thousand pounds of bombs. It was a good plane made famous by the Doolittle Tokyo raid in April 1942, but it wasn't what Tom and I wanted.

Annie and I set up housekeeping in a little apartment a fellow named Leroy Guthrie and his wife had converted from a sun porch on the side of their house in Morehead City, about twenty miles from Cherry Point. People were converting everything from attics to potting sheds into military housing. We were their first tenants. The place was just across a causeway from Atlantic Beach, and when we first got there we spent a day on the sand dunes. It was a dull day, or so we thought. The sun behind the clouds was powerful, and we both got the worst sunburns we had ever had. That was frustrating for a pair of newlyweds. We could hardly touch each other for about a week. We'd look at each other, both bright red, and burst out laughing at the expression on the other's face.

Cherry Point was awash in second lieutenants. The United States was throwing new forces into the war at a high rate. Navy and Marine losses in the Pacific had been high. Adding new squadrons and replacing casualties in existing ones caused some training bottlenecks. The squadron commander was Colonel Karl S. Day, who had a reputation in the flying world from being the chief pilot for one of the airlines before the war. Under him was a Captain Danco. "Daddy" Danco was reputed to have been a member of the French Foreign Legion, but now he was in charge of keeping 120 fresh new pilots busy without enough planes to give them flying time. He operated what we called "Daddy Danco U."

Training filled every day. We spent mornings in classes on weapons, engines, navigation, combat tactics, and Marine history, worked in special trainers designed for aircraft recognition, and sat in blacked-out rooms for night-vision training. In the afternoons he sent us running in the steamy North Carolina lowlands. But the one thing we didn't train in was flying. Because of a shortage of aircraft, we had trouble even getting the four hours a month required to earn flight pay. Only three weeks after Annie and I arrived in Cherry Point, the Marines decided the solution was to send some of us to California.

We headed west in a two-car caravan, the '34 Chevy and a spiffy '39 with practically new tires we bought for the long trip. It cost $450 that we didn't have, but the First National Bank in New Concord wired the money as a loan, and we hit the road deep in debt.

We spent one night in New Concord, left the '34 for Dad to put back on his lot, and proceeded west. We had a certain amount of travel time, plus what the Marines called proceed time, to make the trip. We planned our route stop by stop, which was a necessity in those days. You could go a long distance between gas stations in the more remote sections of the West, and finding a place to stay wasn't guaranteed. One place we stopped, we had to set the legs of the bed in cans of water to keep a colony of ants from crawling in with us.

Annie and I made a grand adventure of that trip, heading each day into the setting sun, seeing parts of the country neither of us had ever seen before. Often, as we drove, the sheer splendor of America was overwhelming, and it was impossible to believe the country was at war. The West presented vistas more magnificent than anything we had imagined. The horizon stretched forever over the Great Plains. Then the mountains loomed up in New Mexico larger and more raw than the eastern mountains, even the rocky peaks I had climbed in New Hampshire. We spent half a day at the Grand Canyon, and stopped in Las Vegas, where Annie hit a jackpot with one quarter in a slot machine and retired with her winnings. We rolled into San Diego for our first view of the Pacific Ocean without a single flat. Twelve hours later I reported for duty at the Naval Auxiliary Air Station at Kearney Mesa, just above La Jolla.

San Diego was the usual scramble for a place to live, but Annie was resourceful. Carrying a notepad and a pencil to communicate with, she went to the gas company and the electric company in San Diego to find out where service had been canceled. Then she rushed out to see what kind of places they were and if they were still available.

She soon found us a one-room efficiency apartment on a little sand alleyway in Mission Beach, a San Diego suburb. It was two blocks from the ocean. Each day when I drove to the base she would head to the beach.

At the base, Tom Miller and I were wondering how to get ourselves out of a fix. Choosing multi-engine training had not turned out the way we planned. We were assigned at Kearney Mesa to a transport squadron flying R4Ds, the Navy and Marine version of the commercial workhorse DC-3s.

The R4D was a fine, sturdy plane. But like the PBY flying boats back at Corpus Christi and the PBJs at Cherry Point, it left a lot to be desired compared with lighter and more maneuverable fighters. The transport's nickname, "flying boxcar," said it all. The kinesthetics of flying didn't quite apply. You would move your hand on the stick or your feet on the rudder, but where a fighter would respond instantly, the R4D, with wheels that protruded from the wings even when retracted (the tail wheel didn't retract at all), would get around to it eventually.

Across the field from the R4Ds we had been assigned to fly was an observation squadron of Grumman Wildcats. The F4F Wildcats were fighters that had done well for the Navy in the battle of Midway and for the Marines at Guadalcanal. VMO-155—the V stood for "heavier than air" in Marine parlance, with M for "Marine" and O for "observation"—had been in Samoa in the early days of the war. The action had pushed farther west, and the observation squadron had been transferred back to the United States to be re-formed into a fighter squadron that, according to the ever-present buzz of rumor, would fly Chance Vought Corsairs.

That squadron beckoned like a siren song from across the field. Pretty soon Tom and I were angling around to see what we could see. The view was encouraging. We found the squadron's skipper, Major J. P. Haines, sitting alone in the mess hall one day. He was a rosy-cheeked guy with a friendly, approachable manner, and we asked if we could join him. He waved us to a seat and we introduced ourselves. "Call me Pete," he said.

Tom and I went into full wheedle mode. We moaned and groaned about how we wanted to get back into fighters.

"I've got slots for a couple of pilots," he said. "Tell you what. Go through channels, but if you can get the okay from Colonel Roberts for a transfer, I'll take you on."

That's when the trouble started. I'd been pretty persuasive all my life, and I was used to going straight to the person I knew could get the job done. That led me to commit one of the cardinal sins of military discipline: I ignored the chain of command.

Lieutenant Colonel Dean C. Roberts was our group commander. Under him, a Major Zoney was the commander of our transport squadron, VMJ-353. I should have gone to Zoney to ask permission to request the transfer from Colonel Roberts. Instead, I went straight to the colonel's adjutant, L. H. Buss. Buss got the colonel to approve our transfers. What I didn't know was that Tom had done it right. He had talked to Zoney. When the major found out the transfers had already been approved, he called us on the carpet.

I've never been dressed down like that before or since. Zoney was a tall blond man who stood us at attention in front of his desk and chewed us out royally. Red-faced, his voice raised, he told us we were under his command. He said, "You're mighty cocky, being brand-new pilots. But in this man's Marine Corps, you need to learn where orders come from. They don't come from you. They don't come from you going to the group commander. They come from your immediate superior. That's me. I'm running your lives. I'm in charge of your careers. What you did was insubordination and I won't tolerate it. So let me tell you what I'm going to do about those transfers. I'm canceling them. You boys are going to fly transports until you muster out, and you're going to love every minute of it." It was the worst reaming-out I got in twenty-three years in the Marine Corps.

I felt terrible when we slunk out of his office. Not only had I squandered my own chances, I'd let Tom down as well. One of the worst things a Marine can do is let his buddy down. But I wasn't ready to give up. I apologized profusely and did everything I could to smooth things over. Zoney finally relented and gave me his permission to reapproach Colonel Roberts. I made my pitch, but I didn't have high hopes. I vowed that whatever happened, I'd never again bypass authority in the Marine Corps.

A few days later, to my surprise, joy, and great relief, orders came through transferring me and Tom over to the fighter squadron.

CHAPTER 6

Annie and I had enjoyed the ocean breeze during our five or six weeks in San Diego. Two days after my transfer to the fighter squadron, though, the squadron itself was transferred from Kearney Mesa to the Marine Corps Air Station at El Centro. This took us over the Laguna Mountains and into the Imperial Valley between the salt lake called the Salton Sea and the Mexican border. We arrived on July 4, 1943.

El Centro was experiencing a heat wave. It was more than two hundred feet below sea level and one of the hottest places in the United States, and the day that we arrived with our pots and pans and bedding once again packed in the back of the '39 Chevy, it was well over 100 degrees. We checked into the Barbara Worth Hotel. The hotel had air-conditioning on the first three floors, and vacancies only on the fourth. The room we got was like an oven. We opened the window as far as it would go, took off our clothes, and lay on the bed. An announcer in-

formed us on the bedside table radio that the temperature
had broken 120. I think he said it was a record for the
date. Annie was wilted, and I wondered what I had gotten
us into. When we could summon the energy to look at
each other, we had to laugh.

The town at least had some palms and a few shade
trees watered by the irrigation canals that fed the Imperial
Valley farms. The base sprawled on the desert floor a few
miles west. It consisted of two runways in an X, and a row
of adobe-colored hangars, shop buildings, and offices that
ranged from north to south and closed the X's western
end. Here the shade was nonexistent. Heat waves shim-
mered above the desert floor.

Our training schedule at El Centro made allowance
for the heat. We started flying early in the morning and
stopped in the afternoon, when it was so hot on the runway the planes' tires were likely to blow. Guys who didn't
have anything better to do cracked eggs on the ramp just
to watch them fry. We flew again after the sun set.

After fifty hours in SNJs—a retractable-gear plane resembling the Vultee "Vibrator" we'd first seen in Corpus
Christi—to make up for training we lacked, Tom and I
moved to the F4F Wildcats. This was my first experience
with a real fighter. It was a good plane, but newer airplanes had outstripped its technology and military capabilities. The pilot still had to crank the landing gear up and
down by hand, and the Wildcat was slower than the
Japanese Zero, its main rival. But pilots like Joe Foss
proved that even with the Wildcat's shortcomings, pilot
skill and courage could reverse the odds. Joe flew at
Guadalcanal, where he shot down twenty-six confirmed,
plus fourteen "smokers," one of the best records ever, and

all in the F4F. Some of the instructors had flown Wildcats in action at Guadalcanal. They had a lot to teach us about combat tactics and gunnery.

The Chocolate Mountains rose to the northeast of El Centro. The gunnery range was there. We would take turns towing a wire mesh banner about thirty feet long and four feet across behind another Wildcat. You could weight the banner so it towed vertically or horizontally. Half a dozen pilots would take turns firing at it from various approach angles.

The Wildcats had six .50-caliber machine guns. You charged them by hand, lifting and dropping a handle on each gun to load the first round. The gunsight consisted of an angled glass plate just inside the windshield. Looking through the plate, you saw the reflection of a dot—the "pipper"—representing the aim of your guns, in the middle of a series of reflected circles. These helped you measure the distance you'd have to aim ahead of the target in order to hit it. I spent hours memorizing the charts that told you how many millimeters from the pipper on the sight ring you'd have to lead the target, depending on your angle of approach, your range, and how fast the target was flying. But mostly you learned to hit the target the way skeet shooters learn to hit the clay pigeon—by aiming ahead of it and discovering, by intuition and by practice, how much of a lead to give.

When you were the tow pilot with all this going on one thousand feet behind you, you hoped your buddy didn't make a low-angle "flat pass" where the bullets came close to the tow plane.

Sighting in and squeezing the trigger as you rushed at the target at 350 miles an hour was a thrill. Every third round was a tracer round, so you could see if you were hitting and correct your lead if you weren't. But the first rounds were what you tried to hit with. That meant you had your aim down cold and weren't wasting ammunition.

We made run after run, and kept score religiously. The tips of each plane's practice rounds were dipped in a different color mix of ink and paint, and a hit on the target would leave a trace. We couldn't wait to fire our rounds and get back to the base to see who had hit the banner the most times. Real competition developed. Guys would stop speaking to each other over whether a speck of color was blue or green, red or orange.

We made practice bomb runs using two-pound mini-bombs that contained shotgun smoke shells that let us see where they landed. They were effective for training because their trajectories mimicked those of 500-pound combat bombs. We never used live bombs. Those were all going into combat.

The F4F was a squirrelly bird. Its close-set landing gear, combined with a very soft suspension system, made it want to wander on the runway when you landed. Crosswinds gave it fits on landing. The sight of an F4F running off the runway and bouncing out across the sand wasn't at all unusual. The hand-cranked landing gear always contained a potential for disaster. Forgetting to lower the gear was a cardinal sin, resulting in a belly landing. One evening at late dusk, a pilot from a neighboring squadron came in with the gear still up, slid off the runway, and bounced across the desert until he flipped upside down across an irrigation ditch. When the crash crew got there, they saw the pilot hanging from his straps, and the wheels inching up out of the bottom of the fuselage. He was still grinding on the gear crank, struggling to lower the wheels so he wouldn't get blamed for his forgetfulness. But for all the Wildcat's foibles, I developed into a pretty good shot in that plane.

The heat in El Centro increased as the summer deepened. Annie and I eventually found a little apartment, a duplex, with a desert cooler. This wasn't an air-conditioner, but a device that cooled the air with a fan that drew dry desert air through a screen of dripping water. Evaporation cooled the air. It worked surprisingly well.

The late summer brought a cricket infestation. It was like one of the plagues of Egypt. Driving over their bodies on the road sounded like a bowl of Rice Krispies when you pour in the milk. One morning I opened the front door to head over to the base, and dead crickets were piled four inches deep against the door and the walls of the building.

VMO-155 was flying seven days a week. Although we were now a fighter squadron of twenty-four planes and thirty-six pilots, the Marines were slow to change our designation from VMO to VMF; we would keep the observation designation for most of the war. The pilots had a day off now and then. Annie and I usually spent those days driving up to Los Angeles or Hollywood, back to San Diego to the beach, or up to Mt. Laguna, where we rode horses on cool forest trails to beat the heat and came back with sore posteriors.

Tom Miller and his new wife would sometimes join us. Tom had arrived at Kearney Mesa after our transfer from Cherry Point with the news that he had joined the married ranks. He had known Ida Mai Giddings since they were children; she had been studying journalism at Washburn University in Topeka, Kansas, when Tom interrupted his train trip west to visit her. They had gotten married, and she had joined him in El Centro after finishing the semester.

The four of us would get in the car maybe once a week and drive fifteen miles down to Mexicali, Mexico, where meat wasn't rationed and you could get a big sirloin steak at a restaurant called the Golden Lion. We were served horse meat, as it turned out, but we didn't know that at the time.

I could drive all over because I supplemented my gas rations with leftover 100-octane gasoline from the planes. When a plane's tanks were drained for maintenance, the fuel couldn't be replaced because of the danger that it had gotten mixed with sludge or water. The crews were supposed to dump it or let it evaporate, but they put it into fifty-gallon drums out behind the squadron, and it was standard procedure to pull up and add a few gallons. You couldn't add too much, because the high octane would burn the valves in an automobile engine.

Annie and I were at home in front of the desert cooler one Sunday night when the phone rang. Henry Knauth was on the line. Hank and I had been friends since we roomed together at Iowa City. Somehow we had wound up in the same squadron. He wanted me to come down to the Barbara Worth Hotel.

"It's ten o'clock, Hank. I've got to fly early in the morning. What's going on?" I said.

"I'm down here with Dorie, and we've decided to get married," he announced. "We'd like you to run us over to Yuma."

Yuma, in Arizona, was to California what Kentucky was to Ohio in terms of marriage laws. You couldn't get married on a whim in California, but there was a cluster of justices of the peace just across the border, advertising their services with blinking incandescent signs. Dorie was a girl Hank had met just the week before in Los Angeles.

I put my hand over the mouthpiece of the telephone and said to Annie, "It's Hank Knauth. He's been drinking."

Speaking to Hank again, I said, "Let me talk you out of this. Maybe you've had too much to drink."

"It's not just me," he said. "Stan Lutton's here, too, with his girl, Kay. We all want to go. We're not drinking, and if you don't want to take us over, we'll find somebody who will."

"Okay. Okay. Don't do anything until I get there." I hung up the phone and turned to Annie with a wry face and a shrug. "Why don't we drive over to Yuma?" I said. She jumped up, and off we went to the Barbara Worth Hotel.

Hank and Stan and the two girls were all stone sober—or drunk on romance, I suppose. They'd decided they wanted to get married, and nothing would do but that we were going to drive them to Yuma. We piled into the Chevy coupe, three in front and three on the luggage platform in the back, and headed east across the desert. Right after we hit the state line we saw a sign for a justice of the peace. By then it was after midnight. We knocked, and after a few minutes a man in a bathrobe opened the door.

He performed a double wedding in his bathrobe while his wife played a little organ. Annie and I stood up as witnesses. Ten minutes later we were in the car again, driving back to El Centro, with Mexican music playing on the radio. The war was all but forgotten. Annie whispered in my ear, "I told you we should have gotten married in Kentucky."

The squadron's F4U Corsairs arrived in September. The plane at that point was a Navy castoff. It was designed for carrier duty, but the combination of its power and stall characteristics made it very difficult to land on an aircraft carrier. The Navy had sent the designers back to the drawing board and turned the early planes over to the Marines because we were landbased.

I made my first flight in one in late September. You sat low in the cockpit under a "birdcage" canopy with extensive bracing. The bracing limited visibility, and had the added effect of making the pilot feel as if he were in jail.

But even with its drawbacks the Corsair was an impressive plane. Its eighteen-cylinder Pratt & Whitney radial engine displaced twenty-eight hundred cubic inches and produced two thousand horsepower, the most ever in a fighter engine at that time. It took a big propeller to harness that much power. The Corsair's three-bladed prop was thirteen feet six inches in diameter, and that in turn had led its designers to create wings that dipped down and turned upward in an inverted gull wing. It was the only way to get the plane's nose high enough for the propeller to safely clear the runway without the designers having to make the landing gear too long. The Corsair was difficult to handle in a crosswind, and the nose-high attitude, with the cockpit halfway back on the fuselage, meant you took off and landed by feel as well as sight. You had to ease the plane down in any case, because the left wing would stall a little before the right and drop if you weren't careful.

Landing the Corsair was like milking a mouse, we liked to say around the squadron. It was a delicate job.

The air, not the runway, was the Corsair's element. Flying it once, none of us would have traded it for any other fighter. You could take off, point the nose up, climb for the clouds, and not stop until you got past thirty thousand feet, although you'd rarely take it up that high. It could fly at 417 miles an hour, turned on a dime, and was well armed with three .50-caliber machine guns in each wing. Under each wing it had three bomb or rocket racks, and a stub that could be used for a bomb or an additional fuel tank. Some models had a center bomb rack.

Vought was working to smooth out the problems in this very good fighter, as was Goodyear, which was building the plane under a contract with Vought. They produced a number of changes that fall, and enlisted Charles Lindbergh to fly one of the new planes around to the squadrons and talk about its characteristics. Lindbergh brought the new Corsair to El Centro in October.

Word got around that he was coming. A handful of us who weren't flying gathered to watch him land. The new plane was a beauty, painted a glossy, jewel-like navy blue. It had a slightly rounded canopy without the "birdcage" bracing, and was a little less nose-high, with a longer tail wheel. Lindbergh taxied to the ramp. We were as eager to see him as we had been to see the plane. He was a legend and a pioneer and every pilot's hero.

He climbed out of the cockpit and came down the plane's right side, where the step pads and hand grips were, and walked over to where the skipper and the rest of the El Centro hierarchy were waiting to greet him. He had his flight helmet tucked under one arm, and he looked older than the pictures of "Lucky Lindy" I remembered from his flight over the Atlantic, and from his book about the achievement, *We*. His eyes had seen a lot of horizons. I couldn't say why, but he inspired in me a feeling of kinship. It had been eleven years since his son was kidnapped and killed, and the nation had been as transfixed by his tragedy as it had been by his triumph five years before that. He had lived the best and endured the worst. Now here he was, at forty-one, helping the country produce planes that would help us win the war.

Three squadrons were training at El Centro. Lindbergh stayed two or three days and briefed each one. The main change in the Corsair, beyond the canopy and tail wheel, was a spoiler on the leading edge of the right wing that made both wings stall at the same time and made landing easier. The squadron commanders and executive and operations officers were getting to check the plane out in the air. I was one of the few others, having approached Pete Haines ahead of time and persuaded him to put a lowly second lieutenant on the list.

After a half hour in the new Corsair, I loved it that much more. The cockpit was raised a little, and you could see much better out of the bubble canopy. Landing was a lot less touchy. You still couldn't see ahead while you were taxiing, but that was a small thing; Corsair pilots were used to zigzagging when they taxied, looking out one side of the plane and then the other. A whole squadron taxiing looked like some kind of primitive dance performed by crabs scuttling in formation. I climbed out of the cockpit from my test flight reluctantly. It was almost as if my training had ended with that flight. I was ready to go into action.

Annie and I spent a cozy Christmas together, our first as a married couple. We had a good time, although it seemed odd to have Christmas without snow or even a Christmas tree, which proved impossible to find. I gave her a Bulova watch.

Christmas reduced our activity schedule. The squadron crew needed some time off after the hectic training period. We mustered late and spent the day playing softball and basketball. Somebody said we were highly paid softball players. About half of us were married second lieutenants making $392.50 a month. That felt like real money in a place with as few ways to spend it as El Centro. Annie was making money, too. The Methodist church had been paying her to be its organist ever since the pastor listened to her play and proclaimed her the finest organist he'd ever heard. She and the choir had performed Handel's *Messiah* with a large choir before a packed church in the weeks before Christmas. I went on Sundays when I could, with apologies to John Knox and the Presbyterians, and every time I did Annie tried to play "Be Still My Soul" because she knew it was my favorite.

The year turned, and I kept wondering when word would come that we were moving out of training and into the real thing. Annie and I, on one of our trips to Los Angeles, had seen the world premiere run of the great movie *Guadalcanal Diary* and met Joe Foss, the Guadalcanal hero.

Meeting him and having learned from the decorated

combat-wise flying veterans of the Pacific theater increased my pride in being a Marine. I was beginning to learn what that indefinable something was that I hadn't been able to put my finger on when I'd decided in Corpus Christi to switch to the Marines. I was talking with Pete Haines in the ready room one day and the conversation turned, as it often did, to tales of the Marines leading the way across the Pacific.

I said, "The Navy and the Army are out there, too, Skipper. Why are the Marines so much better?"

Up to that point, the conversation had been half joking, half serious, but Pete turned sober, paused a few seconds, and then said, "Marine training makes each Marine more afraid of letting his buddies down than he is of getting hurt himself. And that wins battles."

I've always remembered that, and I believe it's true.

I might not have asked that question if I had been through Marine boot camp, where that feeling of trust and confidence is instilled. I felt it just the same, however. The Marine pride and esprit de corps that came through in the tales of the returning veterans worked its way into the training of VMO-155. It showed in our ground preparation, our flying and gunnery practice, and the competition among us to be the best as we waited to learn our destination in the war.

The orders came on January 6, 1944. We stood down, our equipment was packed, and we started getting ready.

I said goodbye to Annie before dawn on the morning of February 5. She was brave, and we both tried to be cheerful, although neither of us felt much like it. Looking for something to say, I remembered a tagline from a half-forgotten story. "I'm just going down to the corner store to get a pack of gum," I said.

"Don't be long," she whispered, biting her trembling lip.

We kissed goodbye and I got on a Navy bus with the other pilots. The whole squadron headed to San Diego in a convoy with trucks that carried the airplane mechanics, the fuelers, the ordnance and instrument people, the men who worked on airframes, the cooks and their stoves, the doctor and his medical supplies, the scheduling and operations staff, the parachute and flight equipment experts, all the tools and gear, and the supply and logistics people who kept it all moving. The squadron wasn't just the pilots and the planes; it consisted of about 275 men and their equipment—a self-sufficient, special-purpose town on the move.

We boarded the USS *Santa Monica* at two o'clock that afternoon. She sailed at eight the next morning, headed for Hawaii. I watched the coast of California until it slipped below the horizon and wondered what I would do before I saw that land again.

The *Santa Monica* was a converted banana boat, outfitted with five-inch guns and twenty-millimeter cannons. Crowded with eighteen hundred enlisted men and officers plus cargo in the hold, she was short of room. The bunks were stacked several deep, with only about eighteen inches of sleeping space between them, where clusters of bananas had hung. We had a Navy destroyer and a blimp with us for the first day. Then our escorts left us and we were alone in the open sea. The ship's bos'n told us the *Santa Monica* was fast enough to outrun a submarine. I doubted that, and anyway, he didn't say anything about torpedos.

A torpedo could have hit us and I probably wouldn't have noticed. I was so seasick I didn't care if I lived, died, or fell overboard. It was odd because I had never experienced airsickness. I visited the rail twice on the first day, and I wasn't alone. After that my stomach settled down a little bit. My color moderated to a grayish white from a violet shade of green. The sleeping quarters belowdecks were close and foul, so I brought a blanket topside and spent most of the voyage on the deck both for the air and the convenience of the rail, and stumbled through the daily abandon-ship and fire drills.

Nobody aboard was happier than I was when we reached Hawaii on February 12 after six days at sea. We dropped about fifteen hundred officers and men of the Fourth Marine Division on the island of Maui. A day later the *Santa Monica* steamed into Honolulu and Pearl Harbor. I walked down the gangplank onto blessedly stable land, and with the rest of the squadron boarded a convoy of buses that took us to the Marine Corps Air Station at Ewa, nineteen miles from town.

Ewa was a lazy man's paradise. We were supposed to be posted to the Marshall Islands after first testing the new planes we would fly in combat there, but the planes had not arrived. While we waited, we spent our time lying around the swimming pool, watching the movies that were free on the base and changed every day, and taking the public bus into Honolulu.

I had heard about the famous beach at Waikiki ever since I could remember. My first trip there was a disappointment. The beach was narrower than I expected, with coral patches here and there. I warmed to it quickly, though. The Navy had taken over the Royal Hawaiian Hotel on the beach, and the United Service Organizations—the USO—operated a stand there. Servicemen could rent swimming trunks and towels for a quarter. The waves were everything I had heard about and more. Several of us hired an outrigger canoe and rode in on a long series of breakers. Then we rented surfboards. I loved riding the waves, but standing up on the board usually was a disaster.

I missed Annie terribly. There had been no reason for her to stay on in El Centro after I shipped out. She had packed the car again and headed east with Ida Mai Miller, who was eight months pregnant, and two other squadron wives driving in another car. She dropped Ida Mai in Fort Worth, Texas, and the convoy turned north. Annie wrote to tell me that one of the other women had a dog that had to have its own chair in the restaurants where they stopped to eat. My mother met Annie in St. Louis for the remainder of the trip back to New Concord. She planned to stay with her parents until I came home.

Annie would really have enjoyed the beach at Waikiki, and I would have enjoyed it more if she'd been there. It had been only a couple of weeks by then, but it felt longer. I passed a place near the Royal Hawaiian that was selling full-blown hula outfits—leis, bra, flower bracelets, and grass skirt. I immediately put that outfit on Annie in my mind's eye. It looked mighty good on her, so I bought one and put it in the mail to her.

"I want you to send me a picture of you in this," I wrote. "As for the grass skirt, I'll bring the lawn mower with me when I come home."

The profusion of Japanese in Honolulu confused me

at first. We were at war with Japan. I had never been around large numbers of Japanese-Americans. Japan's imperial forces were known by then not only for fighting to the death, but also for torturing and killing their prisoners. Survivors from ships sunk by Japanese torpedos had told stories of submarines surfacing and their crews machine-gunning sailors clinging to the wreckage. But after first being uncomfortable seeing Japanese "running around loose on the streets," as I wrote in the diary I kept, or sitting next to a Japanese person on the bus, I decided that they were just as loyal to the United States as I was or they wouldn't be there. I realized it was too easy to discriminate against the Japanese based on their appearance. Japanese-Americans had been interned on the mainland after Pearl Harbor. As far as I knew, no Americans of German descent had been taken to camps when the United States entered the war against the Nazis.

After about a week of recreation on Oahu, our planes arrived and VMO-155 received orders to proceed to the Midway Islands. Our posting to the Marshalls had been delayed for some reason, and we were being sent to Midway to relieve a squadron that had been out there about two months.

The action in the Pacific had moved well beyond Midway in February 1944, but it still filled a strategic gap in the ocean between Hawaii and the Aleutian islands of Alaska. At that stage of the war the Navy used it as a submarine base. Subs in the Pacific came to Midway to replenish their supplies and have the crews there repair their battle damage. Our job was to protect the sub base. We would also bring ourselves back up to "fighting trim," with intense gunnery training the likes of which we hadn't done since we stood down at El Centro.

We flew the twelve hundred miles out to Midway on February 21 in groups of officers and enlisted men aboard a string of Curtiss Commandos. Tom Miller, Bill Blades, and Monty Goodman were the officers with me. Monty's real name was Miles, but nobody ever called him that. He was a lively kid from central Pennsylvania. He had become one of my closest friends starting back at El Centro. Monty had black hair and freckles, and he loved music as much as I did. He was a big fan of Frank Sinatra, and he did a great Sinatra imitation, singing into the end of a broomstick that he caressed as if it were a microphone on a stand. Sometimes I sang harmony behind him. Monty was one of the most likable and funniest pilots in the squadron. He tried to keep us entertained, but the roar of the engines was too much and it often drowned him out.

The squadron we flew with was VMJ-353, the outfit Tom and I had left so we could fly fighters. Chugging along over the waves in the Commando at 215 miles an hour didn't make either one of us regret the choice we'd made.

Our planes followed us, half the complement of twenty-four new Corsairs by ship, the others being flown from Ewa and finally arriving after setting down at French Frigate Shoals to wait out five days of bad weather. When they reached Midway they landed in a stiff crosswind without problems, indicating that the changes Lindbergh had briefed us on at El Centro had done the trick.

Midway is actually two islands, Sand Island and Eastern Island, in a lagoon ringed by a coral atoll. It is a dot in the ocean, not much more than two square miles of land taking both islands together, and a bump on the horizon not twenty-five feet high. We shared the islands with tens of thousands of sea birds that made a racket day and night. The gooney birds—black-footed albatrosses—provided slapstick entertainment. Watching them was like watching the Three Stooges, a pratfall every minute. They spent months at sea out of sight of land, and returned to Midway to mate and nest, but they weren't used to land-based operations and more often than not, when they tried to land they tumbled beak over webbed feet in a cloud of dust.

Watching the young birds learn to fly was even better. They would stand on top of a dune, stretch their wings to get a feel of the air, run a couple of steps, and launch into an unstable glide, heads swiveling from side to side, that usually ended when they hit the sand in a rolling pile of feathers. They would stand up, shake off, and look around as if to say, "That's the way I planned it all along." Then they would waddle up the dune and start over.

We had a gunnery and navigation training syllabus to follow, but the weather kept us on the deck for most of the first two weeks. When we could get up, we often flew the SNJ trainers we had come to know at El Centro. When we couldn't, we hung around quarters or the ready room and listened to the Japanese propagandist, Tokyo Rose, on Radio Tokyo. One of the mysteries of life on Midway was how the Japanese knew chapter and verse about what was going on there. When the Seabees finished one of the runways, Tokyo Rose was on the air the next day congratulating them on their good work and saying the Japanese would be along in a few days to take over.

We wished they would try. We were ready to give them a warm reception. We had four Corsairs on alert status twenty-four hours a day, fully armed and ready to go. They could be in the air within two minutes. We could scramble the rest right on their heels. Nothing the Japanese had could handle the Corsairs in the air, and the gunnery practice we managed to get in showed that the squadron was hitting pretty well.

One day in March Tom sauntered in from mail call with a cigar stuck in the corner of his mouth. He handed one to me along with a letter he'd received. "What do you think of that, Jughead?" he said with a grin on his face.

He called me Jughead because *J* and *H* were the initials of my first and middle names. Tom's middle initial was *H* also. He was Tithead.

The letter from Ida Mai said that she had delivered a son, Tom H. Miller III. Mother and baby were doing fine. Tom was on his own little cloud for several days.

The weather broke toward the end of March, and we started to fly every day. One of our rare exercises was an "oxygen hop"—a flight close to the Corsair's altitude limit of around thirty-two thousand feet. These flights were aimed at improving our ability to attack Japanese bombers that were flying at higher altitudes. They also let us practice flying wearing oxygen masks, and testing the handling characteristics of the plane in thin air. Coming down, we had to be careful of something new to us, high Mach number, or fraction of the speed of sound, where control of the Corsair would become difficult. I didn't want to repeat the free-fall I'd been in back in El Centro, when the left elevator came off the Corsair in a 460-mile-per-hour dive and I barely regained control.

The Navy submariners whose boats were in for repair loved to fly with us when we went up to do acrobatics in the SNJs. I imagine they liked the lack of confinement, the freedom of movement, and the chance to see as far as the horizon. In exchange, they sometimes took pilots on the one-day shakedown cruises they made after receiving repairs.

The USS *Barb* came into Midway for some supplies and work. The *Barb* and her crew were later famous for sitting on the bottom of Tokyo Harbor and torpedoing a new Japanese carrier on its launching day, then escaping through open submarine nets. This trip into Midway was occasioned by a periscope problem, and Tom, Monty, and I were aboard on the shakedown cruise after the repair. We were down at maximum depth when the packing around the periscope blew. Water came in as if somebody had turned on a fire hose. I'm sure we had the fear of God in our eyes, but the crew all thought it was pretty funny because they knew that water could be pumped out a lot faster than this was coming in and that now they were going to get a few extra days on Midway. That was my only dive in a submarine. The chow was as good as it was cracked up to be, but I wasn't sorry I had chosen flying.

The dangers of flying were always there, too. On Easter morning, 1944, we were going out on a gunnery mission. The first hop, led by our operations officer, Major John "Drifty" Reynolds, was ready to take off. Drifty had started down the runway on a full-power takeoff roll when Joe Johnson decided to taxi across. Drifty couldn't see him until the Corsair's tail wheel came up. When he

did, he horsed back on the stick. He had enough speed to
get airborne but not to clear Joe's plane. The impact
flipped Joe's plane over and it burned, while Drifty's plane
crashed on the runway, wiping out the landing gear. Drifty
only needed a few stitches, but Joe was killed. He was one
of the best-liked members of the squadron, always the life
of the party. We buried him at sea from a PT boat the next
day, and I heard the familiar sad notes of taps again as his
body slid from under the flag into the deep water.

Joe's was the only death while VMO-155 was at
Midway, but there were a number of accidents. Stan
Lutton, from the Yuma midnight wedding party, spun out
of control and fell from twenty thousand feet before bailing out at fifteen hundred. Landing accidents were not un-
common. "Danny" Danitschek had one of the worst. He
caught a wing on a sand dune after going out of control
and tore off both wings and part of the tail. The impact
ripped the engine off the plane. Danny ended up in what
was left, upside-down in a dune, with sand in his eyes—
miraculously, his only injury.

I thought flying was not much more dangerous than
any other type of work, with the exception that one slip, a
lapse of just a second at a crucial time, could make a lot
more difference. I tried to eliminate carelessness from my
flying habits. Things happened so quickly in fighters that
the smallest oversight could have disastrous results. I
checked, rechecked, and tested every gadget on the plane
before I took off. Flying was intolerant of human error,
and I considered being check-happy a good and inexpensive form of life insurance.

Another way to make sure you stayed alive in a fighter
was to be a better shot than the other guy. We spent a lot

of time in gunnery practice, since we'd be doing the real thing pretty soon now. Tom and I had a competition going between us. We had the best hit averages in the squadron, about 6 or 7 percent, and I figured it would pay off when we were shooting at something other than a tow target.

Our relief squadron, VMF-313, arrived at Midway at the beginning of May. A few weeks later we left aboard a convoy of Commandos headed for Ewa, leaving behind our Corsairs for the new squadron.

At Ewa we checked out the new planes we'd fly in the Marshalls and waited for transportation. Finally, on June 28, we boarded an escort carrier, the USS *Makin Island*, and set sail for the Marshalls.

The trip passed easily. The escort carriers were called "Kaiser Coffins," a nickname that owed itself to their light construction, small size, and the fact that they were made by Kaiser, the firm that had brought the mass-production methods of the auto assembly line to wartime shipbuilding. But the *Makin Island* was a lot more stable than the *Santa Monica* had been sailing to Hawaii. I never felt seasick. My cabin in the junior officers' quarters was only a bulkhead away from the forward engine room, however, and to escape the heat I took to sleeping topside under the SB2C dive-bombers and the TBF torpedo bombers on the flight deck. Tom did the same. The moon was out every night, the breeze was cool, and sleep came to the sound of waves against the ship's bow. We woke early and spent the days reading, shooting the bull, and playing volleyball on the forward elevator. I had a dog-eared copy of Robert W. Service's *Tales of the Yukon,* rollicking poems that I found fun to read. With very little coaxing, Monty entertained with his Sinatra imitations. He sprained his ankle playing volleyball and was on crutches for a while. Sometimes it was easier to carry him, and he'd pile on my back and away we'd go. I spent part of each day writing to Annie.

We sailed into Majuro Harbor in the Marshalls on the afternoon of July 3. The next morning, with the ship at anchor, the crew brought our Corsairs up from the hangar deck, lowered their folding wings, and prepared to catapult us into the next phase of the war.

CHAPTER 7

The Corsair strained at its cable on the catapult track. I rechecked to see that the flaps were down, pushed the throttle to full power, sat back in the seat with my head against the headrest, and gave the ready-to-launch salute to the catapult officer on deck. Two seconds later the plane and I were off the deck going ninety knots. Afterward I wrote in my diary that I was "out in space flying and not too sure about how I got there. That is really a boot in the tail."

I followed the parade of squadron planes being launched from the *Makin Island,* circling Majuro and landing at the airfield. From the air it looked peaceful and calm, a little piece of paradise with clear water lapping against white beaches fringed with coconut palms.

As recently as February, though, the Marshalls had been the scene of vicious fighting. The cluster of islands and atolls, scattered across a vast patch of the central Pacific more than 2,000 miles southwest of Hawaii, was important to the Japanese. American landing forces had encountered fierce resistance on Kwajalein, Roi-Namur, and Namu, where 11,612 Japanese were killed of a total force of 11,841. Fewer than 600 American soldiers and Marines of a force of more than 40,000 died in the assaults. One hundred and ninety-five Marines died taking Eniwetok, and 2,612 of 2,677 Japanese defenders.

Once American forces effectively controlled the Marshalls, they moved west, bypassing a number of the islands still held by the Japanese but cut off from support. Majuro lay near the center of a quadrangle formed by the four main atolls still under Japanese control—Maloelap and Wotje to the north, Mili to the south, and Jaluit to the southwest. The Japanese there could not do anything offensively. Despite thousands of troops and a multitude of antiaircraft weapons, control of some ports and airfields, and large caches of supplies, they had no planes or ships. Our assignment was to keep them suppressed, to prevent the imperial forces from sneaking back onto the islands by air or submarine.

The first thing we did on Majuro was check out on the F6F fighters that VF-39, the Navy squadron that preceded us, had been flying. These were Grumman Hellcats, a successor to the F4F Wildcats we'd trained in back at El Centro. The Hellcat had the same 2,000-horsepower engine as the Corsair, but to my mind that was as far as the comparison went. I found the plane sluggish and slow in a dive. It did land easily, and handled well on the ground.

In our flight setup now, I led a division of four planes. Pete Haines had irked some of the more senior men when he chose me as a division leader, but he had confidence in me and that's what mattered. Crooning Monty Goodman was my wingman. Tom Miller led the second section, with a pilot we had picked up at Ewa, Ed "Tyrone" Powers, as his wingman. The squadron had three eight-plane "flights," under the command of the squadron commander, the executive officer, and the operations officer. Mine was the second division of the skipper's flight.

We got our first combat assignment a few days after arriving at Majuro. The fighters were part of a team ordered to strike Taroa, one of the islands of Maloelap Atoll, about 110 nautical miles north of Majuro. Our mission was flak suppression. We would go across the island ahead of the dive-bombers and strafe to keep the antiaircraft fire down. We took off early in the morning. It was a fine day with scattered clouds.

We came over the island, and at a signal from the leader of the dive-bomber group, the skipper rolled his division off to the left and started a shallow dive for raking fire across the island. My division was next. High overhead, the dive-bombers were rolling into their steeper bombing runs. The attack was an aerial ballet choreographed so our fighter runs would end just before the dive-bombers dropped their bombs, but too late for the antiaircraft gunners to sit up and take aim on them. We maintained radio silence for maximum surprise, and if we could come at them out of the sun, we did.

Before I started my dive, I looked back and saw Monty dropping back slightly, as he should. He was grinning, and he nodded and pointed down, a sign for "Let's go get 'em." I rolled into my dive, and when I was sighted on the first antiaircraft position on the near side of the island, I squeezed the trigger on the six .50s, stopping only long enough to avoid overheating the guns, which would make the shells tumble instead of converging at the bore sight point a thousand or twelve hundred feet ahead and then fanning out again. Monty was in good position to one side and back a little.

I hit several targets as I came across the island from the south. Streams of tracer fire came at us from the ground. I pulled out very low on the treetops, banked to the east, and climbed to the rendezvous point. As far as I could tell, I had not been hit. Looking back at the island, I saw a series of explosions as the dive-bombers dropped their loads and pulled out to clear the island. Up high, black puffs of smoke hung in the air from the large-caliber antiaircraft fire directed at the bombers.

We had to go to eight thousand feet to find a clear spot in the clouds to rendezvous. I heard one of the dive-bomber pilots say on the radio that he had seen a plane go down. Seven of us joined up at the rendezvous. Monty was missing.

Four of us flew to the spot where Monty's flight path would have carried him. There was an oil slick on the water a mile offshore, and some yellow dye marker from a Mae West, a pilot's life preserver. We crisscrossed the area for two hours. So did one of the PBY flying boats that we called "Dumbos"—likening them to flying elephants—that accompanied missions such as ours in case a pilot had to ditch in the water. None of us saw anything more. A rain squall out to the west produced a rainbow that was all too fitting a symbol. Finally, low on fuel, we flew back to Majuro.

After I landed, I went to the Quonset ready room. The skipper was already there and had passed on the bad news.

The division's other pilots all had waited, and you could see it in their eyes. I went to stow my parachute. We had individual parachute bins in an area behind the ready room. My division's four were right together, just as its pilots had tried to be. The bin with Monty's name on it was empty. I had never felt so helpless. The pent-up feelings of the last few hours let loose, and I stood there and sobbed. We all knew what could happen when we flew into combat, although we rarely talked about it. The war of waving flags and musical send-offs was gone, and I had seen my first day of the real thing. I had lost one of my closest friends, and war had suddenly become very, very personal.

We gathered Monty's things and sent them to his parents with a letter from the skipper. I also felt a need to write them. I stared and stared at the blank page, not knowing what to say. I thought about a hundred things, about the brotherlike relationships that grow from the unlikely circumstances of war. Monty was Jewish, and his family owned a furniture store in Harrisburg, Pennsylvania. I was a Protestant plumber's son from a state away. I had depended on him as a friend. He was my bunkmate on the ship out from San Diego. We had gone on liberty together in Honolulu. We had roomed together, with Tom Miller, at Ewa and at Midway. We had harmonized on "You'll Never Know" at midnight in the shower until the skipper had stumbled in sleepy-eyed and ordered us to shut up and go to bed. We'd done a corny dance routine from *Coney Island,* a show we'd seen at Midway.

I ended up telling them what a good pilot Monty was, how proud he was of his ability to fly the Corsair, how he always hung in there no matter what. I wanted to let them

know how much he meant to us, and that now we were
fighting for him, too.

The squadron flew missions almost daily. Sometimes we
flew two missions in a day. All twenty-four planes flew if
they were in commission; usually two or three were out
being repaired. Pilots rotated, and averaged flying every
other mission.

We attacked the Japanese-held islands all the time.
We usually dropped bombs. Occasionally we did flak
suppression for dive-bombers, as we had on the first day.
Sometimes we kept it all within the squadron, a few of us
doing flak suppression ahead of our own bombing runs
with the Corsairs.

Sometimes we were called out for reconnaissance
follow-up missions. These were fun because they were less
structured than the attack missions.

Late one evening a month or two into our stay on
Majuro, a recon plane reported it had seen a Japanese seaplane in the lagoon near Emidj, the main island of Jaluit
Atoll. Four of us, including Tom and me, with Pete Haines
leading, took off before dawn the next morning. We
reached Emidj at first light, broke through a cover of
heavy clouds at a few hundred feet, and skimmed the island, just clearing the palm stumps and buildings. No tracers came up at us; it must have been too early. We saw no
planes other than the burned-out wrecks on the airfield,
so we kept going around the entire atoll, which was about
thirty miles from north to south and fifteen miles across.
Away from the Japanese-occupied areas, native huts and
gardens looked like scenes from a South Pacific travel-

ogue. We strafed a building at the base of a radio tower and then headed back to Emidj to do some strafing there.

This time we came in fast and low. In a fast dive, the air rushing through the oil cooler baffles in the leading edge of the Corsair's wings made a whistling sound. You couldn't hear it from the cockpit, but the sound struck fear into anyone who heard it down below. The Japanese had learned to call the Corsairs "whistling death."

They were ready for us this time. They were shooting antiaircraft "in copious quantities," as my dad would have said. The overlapping streams of tracers coming past us looked like flaming golf balls. But the antiaircraft gunners got no hits this time, and we all flew safely back to Majuro.

The Corsair's power gave it an edge on other fighters because it could carry an effective bomb load easily. Our standard load was three one-thousand-pound bombs, one under each wing and one on a rack on the centerline of the plane unless we needed extra fuel and had to replace that with a drop tank. By comparison, the PBJs (B-25s) the Marines flew in the Marshalls carried a standard twenty-five-hundred-pound bomb load, albeit for far longer distances. Lindbergh came out and joined our squadron for a couple of missions when we carried a pair of two-thousand-pounders, but that meant flying close to the performance limits of the Corsair, and we didn't do it often.

Depending on our targets, we sometimes carried five-hundred-pound armor-piercing bombs with long needle-shaped steel noses designed to penetrate the concrete bunkers the Japanese used for storing fuel and ammunition. You knew you were on target if you got a secondary explosion after the bomb hit. The secondaries could be spectacular, as well as good news to our intelligence officers in the debriefings at the end of every flight.

The dive-bombers had dive brakes. These were large, sometimes split flaps that came out into the airstream to slow the dive. With the Corsair, we had a different kind of dive brake. Our technique was to come down from high above a target and drop the landing gear. The gear and their flat plate fairings acted as an air brake, making a bombing run slower, more stable, and more accurate. Our bomb runs hit a higher percentage of their targets than they would if we had come in at full speed. You would drop your bomb load, pull out and raise the gear, then jink away in a series of quick maneuvers that made it hard for the antiaircraft gunners to hold an aim point ahead of you that would get a hit. You jinked up and down if you were low and they were firing at you horizontally, and from side to side if you were high and the flak was coming up.

We would go through spells of several days in which hardly any antiaircraft fire came at us. Then one day we would fly over the same island and they would throw everything at us but the kitchen sink. We would fly through clouds of tracers so thick it was hard to see how a plane could make it through without getting shot full of holes, but usually we did.

The danger of combat flying did nothing to diminish my love of flying in general. If anything, it enhanced it. This was flying with a purpose. We had been in no air-to-air combat, which is what every fighter pilot wants to do. But bombing runs into antiaircraft fire were a test of skill, nerve, preparation, and focus that I relished. Nothing gave me more pleasure than to be flying the Corsair. At the controls of a small, high-performance aircraft day in and day out, you reach a point of oneness with the plane. Some people would call that the zen of flying, but that's too deep for me. You're just a part of the plane, not separate from it, as if you are the brain and it is the body, doing some things you don't have to think about, like breathing, and other things that are converted from thought to action in an instant.

Pete Haines was aware of my feelings about airplanes, flying, and the Marine Corps, which felt like a second family. One day not long after we got to Majuro he called me and Tom Miller into his office next to the ready room. "You two should think about applying for regular commissions," he said.

Coming out of Navy cadet flight training and switching into the Marines, we were officially reserves. I wasn't sure what I wanted to do when the war was over, and neither did Tom. I certainly wasn't sure I wanted to stay in the Marines. But the skipper pointed out that the applications would take a long time to process, and we could always decide, if we were accepted, to opt out when the time came.

Tom and I figured we had nothing to lose, and so we both put in for regular commissions. I didn't know if Annie wanted a Marine life, but I wrote to tell her I had applied and said we could talk about it and decide when I got home.

VMO-155 flew out of Majuro for four months. I was the squadron engineering officer for part of that time, owing to a shortage of qualified personnel. Engineering was the largest department in the squadron. It kept the airplanes in commission and flying, and repaired combat damage. I tried to set it up to run properly and stay out of the way of the NCOs who took care of the work.

We lived fairly well at Majuro, on a coral strip between the ocean and lagoon. We slept in Quonset huts we built ourselves, from kits, and set up on sawed-off coconut logs. Each accommodated twelve to fifteen men. We even had electricity, from generators. We spent hours snorkeling in the lagoon or over the tidal shelf on the ocean side. The greatest dangers we faced when we weren't flying were sunburn and falling coconuts, which we knocked off the trees as a precaution.

The food was another matter. For one two-month period we had no fresh meat, vegetables, or fruit. The dehydrated foods we ate had as much taste as the cardboard they came in, and we relied on canned fruits and vegetables plus peanut butter and jelly. When weevils invaded the flour supply, at first we would hold the bread the cook baked against a light and pick them out. After a while it didn't seem to matter. The cook made valiant efforts. He was a Spam genius. He made Spam patties, Spam loaf, Spam stew, and Spam in gravy. I liked Spam well enough, but I decided that if I survived the war, I would never again eat a Vienna sausage.

Near the end of our stay on Majuro, we received orders to attack Nauru, an island in the open ocean far to the southwest. Nauru had rich phosphate deposits and a busy processing plant. Japan had taken the island from Australia early in the war, and although the conflict in the Pacific had pushed past the island and into the Philippines, Nauru was still producing phosphate for the Japanese war effort.

Nauru was too far to reach in a flight from Majuro. We needed a staging area closer, and for this we flew south to

Tarawa Atoll, in the Gilbert Islands. Japan had occupied Tarawa two days after the attack on Pearl Harbor. The Marines took it back almost two years later in one of the most savage and costly battles of the war in the Pacific. The fighting raged for more than three days. When it was over, the Marines had lost a fifth of their five-thousand-man invasion force, and only seventeen of five thousand Japanese remained alive. Photographs of the carnage helped make the battle for Tarawa one of the most famous of the war.

We were on Tarawa for several days, waiting out bad weather. Our planes were tied down, but we ran them every day to keep the engines in good shape. The big Corsair props always caused a lot of prop wash; it threw up coral sand and dust behind the plane. One day I shut the engine down, and when I climbed out of the cockpit and walked around behind the plane I saw that the prop wash had scoured the sand from around a human skull.

I dug it up and took it to Dr. Roy Johnson, the flight surgeon. Doc Johnson said it was almost certainly Japanese, a remnant of the battle.

There were many other remnants, as I discovered exploring some of the bombed-out fortifications on Tarawa. They were littered with bones. Disposing of corpses was a big problem for American forces in the Pacific, because the Japanese preferred death to what they considered the dishonor of defeat.

Even from Tarawa, the flight to Nauru was a long one, around four hundred nautical miles with no islands to serve as navigational guideposts. We carried 150-gallon drop tanks. The trip required absolutely accurate dead reckoning—adjusting course to account for wind speed and direction in order to end up in the place you want to go. By this time, we were experienced dead-reckoning navigators. Even so, we took off with a PV-2, a military version of a Lockheed Lodestar, giving us a navigation lead. A Dumbo trailed along for search and rescue. With the distance of the journey, it gave us an extra level of comfort knowing they were there.

Nauru was a good-sized island of about eight square miles that rose higher above the water than the atolls we were used to striking. It was just south of the equator, and this was my first time crossing into the Southern Hemisphere. We saw the island in the distance and set up our attack. The hills and the terrain dictated that there was only one way to hit the phosphate plant. We came down in a line into the heaviest antiaircraft barrage I had seen thus far, all concentrating on our approach route. Black bursts of it dotted the sky from ten thousand feet down to two thousand. We flew into it because we didn't have a choice, each dropped our two thousand-pounders, and jinked out of there. It was a miracle that nobody went down.

We had seen no women for several months when Bob Hope brought his USO show to the Marshalls. Hope and his bevy of Hollywood starlets provided a break in the routine. You could watch that show and remember the people at home, but it also made you remember just how far away they were.

I wrote Annie almost every day. All our letters home were read by a censor before they were sent on. We could refer to palm trees and lagoons, but not specific places. I found a hundred ways of describing Majuro and the islands we saw from the air, telling her of the beauty that we were having to destroy to win the war. I found a thousand ways of telling Annie that I loved her, and of the longing that grew greater the more time I was away from her.

I described in my letters a war that was horrible and beautiful at the same time, and a routine that for the pilots provided a similar contrast. The squadron spent mornings strafing and bombing, and afternoons playing volleyball on the beach and snorkeling. At night we sat on coconut logs and watched movies on an outdoor screen, with our ponchos and sun helmets handy for the rain showers that came like clockwork.

Around November 1, after four months of attack flights out of Majuro, including two trips to Tarawa for strikes on the phosphate plant at Nauru, we got orders to move 275 miles northwest to Kwajalein Atoll. We flew the planes up while the rest of the squadron packed the gear and followed on a transport vessel called the *Hawaiian Shipper*.

Kwajalein was the largest atoll in the Marshalls, spanning almost eighty miles from northwest to southeast. It looked like a comma wrapped around a huge lagoon.

For several days we flew out of Roi-Namur, an island at the atoll's northern point that was our air group's headquarters. One of our assignments was a bombing run on Jaluit Atoll. The skipper had some squadron business that kept him from flying, and "Drifty" Reynolds, now the exec, hadn't yet arrived from Majuro, so Major Haines told me to lead the strike.

It was a long run, about five hundred nautical miles to Jaluit and back. Several of us carried two-thousand-pound bombs. We came in high with our gear down, and the two-thousand-pounders exploding at close intervals sent out shock waves briefly visible as vapor rings. The explosions seemed to radiate out to cover half the island, and the ground seemed to rise, shake, and settle down again.

I was climbing from my run when the plane shook as if somebody had hit it with a brick. I looked for the damage, and saw the plane had taken a hit on the left wing. A 20-millimeter explosive shell or its equivalent had taken a chunk the size of a man's head out of the wing's leading edge. I tried the controls and found that the cable leading to the aileron trim tab on the back edge of the wing had been cut. That made it hard, but not impossible, to fly. But the hit was dangerously close to the oil cooler baffles; if the oil cooler was damaged or if the engine lost its oil, the engine would soon seize up and I'd have to bail out over the ocean or make a water landing.

I didn't pray much when I was flying. I never believed in foxhole conversions, praying just when you were in trouble and making promises you'd never keep. But I'm not ashamed to say I prayed on that trip back. The Mae Wests we wore contained dye marker and shark repellant, the rescue crews were good, but I didn't want to go down in the Pacific.

The oil cooler was intact, and I made it back to Roi-Namur, where the wing had to be replaced.

The bombing run went well enough that the area we hit on Jaluit was removed from our target list. It felt good to know the skipper had trusted me enough to lead the squadron, and that the attack had been successful.

We moved from Roi-Namur to Kwajalein Island when the rest of the squadron set up camp there. It had been devastated in the previous February's fighting. Majuro had been lush with foliage, but on Kwajalein the only trees standing were the dead stumps of coconut palms lopped off by shellfire. The Kwajalein airstrip ran parallel to the beach on the island's southern shore, and started and stopped at the ocean on both ends. The entire squadron lived in tents between the runway and the water's edge.

Tom, Dick Rainforth, and I set up our tent on a wooden platform eighteen inches above the sand, with the tent flaps raised to let the air in. At high tide the water flowed under it to a low spot behind. Along the shore, you still could find bullets, shell fragments, and hulks of equipment left from the battle for the island. The wall of a blasted-out blockhouse stood at the corner of our porch, and at night we built fires against it to roast marshmallows that Annie sent me in her care packages from home.

We slept on "rubber-band bunks" hammered together out of two-by-fours, wrapped around with rubber strips we cut from worn-out Corsair inner tubes. They were comfortable, once you got the tension right.

Tropical rain showers blew in every afternoon. We rigged a trough and collected the fresh water in a barrel. There was enough for cooking and bathing, but only occasionally for laundry. We rigged up a wind-driven washing machine to keep our clothes clean, just like in the movie *South Pacific*. This was a fifty-gallon barrel fixed with a pole, a little windmill propeller, and a pulley that moved an agitator in the barrel up and down. We would build a fire under the barrel to heat the water, and with salt-water soap your clothes got reasonably clean. Or we would just bundle them and tie the bundle to a coral head and let the waves wash them for a couple of days. Then we'd hang them up and let the rainstorms rinse out the salt water. As long as they were clean, what they looked like didn't matter.

This primitive, almost idyllic existence stood in stark contrast to the new weapon we were using against the Japanese. A thickener for jelling gasoline or kerosene had been developed a year or two before. Its name was napalm. Put into a 150-gallon drop tank and fused with white phosphorus in the tank cap that would flame when the tank shattered and ignite the jelled mixture, it made a devastating firebomb.

We first used napalm bombs in an attack on Jabortown on Jaluit Atoll, which had been the seat of the Japanese occupying forces when they controlled the Marshalls. We flew in behind another squadron. Tom Miller and I were not in our usual places near the head of the squadron, but were the last two planes in to attack. The skipper and the planes ahead of us dropped a total of sixty-two napalms and blanketed the town. When Tom and I came in, the whole island was a mass of smoke and flames. We flew in low, through the smoke, and there were only two areas not burning. He took one and I took the other. Nine hours later reconnaissance reported seven large fires still burning.

The destruction was total. Later, when we flew over Jabortown, there was nothing left.

The napalm was so dangerous that pilots were instructed to bail out and ditch their planes if their napalm bombs did not release. Nobody wanted a plane to go out of control on landing and firebomb our own planes and crews.

We made several napalm runs. Each plane carried

three bombs. You couldn't target them the way you could a regular bomb. We called the runs "nape scrapes" because you had to come in low, right at the target, drop the napalm, and pull up quickly as the inferno spread out underneath.

A napalm attack had a horrible beauty. The bright reddish orange of the flames and billowing clouds of black smoke against the green of the trees, the light green of the ocean water on the reefs, and the dark blue deep water made one of the most eerie, awesome, and sobering sights I had ever seen. We routinely used napalm in areas where intelligence thought there were a lot of people. It was terrible to think what it was like on the ground in the middle of those flames.

Flying in combat, you don't look into the eyes of the enemy you try to kill. You attack his machine with yours, or attack the area he occupies. But napalm was a hideous weapon and it made you think. Then the psychology of war took over. We were fighting in a war we hadn't started, for the survival of our country, our families, our heritage of freedom. We were there to do what had to be done. Using napalm didn't fit with peacetime sensitivities, but peace and a return to the sensitivities that it permitted were what we were fighting to achieve.

You don't think about that at the time, of course. You inhabit a small universe of you, your unit, and the enemy, and you're fighting for your own survival and that of the men you're with. It is destroy or be destroyed.

Sometime before Christmas, I came into the tent after mail call and saw Tom with his head down on his arm on the little desk he had made out of two-by-fours and food crates. I let him alone because I thought he was just tired, but when he looked up his eyes were red.

"What's wrong?" I asked.

He handed me the letter he'd just received from Ida Mai. Their ten-month-old son, Tom junior, had died. She described a rare intestinal blockage in which a part of the intestine telescopes into another. He had had corrective surgery and was recovering, but then something had gone wrong. I remembered how happy Tom had been back in March when he'd gotten Ida Mai's letter, and the several photographs since that he kept propped up against the kerosene lantern on his makeshift desk. He had never held his son.

We were close to the end of our year's tour by then. His son's death hastened the end of Tom's stay. He left the squadron on Christmas Eve for a sad trip home to Ida Mai in San Antonio.

I left Kwajalein at the end of my overseas tour six weeks later. I had flown fifty-nine missions, been hit by antiaircraft fire five times, fired thousands of rounds of .50s, and dropped countless general-purpose, incendiary, and napalm bombs. I had been awarded two Distinguished Flying Crosses and ten Air Medals, and had written at least two hundred letters to Annie. My role in the war to that point had been a relatively minor one, but I was proud that I had done what was asked of me as best I could. And while the Allies continued to advance on both fronts, the war was by no means over. The buildup for a final assault on Japan was well under way. I expected to be rotated back overseas in a matter of months.

In Hawaii, the pilots with starched shirts and pants

looked strange to me after my sea- and rain-washed khakis. Dick Rainforth and I were traveling together. Despite the long journey from Kwajalein, we didn't spend a night in Hawaii but got right on a military version of a DC-6 and flew to San Francisco.

I was eager to get home to Annie, but Rainy and I spent a couple of days in San Francisco for processing. After more than six months of showering in rainwater collected in canvas bags, we took pleasure in the luxury of stateside life. We went to the barber shop in the Fairmont Hotel on the top of Nob Hill and ordered the works. I had a haircut, a shave, a mud pack—everything they offered.

As I headed back across the country in my winter greens, by air this time, I was moved by the looks and words of gratitude the uniform attracted. I felt all over again the patriotic pride that I had felt as a boy playing echo taps with Dad at the cemetery in New Concord on Decoration Day. And there was the added sense that I was paying my dues, as so many others had done ahead of me.

Everywhere was a swelling expectation of an Allied victory. The Germans and Japanese were being pushed steadily, albeit at terrible cost, back to the countries from which they had launched the war that had engulfed the world. Desperate tactics and horrible atrocities marked their retreats, but neither stopped fighting. Japanese kamikaze pilots were killing themselves in attacks on American ships, while Japanese soldiers savaged civilian populations in Manila and elsewhere in the Philippines. The Germans hurled rocket bombs at Allied cities, marched inmates away from death camps, and tried to destroy evidence of their extermination factories. There was more to do before victory was ensured. This the people also seemed to know.

New Concord looked the same as when I left it. It was the quintessential American small town. Annie was even more beautiful than I remembered her. We had decided in the letters we had written back and forth that as soon as I got home we would try to start a family.

When my leave ended, I was ordered back to Cherry Point in North Carolina, then assigned a stint at the Naval Air Test Center at Patuxent River, Maryland. Annie and I lived in an old hotel called the Seven Gables. Down at the heels and not air-conditioned, it sat at the edge of the river, which is an arm of the Chesapeake Bay on the western shore of southern Maryland, just a few miles from the base. We called the place "Seven Stables."

My job at Pax River, with a number of other experienced combat pilots just back from overseas, was to put new fighter and attack planes through their expected hours of service life as fast as possible. We were not evaluating the planes, we were flying them—hard.

Aircraft development had been accelerated by the war. Grumman was producing two new planes. I flew various types, but the one I flew the most, the F8F Bearcat, used a more powerful version of the Pratt & Whitney engine that powered the Corsair, on a smaller and lighter airframe. They were also producing a twin-engine night fighter, the F7F Tigercat. The idea was to put these planes through everything they would go through in combat, other than being shot at, so as to work all the kinks out ahead of time. It was called Accelerated Service Test.

Service Test called for a certain number of takeoffs and landings, so many hours under wide-open full power, or combat power, time at maximum cruising speed, and a prescribed number of hours of acrobatic time, jinking and dodging and diving as if you were in aerial combat. You dropped bombs and fired the guns. We worked eight-hour shifts, and of that you would put in two three- or three-and-a-half-hour flights if the planes were in commission.

There was less air traffic then, and you could go anywhere you wanted. When I was flying the midnight-to-eight-A.M. shift, I liked to fly down to Cherry Point, where Tom Miller was stationed at the time. He and Ida Mai had rented a house outside of New Bern, a few miles from the base, and I'd get down low and give them a wake-up call at six in the morning, flat-hatting a few feet off their roof. In the daytime, pilots would practice dogfighting with each other.

One of the planes I flew at Pax River was the FR-1, the Ryan Fireball. Jet engines were just finding their way into Navy fighters, not as main engines but as boosters for use in combat situations. The Fireball was a single-engine prop plane with a jet engine hidden in the fuselage, and twenty minutes' worth of jet fuel. Flying the Fireball up to the P-47 base at Dover, Delaware, was great sport. You'd hook up with a P-47 pilot and get beside him, cut the main engine while you were flying with the jet, and feather the prop so it stood still. He'd almost spin in, trying to figure out how you were flying.

But at any second flying might serve you with a stark reminder that it could be dangerous as well as fun.

Fred Ochoa had been in VMO-155 in the Marshalls. "Bubba" Ochoa was a South Texas boy like Tom Miller. He was also at Service Test flying the F7F Tigercat. One night my plane was down for repairs and I was asleep on a couch in the ready room in the middle of my shift when a voice came on the squawk box. It was the operations office, wanting to know if Glenn was there.

"Yeah, this is Glenn," I said to the disembodied voice.

"Are you sure?" the voice came back.

"Last I checked," I said.

"Come to operations, on the double."

I hustled over there to learn that a plane had crashed in some bad weather in the Blue Ridge Mountains of Virginia. The pilot was killed, and the local sheriff had called to say he had found a dog tag with the name Glenn. The duty officer was getting ready to call Annie.

But the sheriff had made a mistake. What he had thought was a dog tag was a metal-rimmed cardboard tag on the parachute. Bubba Ochoa had borrowed my parachute that night because his was being repacked.

Bubba was a good friend and a good pilot, and his death struck me as particularly tragic. He had survived more than six months of being shot at in the Marshalls, only to be killed testing planes for combat after he got home. I didn't know yet what I was going to do with the rest of my life, but that was my stark introduction to test flying.

The war in Europe ended while we were at Pax River. The celebrations were subdued around the base, because the Japanese were still fighting hard on Luzon, in the Philippines, and Okinawa, and American soldiers, sailors, and Marines were still dying. The pilots at Pax River knew the war wasn't over for us, either. We felt a responsibility for the new crop of planes that was going out to the Pacific.

As the noose tightened around Japan, I received orders sending me back to Cherry Point. Annie was pregnant by then. At Cherry Point, I was assigned to VMF-913, a Corsair squadron of about 150 pilots. We were there on hold, expecting to be sent back to the Pacific for the invasion of Japan. But then the United States dropped the atomic bombs on Hiroshima and Nagasaki. Horrible as that weapon was, it would have been far more horrible to have the war go on; military planners had predicted as many as two million deaths if we had had to invade Japan.

On August 15, Japan surrendered at last, and America could breathe again.

CHAPTER 8

The end of the war left me uncertain about what to do. I had been promoted to captain in July, but in the fall of 1945 I was still waiting to learn if I would be offered a regular commission. That was just as well, because while I liked the Marines and the esprit de corps, I still didn't know if I wanted to make a career of it. What I did know was that I loved flying and wanted to keep doing it somehow.

That decision had already disappointed my father. When I first returned to New Concord from the Pacific, he had taken me aside and asked me to think about joining the plumbing business he had built up from nothing. He was almost fifty then. His thick, stout, strong body was showing—and feeling, he said—signs of wear from twenty-four years of digging pipe trenches and septic tank holes, wrestling boilers into basements, and twisting wrenches against stubborn pipes.

"It's a good business, Bud, and you could do it proud,"

he said. "You're known and liked here, and it's a foundation you and Annie can build a family on."

This was certainly true. But as proud as I was of my dad for building a successful business from scratch, I didn't see that as my life's work.

Doc Castor, it turned out, also had plans for me. He, too, took me aside during that leave.

"Johnny," he said, "I know you've had an interest in chemistry and science. Have you ever considered dental school? I've got a fine practice here, and it would be a fine thing for you and Annie if you were to take it over. Annie could work in the office. You could keep it all in the family."

I was grateful to both of them. Their confidence in me to carry on the careers they had built their lives around made me feel proud. But the war had changed me more than my parents realized. It had taken me away from the confines of New Concord. My hometown was a wonderful and cherished place, but I had my sights set elsewhere. I told them I wanted to keep my options open.

At Cherry Point, still waiting for word on my commission, I talked with another pilot in the squadron about buying some war surplus transports at giveaway prices and going into the shipping business. The North Carolina coast around Cherry Point was rich with seafood. We thought we might fly fresh seafood from the coast to the Midwest. That idea lasted for all of fifteen minutes.

I thought about going to work for the airlines. But I knew down deep that I had given that up when I had opted out of twin-engine flying in favor of fighters back at Kearney Mesa. There were plenty of experienced transport pilots vying for those jobs anyway.

Then my commission came through. As soon as I got the word, I knew what I wanted. I liked the Marine Corps, I liked the life, and I particularly liked that kind of flying. Annie and I talked it over. I think in the back of her mind she had never really expected we would settle down in New Concord after the war. She knew how much I loved flying, not just from me but from pilots I had served with. She had learned what I would never tell her in so many words—that I was proud of my ability as a pilot. Part of my love for her grew from knowing that she understood me so perfectly. She knew that flying had become part of who I was, and she was proud of me.

"Honey," she said, "I know flying makes you happy. When you're happy, I'm happy. Let's join the Marines."

We still had the little '39 Chevy coupe we had bought the first time we were in North Carolina. As Annie's due date grew closer, I worked on that car obsessively, making sure it ran perfectly and always had a full tank of gas.

Cherry Point had no military hospital. The closest one was at Camp LeJeune, about fifty miles away, and that was where the Marine wives at Cherry Point went to deliver their babies. I didn't want Annie delivering by the side of the road because I'd forgotten to keep the car gassed up and tuned.

Thanksgiving passed, and we counted the days into December. She assured me that everything was going to be fine and not to worry, but I was not that certain. She twice started feeling labor pains. Each time I was up, dressed, had her hospital bag in the car, and was ready to go when the pains subsided. The third time it was about two o'clock in the morning of December 13, in the middle of a rainstorm. Her pains started coming closer together, and we knew we were headed for the hospital. I helped her to the car in the pouring rain, and we started down the road to Camp LeJeune.

The wipers in the Chevy were the one thing I hadn't thought to check. The rain was blinding. Every minute or so we'd be rattled by a thunderclap. Every jagged flash of lightning showed an expanse of rain-soaked swampy countryside with no houses and no lights. Annie's pains kept getting closer together, the rain kept pouring down, and I tried to remember what little I knew about being a midwife. The cries "Towels! Hot water!" came to me from half-remembered movies, but the only hot water was in the radiator and we had no towels, only a bundle of baby blankets.

We finally got to the hospital at about four-thirty in the morning. Annie delivered our son, John David, only two hours later. He weighed seven pounds thirteen ounces, and had Annie's dark hair and a strong, insistent voice. He was beautiful, and we loved him from the moment we saw him.

Dave was not a Tar Heel for long. There was little flying going on at Cherry Point. Fighter production had slowed with the end of the war. People were leaving the services and we had both fewer planes and fewer crews to maintain the planes we did have. That, combined with a surplus of pilots, led to my second transfer from Cherry Point to the West Coast. Annie, Dave, and I left for El Toro, California, in March 1946.

The squadron to which I was assigned at the Marine Corps Air Station at El Toro, VMF-323, was known as the "Death Rattlers." The nickname had followed the squadron home from the war. I was its operations officer.

Demobilization had its effects at El Toro, too. Mechanics and maintenance personnel were in short supply. In order to fly, the pilots had to work on the planes themselves. I could tune a six-cylinder Chevy and four-cylinder Model A and Model T Fords, but the 2,000-horsepower Pratt & Whitney Double Wasp radial engine with its eighteen cylinders presented different problems. Plus the need for roadside assistance could be a lot more critical. I was more comfortable flying planes that had been worked on by qualified mechanics. Nevertheless, we kept as many planes in commission as we could.

The Corsair had acquired an impressive reputation as a fighting plane during the war, and there was a great demand on Corsair squadrons such as ours for flyovers and for demonstrations at air shows. We performed up and down California from San Francisco to San Diego. The crowds always reacted with great enthusiasm, expressing postwar thanks to the pilots and an appreciation for the military technology and manufacturing prowess that had served America so well. And we pilots at least got to practice our formation flying.

This was the first time we had lived on the West Coast in peacetime, and Annie and I both loved it. Every time we could, we headed for Laguna Beach, Corona Del Mar, or the other ocean beaches between Los Angeles and San Diego. We played in the sand with Dave and dived for abalone and took them home to cook. In the heat of summer, we sometimes drove to the Laguna Mountains between San Diego and El Centro. Annie and I had always enjoyed horseback riding together in the cool mountains. Now we hiked, since Dave was too small to be on horseback.

Life in El Toro was fine, but the flyovers soon paled as an exercise in flying. I didn't seem to be learning anything new as a pilot. It felt as if I were treading water while the postwar military was trying to figure out what to do with itself.

About this time, word came down that the Marines were looking for volunteers for China duty. Rumor had it the tour would be a short one, six months at the most but more likely three or four, with credit given for a full year's tour. The squadron's job would be supporting General George C. Marshall's negotiation teams that were trying to broker a peace between the Nationalist Chinese, led by General Chiang Kai-shek, and the Communists of Mao Tse-tung. I had been home for over a year and was due for another overseas assignment. Annie was five months pregnant that November. With Dave still a baby, I thought that if one short tour could take care of my overseas duty for several years to come, while Annie waited in New Concord, this would be the thing to do. Annie agreed. I volunteered, assuming I could be back home before our second child was born.

VMF-218, a Corsair squadron of Marine Aircraft Group Twenty-Four, First Marine Aircraft Wing, was based at Nan Yuan Field, just south of the Chinese capital, known then as Peiping. I arrived in December 1946, at the beginning of the Chinese winter, as the squadron's acting operations officer.

Peiping was a strange and exotic place. There were very few Caucasians other than those of us in the military. People on the street stared at us with open curiosity. The Communists directed threats at us; they saw us as supporting Chiang Kai-shek. But the corruption of at least some of those in the Nationalist regime was evident even to those who were supporting it. From our base at Nan Yuan, United Nations relief teams were air-dropping padded containers of flour and other foodstuffs into areas within fifty miles of Peiping where people were starving, and two or three days later those same bags showed up for sale in the big open market behind the Peiping Hotel in the heart of the city.

The hotel was the tallest building in the city then, at seven or eight stories. Our air base, which we shared with the Nationalists, was six miles from the South Gate of the Forbidden City. Now the gate opens onto Tiananmen Square, but then the road from the gate ran through a crowded area of shops and narrow streets. Evidence of poverty was everywhere. The rickshaw drivers ate rice with weak soup they bought from vendors, who boiled the soup from animal entrails in large barrels on the street corners. Many of them slept next to or in their rickshaws, pulled up under the eaves of buildings.

The food at our base was generous by contrast. We ate well, and the garbage from our mess hall was dumped into a ravine outside the electrified security fence. Hungry Chinese would walk up to six miles from Peiping carrying buckets and pick through the garbage looking for something edible. The word spread until so many people came that the Nationalist troops had to put down a near riot. After that the garbage was sold to a man from Peiping who

My father, John Herschel Glenn, served in World War I. He taught me the bugle and military calls.

I'm riding a pretend hobbyhorse in this picture, which was taken when I was 2 or 3.

I'm with Mother, Clara Sproat Glenn, on a summer afternoon.

My sister, Jean, with my father and me. I'm still in the knickers stage.

Annie Castor and I met in a playpen. By the time we were teenagers, we were a couple.

After Pearl Harbor the Navy sent me to Olathe, Kansas, for primary flight training. [U.S. NAVY]

Annie and I were married in New Concord on April 6, 1943.

VMO-155, in training at El Centro, California. *Back row, from the left:* Tom Miller, me, Monty Goodman, Bill Blades. *Front row, from the left:* John Griffith, Pete Haines, Dick Rainforth. [COURTESY TOM MILLER]

Tom Miller and me on the beach at Waikiki, waiting for our combat assignments. [COURTESY TOM MILLER]

I sent Annie this hula outfit from Honolulu.

The pilots of VMO-155, on Majuro in the Marshall Islands.
Boy, we were all so skinny then!

Home on leave in spring 1945, with
Annie and my parents.

The Chance Vought F4U Corsair was the plane I flew during the war and in China afterward. [U.S. NAVY]

With other Marine families on Guam in 1948. Annie's on the right and Dave and Lyn are on the blanket.

I was training
pilots in
Corpus Christi
in 1949, where
this picture was
taken after
church one
Sunday.

In 1953 I'm at K-3, an air base
near P'ohang, South Korea,
where I seem to be enjoying
the food.

Our squadron wore silk
scarves with hearts, and our
nickname was "Willing
Lovers." It came from WL,
our planes' tail letters.
[U.S. MARINE CORPS PHOTO]

My closest call. I got hit with antiaircraft fire after taking out a gun emplacement in Korea. [DEFENSE DEPARTMENT PHOTO]

Ted Williams flew with VMF-311. We called him "Bush" to get his goat.

As an exchange pilot with the Air Force, I shot down three MiGs near the end of the war. [U.S. AIR FORCE PHOTO]

Practicing refueling for my supersonic cross-country speed run, 1957.
[U.S. NAVY PHOTO]

After Project Bullet, I appeared on *Name That Tune* with Eddie Hodges.
[CBS PHOTO ARCHIVE]

Setting the coast-to-coast record was a high point of my test pilot years.
[U.S. NAVY PHOTO]

had a truck. He put it in barrels, took it to the city, and resold it.

I shared quarters in a Quonset hut with several other pilots. We paid a houseboy named Yon to keep things neat and clean. Each day, if he couldn't hitch his usual ride, he walked the six miles out from Peiping and home again. We gave him oranges and cans of Spam and other things. He invited me to visit him, and I accepted. Yon lived in two tiny rooms that he shared with his wife, mother, and two children. He was considered fortunate because he had a job.

We were there to support the peace teams and encourage peace, but even as much against Communism as I was, I couldn't help but wonder why the Chinese people hadn't revolted sooner. I didn't see that they had a whole lot to lose. In retrospect, you could say they had their freedom to lose, but they had no freedom as we know it, and for a great many people it was a choice between Communism and starvation.

We went out on what was called North China patrol every day. We flew a grid-type pattern at low altitude, five hundred to a thousand feet, forty or fifty miles in one direction, then a ninety-degree turn for a few miles, then another ninety and back the other way. We were looking for cuts in roads or railway lines, bridges blown up, power lines down. Our reports let intelligence and the Nationalists know where the Communists were operating. As the squadron's ops, I had people assigned to sectors that covered most of North China, from Peiping and the Great Wall down to Tientsin—now Tianjin—every day. We weren't supposed to fly beyond the Wall, but most of us managed to view it from the other side. The country was stark, rugged, and largely deserted, an inhospitable place to be if you were forced to go down, so we didn't bend the rules too much.

Although officially noncombatants, we flew well-armed. Our guns were loaded and there were bombs on the bomb racks, so that if we came under fire, we could fire back. Intelligence told us the Communists were operating in the area. There were always threats that they were going to come down and attack the base, but they never did. At night we sometimes heard mortar fire in the distance.

Over time, the Communist threats became quite specific. At one of our intelligence briefings, the officer told us the Communists had warned that one Marine a day would be killed as long as we were in Peiping and continued supporting General Marshall's peace teams. We had heard similar threats before, but this time, two days running, Marines were attacked and killed in back-street areas of Peiping. All military personnel were put under base restriction for ten days. When the restriction was lifted, we were allowed into Peiping only in pairs, and only if one person was armed. We kept largely to the corridor between the base and the city.

Pilots routinely carried a .38 pistol in a shoulder holster when we flew. When I went into Peiping I carried my .38 in a pocket of my heavy winter parka. I also carried a blackjack that I had bought in a shop in one of the streets outside the gate to the Forbidden City. It was eight inches long and hard as a rock.

Despite the threats, we visited the famous locations of Peiping, such as the Summer Palace, the Temple of Heaven, the Ming tombs, and the flat stone temple area that ancient

Chinese astronomers had determined was the exact center of the universe around which everything revolved. It wasn't just these destinations, but China in totality that fascinated me. The culture and customs, the tonal variations of Mandarin, the food, the music with only five tones, the Buddhist priests in their robes, all created a core of interest that would remain with me for the rest of my life.

China was so alien, and so utterly different, that it made me think even more than I usually did of home and Annie. No one was more committed to his country and his duty, but I was a happier person when we were together. The days stretched longer without her, and one night I found words to express the way I felt and put them on the back of some old orders:

It's funny how a lonesome feeling—a feeling of utter boredom and futility—comes over me when I am away from Annie. It seems as though a part of me was actually somewhere else. And it's always the part that has a good time and enjoys life. The only remaining part is that which sees things as being rather gloomy and ill-tempered. This feeling certainly is one of the mysteries of life to me. I cannot see it, touch it, smell it, or hear it, but I feel it inside of me as surely as though it were something I could lay my hand on. When I'm alone, a new morning, the sun coming up, white clouds, a rested feeling following a good night's sleep, are all just the beginning of another day that must be hurried through so I can get on to something else. With her, a new day is something to enjoy together and a new opportunity to appreciate all the good things in life. The sun coming up starts the day with a radiance of color that shames a Michelangelo and makes any man-made attempt at beauty a mere insignificance. Can we hurry on and ignore such a thing? Too often we do. It's a wonderful thing to enjoy such beauties with one whose love means more than life itself.

There was more, but I essentially defined myself and Annie as being parts of the same whole. I thought a part of her came with me and a part of me stayed with her when we were apart, and those parts rejoined when I was with her, restoring life's richness and vitality.

After I wrote all this, I slipped the pages in with some other papers. I didn't send them to her, I only wanted to clarify my thoughts. I found them only by accident over fifty years later, and when Annie saw them at last, she was moved by the feelings that haven't changed in all that time.

For the Communists, Marshall's peace initiatives fell on deaf ears. Late in the winter of 1946–47, his teams pulled out. The squadron covered their evacuation by train from Peiping along a southerly course to Tientsin, then north following the coast to the port city of Ch'inhuangtao (now called Quinhuangdao) where the Great Wall meets the ocean. It was a route made familiar to us by almost daily reconnaissance flights. We flew back and forth over the train throughout its two-day journey.

•   •   •

All of us were relieved when the train reached the port and the Army and Navy took over the job of moving the negotiators and their families out of China. If the Communists had tried to attack Peiping while the peace initiative was going on, or if they had attacked the peace teams while they were withdrawing, we would have been in active combat. But Mao and the Communists would not complete their takeover for two more years.

With the peace teams safely evacuated, we were starting to fold up the squadron to move to Guam when I got wonderful news from the Red Cross. Annie had delivered our second child, a girl, on March 19 at Good Samaritan Hospital in Zanesville. We had agreed on the name Carolyn Ann. Now we were four. I thought back to the hope that my tour in China would end before she was born. It hurt not to be there. I hadn't envisioned our life being like that, but I was buoyed by the news that they both were doing fine, and I passed out cigars in the ready room.

While some of the officers and men stayed behind to pack equipment for shipment out through the Marine Division Headquarters at Tientsin, I was part of the flight echelon that flew the planes to Tsingtao (now Qingdao) on the Yellow Sea, and then to Shanghai.

We spent a few days in Shanghai waiting for weather good enough to fly to Okinawa, a way station in the shift of the squadron to Guam. Shanghai was very much old China. We stayed at what had been a German school, sharing quarters with the Navy's criminal investigation unit. Their job was to keep Navy personnel from wandering into areas of the Shanghai waterfront where sailors had turned up missing. One night I asked if I could go with them on their rounds.

I followed two Navy petty officers carrying submachine guns on a round of the opium dens near the harbor in Shanghai. We walked into rooms where the smoke hung so heavy near the ceiling that you had to bend double to see across the room. The places were all very quiet, with men sitting or lying on bunk beds and pallets smoking pipes of opium. They were in such a daze they hardly looked at us. All were Chinese; we saw no Americans in the rounds we made that night.

The weather broke a few days later, and we flew the planes the five hundred miles out to Yonabaru Airfield, on the southeastern shore of Okinawa.

The squadron had been on Okinawa about a week when I received another message from the Red Cross that shattered the euphoria the first had brought. Annie was back in the hospital with a serious infection. The message advised me to get an emergency leave and go home, if I could.

Worry ate at my heart as I flew across the Pacific in a C-54. I'd been able to get no word on Annie's condition since the message, and in the absence of clear information my mind conjured up the worst possible scenarios. What if I lost Annie? I didn't think I could take that. I had missed her in China, but thoughts of her helped me go on. And now that she was the mother of our two beautiful children—I was sure Lyn was gorgeous, even though I'd never seen her—her life to me was that much more important.

Five days of travel and worry took me home at last. I arrived to the joyous news that Annie had been released from the hospital that same morning. She was recovering at her parents' house in New Concord, very weak but free of infection. And Lyn was as beautiful as I had imagined her.

I was home too few days before I had to return to the Pacific. I rejoined the squadron in Guam, where it had gone by aircraft carrier and was setting up shop at Naval Air Station Orote.

As weeks and then months passed on Guam, it grew even clearer that I had miscalculated in believing the sales pitch that VMF-218's overseas tour would be a short one. The Marines at Orote starting casting envious eyes on the Quonset hut village the Seabees had built for Navy personnel and their dependents. Navy men were allowed to have their families with them; Marines were not.

In time, the base CO told the Marines that if we wanted to build our own Quonset huts from surplus kits that were sitting in a warehouse, we could bring our families out. This set off a flurry of building. Instead of the usual volleyball games and swimming and lounging on the beach after our mornings flying, we worked on the Quonset huts. I didn't plan to apply for Annie, Dave, and Lyn to come to Guam, because I thought I was near the end of my tour and expected to be sent home. But I helped with the building, and soon we had about thirty Quonset huts set up ready for occupancy.

As soon as the Marines started the process that would bring their families out, the CO decreed that Navy personnel needed more Quonset huts and that they would get half of those we'd just built. It was one of the most high-handed and unfair moves I saw in my twenty-three-year Marine career, and it set off a near mutiny among the Marines. His edict was overturned, however, by a Marine general who arrived to head the brigade on the far side of the island and became the senior officer on Guam.

More months passed. I watched the families of my squadron mates arrive and set up housekeeping in the Quonset huts I'd helped them build. Finally, in mid-1948, after my tour had stretched to eighteen months, I decided I couldn't be sure when I was going to get orders home, and put in for Annie and the kids to come out to Guam.

The journey that faced her, halfway around the world with Lyn a little over a year old and Dave less than three, would have been daunting to someone who had no difficulty speaking. For Annie, with the elaborate strategies she had to use in order to communicate with people, it was a mountain to climb. But she was determined. She got them all on a plane to San Francisco, reported in to the Alameda Naval Air Station there, and flew with the kids on a big four-engine Navy seaplane, the Martin Mars, to Hawaii. Tom Miller, who also had decided to accept a regular commission and stay in the Marines, was stationed there with Ida Mai. They met my three travelers and put them on a C-54 to Guam.

I waited in the big Quonset "terminal" at Naval Air Station Agana for the flight that was bringing them in. I wasn't alone; several other men were waiting for their families, too, and we cracked bad jokes to hide the nervousness we felt and took turns scanning the horizon for the plane. It finally appeared, landed, and taxied to a halt. When I saw Annie at the top of the stairs, holding little Lyn in one arm and Dave by the hand, I was the happiest man west of San Francisco. The kids really knew me only from photographs Annie had shown them, and I had to be careful not to scare them half to death by hugging them too hard. With Annie, well, that was different. We had a lot of time to make up.

We moved into a Quonset hut that the squadron exec had lived in with his family until they rotated home. It was one of those I had helped build. We had a car shipped out, too, a big used Buick Roadmaster I planned to sell when we left, so when I wasn't flying, the four of us could go to the beach and explore the island that the United States had retaken less than four years earlier at the height of the war in the Pacific. It had cost 18,500 Japanese lives and 2,145 American ones, but already the tropical growth was hiding the scars of war.

Naturally, three months after Annie and the kids arrived, I got orders home. I was happy that we were heading back to the States, but the orders specified that we travel by MSTS—Military Sea Transport Service. Visions of the voyage from San Diego to Hawaii on the *Santa Monica* disturbed my sleep as the date of our departure neared.

Sure enough, a big storm was moving east across the Pacific, and we moved with it. The ship rolled a lot. To get air and relieve the heat, we had a wind-catcher fixed in our stateroom's one porthole. It scooped the air into the room. And that wasn't all, as we learned one night when the ship's bow plunged into a wave and some fifty gallons of cold sea water drenched the bunks where we were sleeping. We crossed the International Date Line, and so had two Thanksgiving dinners while en route.

The coast of California came into sight after two weeks at sea, in December 1948. I said my prayers of thanksgiving then, and renewed them as the transport steamed under the Golden Gate Bridge into San Francisco

Bay with Annie and the kids and me all watching from the deck and tossing coins into the water for good luck, two years after my "three-month tour" had begun.

I had been assigned to the Naval Air Training Center at Corpus Christi as a flight instructor. It wasn't what I wanted.

Most fighter pilots viewed an instruction assignment as a diversion from the active squadron-type flying they preferred. In fact, I had asked for a squadron assignment, and was turned down. I hoped that at least I'd be training fighter pilots. No such luck. I reached Corpus Christi with the family in early 1949, only to learn that, for ninety days at least, I wasn't going to be flying at all. The people who assigned personnel had been waiting for the next Marine junior officer, assuming whoever it was would be tough enough to run a tight ship. Suddenly I was in charge of the downtown headquarters of the shore patrol from six P.M. to two A.M. in a major liberty port for the Navy. My domain included two small jail cells, a tiny office, and a staff of five, including a Navy medic. Sometimes I got extra men for foot patrol.

I had no training whatsoever in police work. Fortunately, I had Chief Botterman. Botterman was a Navy lifer, a chief petty officer whose rank was the equivalent of a master sergeant in the Marine Corps. He stood about six-four and weighed around 240 pounds, most of which was muscle, and he had been on shore patrol duty in Corpus Christi for some time. He knew the back alleys, the red-light district, the Mexican bars where sailors would go to drink tequila with the worm, and the local police officers who worked those beats and called us when sailors were causing trouble, or were in trouble and needed help.

Most of the calls we got had to do with drinking, fist-fights, or a combination of the two. Guns weren't usually involved. Knives and razor blades often were, however. I saw more blood on shore patrol duty in Corpus Christi than I had during the war. Botterman and I broke up a lot of fights after I learned his way of doing things. That occurred during my first week. We got a call about a fight going on outside a bar. I jumped into a shore patrol truck with Botterman. Two other patrolmen followed in another truck. We pulled up at the edge of a big crowd. In the middle, a good-sized scrum of sailors wearing Navy whites was going at it with some locals. The crowd was siding with the locals. I wanted to show how things were going to be done in the shore patrol on my watch, and I jumped out of the truck and waded through the crowd. I was pulling people aside and yelling at them to stand back. I finally reached the middle of the crowd and got ready to break up the fight when I realized I was all alone. Botterman was still back in the truck, and my two other stalwarts were hanging back, waiting to see what he would do. I beat a strategic retreat, and then we got organized and got the sailors out of there. From then on Botterman and I both understood how we would operate. His was the voice of experience.

Shore patrol was like a seamy highlight reel. Corpus Christi's Tuesday night professional wrestling matches always seemed to produce a fighting mood in the many sailors who attended, as if they didn't realize the matches were choreographed and practiced in advance. One night a prostitute stormed into the office, described a sailor, and demanded that we arrest him for nonpayment for the services she had provided a little earlier in the park down the street. Shore patrol wasn't flying, but it sure could be interesting.

Annie and I bought our first house in Corpus Christi. The man who sold it to us owned a lumberyard. He'd built the house for himself and had used all kinds of beautiful wood in the interior. It had three bedrooms, an attic fan, and a huge stone fireplace, and it cost $12,500. We shared it with a nest of scorpions in the pampas grass along the back fence and a lone tarantula that I relocated one cold winter evening from the front step to a mason jar.

Tom and Ida Mai Miller also lived in Corpus Christi. Tom was attached to VF-1, the same training unit I was going into when my career as a shore patrolman ended. Tom Collins, a Marine lieutenant, and his wife, Mary Alice, whom we called Feathers—that was her maiden name—also became our close friends. We did a lot of things together, the kinds of things you do with young children and a Marine salary. We had picnics and cookouts and went to the beach at Padre Island, across a long causeway from the mainland. We played in a summer softball league and a winter bowling league, and ate a lot of meals at the officers club. Annie got a job as the organist at the Methodist Church, and we attended services there on Sunday mornings.

After three months at the shore patrol, orders arrived shifting me to Cabaniss Field in the air training center complex. My first duty as a captain and a new instructor in the Training Command was a six-week tour in the IATU, the Instructors Advanced Training Unit. Ten veteran pilots, most with World War II experience, put the new instructors through the wringer on everything: our ability to fly and teach flying; standardized methods of teaching fighter tactics, gunnery, and dive- and torpedo bombing; and our knowledge of weather and aircraft maintenance. Graduating at the end of six weeks, I was sent over to Advanced Training Unit One to work with cadet student pilots moving from basic flying into fighters. ATU-1 was using Corsairs in the training, and while it wasn't squadron flying, it was the next thing to it.

Three days later the IATU called me back. In less than a week I went from an instructor-trainee to an instructor to an instructor of instructors in the course from which I had just graduated, teaching in the mornings, and flying in the afternoons. I came to appreciate the focus that training instructors required, not just in the techniques of teaching, but in what the most vital aspects of flying and airborne combat were, and how to convey them to students. I could see that teaching flying would make me a better pilot.

Almost as soon as I was called into the IATU, I began corresponding with a similar Air Force training group at Williams Air Force Base in Arizona. The Navy and Air Force occasionally exchanged instructors to work with cadets, but I wanted to know if any of their training methods for instructors were better than ours, and if we had anything to offer them. Several months later I proposed that we send somebody from our unit to Williams to go through their three-month course for new instructors, and that the Air Force send someone to the IATU in Corpus Christi. The idea was approved. I volunteered to go, since it was my idea, and got the assignment.

Williams was twenty miles east of Phoenix, and I flew

my first jets there. The Navy and Marines were just beginning to form a few jet squadrons, but our Training Command was still using prop planes exclusively. I quickly learned that jets changed everything. I'd been used to doing things at three hundred miles per hour, and now suddenly I was doing them at six hundred miles per hour in the Air Force's training jets. They were P-80s, the original jet fighter Lockheed called the Shooting Star, and the T-33, a two-cockpit jet derived from the P-80. The jet engine had a power lag that took some getting used to; prop fighters responded instantly when you pushed the throttle to respond to an emergency, whereas the jet took several seconds to accelerate. It was the same in reverse. You could bring a prop plane close in to the field, cut the power way back, and come in steeply to land. The jets required a longer, flatter landing pattern. But I got used to the new feel of jets quickly.

While the Air Force had a leg up on the Navy in that it was training its new pilots in jets, I felt its training syllabus lacked something essential. They were teaching jet flying, but not the use of the airplane as a combat weapon. I thought we did a better job of training pilots in gunnery, bombing, and the tactics of air-to-air combat.

I was learning to love jet flying when, in June of 1950, the Cold War that had succeeded World War II grew hot. Sixty thousand Communist North Korean troops equipped with Soviet tanks and guns invaded the Republic of Korea, to the south. After the failure of Marshall's peace teams, the Communists had taken over all of China the previous fall. With almost the entire mainland of northern Asia—the Soviet Union, China, and North Korea—already Communist, the invasion solidified fears of a worldwide

Red tide that had to be stopped. The United Nations Security Council condemned the invasion, and President Truman pledged to help South Korea fight back in a "police action." Within days, American troops were on their way from Japan as the major component of a sixteen-nation United Nations force. But five years after World War II, the troops had grown soft and their weapons had deteriorated. The North Koreans pushed them to the southwestern corner of Korea, around Pusan.

But the Communists would get no farther. The Marines and Army landed at Inchon, west of Seoul, in September. Reinforced Army units broke out of their defensive positions around Pusan. By the end of the month the largely American UN force had retaken South Korea up to its border with the north near the thirty-eighth parallel. It crossed the border in October and by the end of the month had pushed the North Koreans back to the Chinese border at the Yalu River.

The Chinese responded by crossing the Yalu with two hundred thousand troops. The next months of the brutal North Korean winter would see both inglorious American withdrawals and remarkable courage and ingenuity in the Marine retreat from the Chosin Reservoir—one of the most notable episodes in Marine history. The infantrymen, aided by Marine pilots in Corsairs, brought out their weapons, their equipment, and their wounded, all while basically surrounded by Chinese. "We are not retreating. We are merely advancing in another direction," said the Marine general, O. P. Smith.

When I returned to Corpus Christi, I applied to the Naval School of All-Weather Flight. This was another three-month course, emphasizing instrument flying, and for me another step up the aviation ladder. In the NSAWF, students came from regular Navy and Marine duty all over the country, with the graduates going back to their duty stations as specially qualified instrument instructors. At the end of that course I had done well enough to be again asked to stay on as an instructor. It was rigorous training, in which vision outside the cockpit was blocked so that you had to fly solely by reference to the instruments, as well as "partial panel" flying, with some instruments blocked out, and flying with intentional instrument failure. I was there long enough to take two groups of students through their three-month courses.

My tour in the Training Command at Corpus Christi ended in the late spring of 1951. Annie and I put the house on the market and made plans to move back east to Virginia. I had been ordered to Quantico and the Marine Combat Development Command for the six-month junior course at the Amphibious Warfare School.

In July, the month I reported at Quantico, the Chinese and North Korean Communists had driven the United Nations forces back to the thirty-eighth parallel. There the battle lines stabilized across a no-man's-land. Ceasefire talks started, but they were hardly worth the name.

The course at the Amphibious Warfare School was another assignment I didn't particularly want but which, as a career Marine officer, I knew was mandatory. The Corps required all junior officers, ground and air, to go through it as a way of familiarizing them with the overall Marine Corps and to teach them how all combat, support, supply, logistics, artillery, air, and command activities dovetail on the battlefield.

For six months we studied every facet of the coordinated warfare that makes the Marines unique among the services. The Amphibious Warfare School was a comprehensive introduction to the big picture of using boats, planes, troops, and guns to win battles. I knew about flying, but the Amphibious Warfare School taught the use of aviation in support of ground troops. With my fellow junior officers, I learned the deployment of artillery and how ground and naval artillery are used in various situations. We learned how to create a plan for taking a particular target—organizing troops to go aboard ship, what supplies and equipment were needed, how long the landing craft would take to get to shore, and the range of naval gunfire it took to support them. We analyzed attack air power for close air support in advance of Marines on the ground, and interdiction deep behind enemy lines. We took turns serving as staff officers in a Marine Corps ground division in simulated combat. I was the commanding general when we took the area around Mt. Vesuvius in Italy. Field exercises with live firing supplemented the classroom work.

The housing situation at Quantico was wretched. All Annie could find for us and the kids was a run-down two-bedroom house just off Route 1 in an area called Triangle Village. Later we moved to a village of two-family Quonset huts on Marine property about eight miles south of Quantico. The fiberboard walls were so thin you could hear the people next door breathing.

It was the presence of friends that made it tolerable. Tom Miller had been transferred from Corpus Christi to the Marine Corps Air Station at Quantico. Dick Rainforth, my squadronmate from the Marshalls, and Tom Collins, from Corpus Christi, were in Quantico as well. Annie and I spent a lot of time with them and their wives and kids, and Annie played the organ at the chapel on the base.

At the end of Amphibious Warfare School, I once again hoped for a squadron assignment. Instead, in December of 1951, I was assigned to the general staff of the commandant, Marine Corps Schools, in the operations section. I was placed in charge of the "E areas," live-fire areas in a vast section of the base called Guadalcanal that was used for exercises. It was my job to schedule what was happening in the live-fire areas each day, making sure that roadblocks were up when they were needed and all of the safety precautions were in place. I was fortunate to again have a Botterman, in the form of a master sergeant who had been doing this scheduling for a long time, and on whom I relied.

I was the only pilot on the staff. Shortly after I reported, my boss, the colonel who was the base operations officer, asked me how much flying time I needed to qualify for flight pay.

"Four hours," I said, but of course that was a minimum, barely enough to maintain proficiency.

"Fine," he replied. "You can have four hours a month away from the office. Don't try to stretch it."

I didn't bother to discuss it. It took an hour just to drive to the field, check out the plane, file a flight plan, and get the wheels up. I took no time off, and did all my flying on weekends. Tom Miller and I would go to the waiting room at the air station on Saturday mornings. There were always a few people who had just gone on leave waiting to see if they could catch a ride somewhere. We'd find out where they wanted to go. Then we'd file a flight plan, load them on an R4D, the military DC-3, and deliver them to their destinations in the discomfort of hard seats and a noisy, unpressurized cabin.

All the while I kept trying to get back to a flying squadron. I took to writing monthly letters to my boss, Colonel Thompson, asking for assignment to Korea. The fighting continued while talks dragged on over the exchange of prisoners. I now had extensive training in all aspects of flying. I had the coveted green card awarded to pilots rated to fly in the most severe weather conditions. To be in a ground post that didn't allow me to use my training just didn't seem right.

But every month I got a letter back stamped "Request denied." The next month I'd write again. Finally the colonel called me into his office and told me to stop writing letters. "You're going to go when the Marine Corps wants you to go, not when you want to go," he said.

In October 1952, the Corps decided it wanted me to go.

CHAPTER 9

Annie had the packing down cold. After nine years as a Marine wife and more moves than anniversaries, she could reduce the household to a stack of boxes and four suitcases in less than half a day. We headed to New Concord, where I would leave her and the kids, then return to Cherry Point in North Carolina for two and a half months of jet training.

I hadn't volunteered for Korea without discussing it with Annie. Dave, Lyn, and Annie, all three of them, entered the equation of duty and risk that had to be considered. But I knew what I had to do.

Annie understood. Once we had decided I would stay in the Marines, she accepted uncomplainingly the life that came with that decision—not only the frequent moves and the scramble for housing that each new move required, but the dangers inherent in flying and the possibility that I would go back into combat.

We didn't talk about how much we would miss each

other or the chance I might not return. Death never entered our conversations. As we drove from the East Coast to New Concord along now-familiar roads, we talked about the prospects for Dave entering elementary school and Lyn kindergarten. The kids counted cows along the road. Then they counted red cars. The four of us sang traveling songs.

I drove back to Cherry Point after they were settled at her parents'. After I was checked out in flying and fighting in the F9F Panthers the Marines were using in Korea, I retraced the roads to New Concord for a short leave before saying goodbye.

Mother and Dad, and Annie's parents, hardly shared my enthusiasm about the assignment to Korea. Of course, they weren't eager to have me shot at, but their attitude was partly due to the nature of the war. Americans were frustrated over the stalemate and a "police action" that had produced over thirty-five thousand American deaths and a hundred thousand casualties, many of them from exposure during winter nights that reached twenty below zero. The South Korean regime of President Syngman Rhee was widely seen as incompetent and corrupt, like much of the South Korean army. It wasn't clear to most Americans what we were doing there, or what we hoped to gain from leading the United Nations' first military action. Dad was as anti-Communist as anyone could be, and even he said that fighting Communism had tucked the United States in with a strange bedfellow in Korea.

Dwight Eisenhower, retired from the Army he had headed as the supreme Allied commander in Europe during World War II, had been elected president in part because of expectations that he would figure out a way to end the war. America wanted out of Korea, at the same time I was wanting in.

Our parents gave me their blessing nevertheless.

One of the last things I did before I left reminded me anew of the unpredictable dangers of war. Tom Collins, my training command friend from Corpus Christi, had been killed in Korea in a takeoff accident. His wife Feathers asked if I would scatter his ashes over his family farm near Amanda, Ohio, as Tom had requested in his will. I borrowed an old F6F from the Marine Reserve base in Columbus. The undertaker walked out to the plane and, with the family watching, opened the urn and handed the small plastic bag containing Tom's remains up to me in the cockpit. I took off and followed Feathers's map to the farm. There, I slid back the canopy, spilled the ashes out into the slipstream, and granted Tom's last wish.

At the beginning of February, I hugged my parents and Annie's goodbye in New Concord, and drove with Annie and the kids to the Columbus airport, where I was catching a plane to Los Angeles. When my flight was called I told Dave and Lyn to take good care of their mom and said I'd be back before they knew it. With Annie, after all this time, I hardly needed words to tell her what I felt.

"I'm just going to the corner store to get a pack of gum," I said.

She smiled, remembering the corny line from the old story that I had used to keep the tears in check when I left for World War II. "Don't be long," she said, giving her same answer.

I felt the kids watching as I kissed her. Then I hugged all three of them at once before I joined the stream of passengers heading to the waiting airliner. Millions of mili-

tary men and women know those kinds of departures;
they're tough, and they never get easier.

I reported to K-3, the First Marine Air Wing's base across
the bay from P'ohang, on Korea's southeastern coast, on
February 3, 1953. I was assigned to one of the two fighter
squadrons there, VMF-311, as operations officer.

The overwhelming impression of Korea in midwin-
ter—even at P'ohang, well away from the northern
mountains—was of cold. The wind off the Sea of Japan
seemed to find every crevice and pinhole in my clothing
and claw its way inside. It ripped the smoke from the
chimneys of the huts where many South Koreans lived.
When I lowered my shoulders and stuck my head up out
of the collar of my flight jacket, I saw a country devastated
and a people numbed by war. The war had surged over
them twice—during the initial North Korean push south,
and again when the mostly American United Nations force
pushed the Communists north again.

Korea in early 1953 was the old Orient of peasants,
low buildings huddled among harsh hills, and a populace
that walked or pedaled to get from here to there and supplemented their meager wardrobes with surplus military
clothing. It was a land of small rice paddies and kimchi, a
Korean staple food consisting of cabbage, onion, radishes,
and a lot of garlic, left to ferment and solidify in large
earthen jugs. Kimchi literally had an air about it; if you
were downwind when someone had the jug open to slice
kimchi for the next meal, it wasn't something you'd forget. The Koreans wasted nothing. In P'ohang each morning, shuffling men in dark clothes hauled barrels of human

excrement for use as fertilizer. About once a week they came to the base to clean out our six-hole outhouse. Officially it was "night soil" that they hauled away, but we called the barrels "honey buckets," as if the stench demanded euphemism.

The great contrast between what I found in Korea, and what I had left behind at home, moved me to write a letter to Lyn that I hoped would explain why I couldn't be there on her sixth birthday in another month:

In our country, Lyn, we all believe that people should be allowed to have their own homes, to work where they want to work, and do the things they want to do without someone always telling them what they must do. But some countries have people in their governments who do not act that way. They want to always be the bosses and tell everyone else how they should live and what the people must do. It's a little like when you play house with some of the other girls. It's no fun if one of the children always insists on being the boss and tells everyone what to do all the time, is it? It's the same way with countries. If one country tries to be the boss and wants to control all the countries, then the countries that want to live their own way must do something about it. The Communist government in Russia wants to control all the countries right now. If we let them do it, it might mean that they wouldn't let us live where we wanted to live, you and Dave couldn't go to a school where you could learn what you should, and they would take a lot of the things we have now. Probably Grandma and Grandpa wouldn't have their nice home. The Communists already did that in some other countries. When they tried to take all of Korea, our country and some others decided it was time to stop them before they could get to the United States and take it, too.

We surely don't want bad people like that coming into our country, do we? So that is why I'm out here, flying against them every day. We go up and drop bombs and shoot at them to make them get out of Korea, and to show them that they can never take our country. The United States is a wonderful country, much more than we think, most of the time, Lyn. People in other countries do not have all the nice things we do, nor can they say and do what they want.

Our flag represents all the things we believe our country should be. It means that children can go to school, the mothers and fathers can have homes and raise their children the way they want, and live the lives they choose. So when you see our flag, think of that. The flag is not just a pretty piece of cloth to look bright and make a nice decoration. It stands for all those things I've been talking about that make our country, the United States, so different and better than the bad countries like the Communists have. Don't you worry about them ever getting to our country, because I, and many other men, are out here to see that they never do.

Does that help explain it a little? Hope so. Children like you and Dave are very lucky to live in the United States. Children out here aren't quite so lucky.

Well, happy birthday, honey. I surely miss you and Dave and Mother, but someone had to stop the Communists and that is one of the jobs the Marines help do for our country, so here I am while your birthday is going on. Will you eat a special piece of cake for me? Here's a big special kiss and a hug for my honey on her birthday—all the way from Korea.

The pilots of VMF-311 were a great bunch. One of them was the Boston Red Sox star Ted Williams. I had just joined the squadron and was sitting in the pilots' ready room one day when he walked in and came over and introduced himself. I had been a baseball fan since I was a boy, and meeting Ted was a thrill. It didn't take a Red Sox fan to know he had batted .406 in 1941 and won the American League Triple Crown the next year before he was called away to World War II, then won another Triple Crown and led the league in batting for two years running after he got back.

Ted had been a Marine pilot at the end of World War II, flying Corsairs, but he had not been in combat. He stayed in the reserves after the war, and was called back to active duty six days into the baseball season in 1952. He had been training in jets, and had reached Korea a week or two before me. We lived in opposite ends of the same Quonset hut.

He was tall, genial, and easy to like, and he developed a voracious taste for the chocolate fudge Annie's sister, Jane, would send from home. His new Marine buddies figured that like every other major leaguer, it would rankle him if somebody said he was "bush league." Given that he was one of the biggest stars the major leagues had ever seen, it did. Naturally, "Bush" became his nickname.

It was only later that I heard him quoted as saying that all he wanted out of life was that when he walked down the street, people would look at him and say, "There goes the greatest hitter who ever lived." In my book he made that hands down. And there was certainly nothing "bush" about him as a Marine combat pilot; he gave flying the same perfectionist's attention he gave to his hitting.

The airfield was on a bluff overlooking P'ohang Bay. We were 180 miles south of the front. Around six o'clock each night we would get our orders for the next day's missions from the Joint Operations Center (JOC) in Seoul. It was my job as ops to prepare a plan that would execute the orders.

The Panther jets we flew were heavily built fighter-bombers with straight wings and tip tanks. They had four twenty-millimeter cannons and could carry a twenty-five-hundred- or three-thousand-pound bomb load, as well as five-inch HVARS, high-velocity aircraft rockets. We could also carry napalm. The Panthers were suited to a variety of missions, all of which came down from the JOC at one time or another.

I flew my first combat mission out of K-3 on February 26, a reconnaissance flight along the main line of resistance, which approximated the thirty-eighth parallel of latitude across Korea. I also flew two instrument training flights that day, and a CAP—combat air patrol—scramble. In these missions, two to four manned planes stood at the end of the runway, hooked up to their starting generators and ready to go the minute we got word of a JOC alert. Sometimes you'd scramble and receive instructions in the air as to where to go along the front, but on this day we were called back to the base.

The next day I flew my first interdiction mission deep behind enemy lines. These flights, a hundred miles or more into the sparsely populated, mountainous North Korean outback we called "Indian country," were aimed at hitting railroads, bridges, and supply and troop staging areas.

We flew constantly. We usually took off early in the morning, armed with what we needed for the day's first mission, and aerial reconnaissance maps and photographs for target and landmark identification. Much of the time we did close support, hitting enemy positions just across the front line from our own dug-in Marine and Army troops. The First Marine Division was dug in around the middle of the front. The First liked to see us coming, because we could get down over the North Korean trenches on the other side and do a lot of damage, particularly with napalm. Accuracy was more important than usual on these close-support flights over the lines. The thing an attack pilot fears most is dropping on "friendlies." The ironclad rule was, "If in doubt, don't drop." Sometimes we'd radio down to the troops below to fire Willy Peter—a white phosperous grenade—toward the enemy lines to help direct our fire.

Tom Miller was also serving in Korea. He had gotten there well before I did, and was flying later-model Corsairs out of K-6, an air base closer to the thirty-eighth parallel. When he found out I was at K-3, my friend flew down to give me some advice. Squadron veterans already had briefed me to avoid "flak traps" the North Koreans maintained at such areas as Pyongyang and Wonsan, in which deceptively attractive targets were ringed by anti-aircraft batteries. Tom added to what they had to say.

"The North Koreans have a lot of antiaircraft, and they're good at using it," he said as we sat over a meal in the mess hall. "Be careful. If you're in a bombing run and you see an antiaircraft gun firing at you, don't be suckered into going after it, because there are six or eight others around, and one of them is going to get you while you're shooting at his buddy. If you're in a hot area, don't make a second run on the same damn target. It just helps the gunners get wound up, and they'll wait for you. And whether you see you're being shot at or not, don't fly in a straight line any longer than you have to. Keep moving until you're well above nine thousand feet."

Tom's advice was sound. We already had orders not to go back in after making an attack. Getting one of their guns wasn't worth risking one of our planes and pilots. But we all hated those antiaircraft gunners with a passion.

It was rare to actually see the antiaircraft guns that were shooting at us. We were looking all the time; our heads swiveled so much we wore silk scarves to keep our necks from chafing. They were sky blue with red hearts, to go with our squadron nickname, "Willing Lovers," from WL, the tail letters on our planes. I'd see the streams of tracers or the AA bursts, but almost never the exact spot the fire came from.

That changed a few weeks after Tom came down to see me.

We were flying near Sinanju, north of the Haeju peninsula along the west coast of North Korea, hitting a complex of buildings the Communists used for staging of equipment and soldiers on the way to the front.

I was making a steep dive-bombing run and was drawing the lead needed to correct for the wind and put the bombs on target when I caught the flash-flash-flash of an antiaircraft gun off to the right. I made a mental note of the position and continued the run, dropped the bombs, and pulled out of the dive. But instead of heading to the rendezvous point, I stayed low on the treetops and made a wide turn heading back toward the gun where the fire had come from. Rendezvous could wait a minute or two. To heck with Tom's advice—this was an opportunity not to be missed.

I flew fast and low directly at the gun emplacement, firing the four twenty-millimeter cannons, and watched the shells tear that emplacement to pieces. That was solid satisfaction.

Just as I pulled up to keep from running into the emplacement, something hit the plane and it suddenly nosed over. I jerked the stick back automatically—by instinct and reaction alone. The plane dipped toward a long rice paddy with a higher embankment at the far end before it reacted to the controls and cleared the embankment by what seemed like inches at close to five hundred miles an hour. In those seconds I expected to crash but didn't. With a lot of back pressure on the stick I could keep the plane climbing out of there. The elevators were working or I would never have pulled out, but the elevator trim that helps equalize control pressures obviously had been shot out. I was going to have to fly home using a lot of muscle power.

But first was the problem of avoiding other antiaircraft fire, so I jinked up and down at low altitude to avoid horizontal fire from the hilltop, and gradually shifted to heading changes as I climbed to where the ground fire would come up vertically.

I flew back to K-3 keeping back pressure on the stick, first with one hand and then the other, because if I'd let it go the nose would have pitched down and I would have gone down like a rock.

All the way back I kept seeing that rice paddy embankment coming at me, with me thinking I'd bought the farm.

The hole in the Panther's tail was big enough to put my head and shoulders through. There were another 250 smaller shrapnel holes around the big one. We figured it was a thirty-seven-millimeter shell that hit me; a larger one would have blown the tail off. Crews replaced the tail, and the Panther flew as good as new.

That night I dusted off an old fondness of mine and wrote a poem about the day's events. In part it went:

> Then off to one side of the tail
> A tracer stream did pass.
> A thought ran flashing through my mind:
> "They're shooting at my [censored]."
>
> If you would bold and older be,
> Be "Tigers" all, 'tis true,
> But on a bombing run just make
> One run, just one, not two.

"You were as right as could be," I told Tom the next time I talked to him. "Hitting the end of that rice paddy would have taken all the fun out of knocking out that gun. From now on, I'm taking your advice."

That was the last time I went back in for a second run. I got hit again a week later. This time it wasn't my fault. We were coming down from altitude in a long glide to start a napalm run. At about eight thousand feet I felt a big explosion and my plane rocked up about ninety degrees to the left. Pilots behind me radioed that I'd been hit, something I was already keenly aware of. I could control the plane, however, and it wasn't on fire, and I made it back to the base okay.

What I saw when I looked the plane over shocked me and made me say a little extra prayer of thanks that night. There was a hole in the wing two feet across. A large anti-aircraft shell—probably their version of our ninety-millimeter—apparently had hit the napalm tank and exploded just off the wing. A direct hit would have taken the wing off. One of the ground crew counted over three hundred shrapnel holes through the wing side of the plane and the cockpit, and I didn't get a scratch. This time the plane couldn't be repaired.

The squadron took to calling me "Old Magnet Ass" after that. But even with all the antiaircraft fire I seemed to be attracting, and a couple of close calls, none was as close as one that almost did in Ted Williams.

Ted got hit on one of his first missions. While the flight controls still worked and he could fly the Panther, antiaircraft and small-arms fire had knocked out almost everything else. He was streaming smoke and fire from around the engine, which in a Panther usually signaled an explosion that would blow the tail off. His radio was out. Other pilots flew close to him, signaling him to eject while he still had control, but he took it into K-13 at Suwon. The landing flaps would not come down, so he would have to land faster than normal. The landing gear wouldn't come down, either. He bellied it in at 150 miles an hour or more, slid up the runway for two thousand feet, came to a stop, jumped out of the cockpit and off the wing, and ran until he was out of danger. Then he turned around and stood there watching the plane burn on the runway.

Ted was known as a pretty fair speedster when he was going after a fly to left at Fenway Park, but people who saw him running from the crash at Suwon say he set a new record that day.

Ted and I flew together a lot. The squadron had a practice of teaming up a regular Marine pilot with a re-servist. The two wouldn't always fly together, but if they were on the same mission, they would be paired. With that arrangement, Ted flew about half his missions as my wingman. He was a fine pilot and I liked to fly with him.

Some of our missions together were "road recces" deep into North Korea. The North Koreans moved at night, bringing troops and equipment down toward the front. We would take off with just a couple of planes early in the morning, so that at first light we would be over one of their supply roads. Then we would drop down and fol-low the road back toward the front, hoping to catch their troops and trucks in the open before they sheltered for the day in tunnels or bunkers or caves. We leapfrogged, with one of us flying at treetop level and the other at a thousand or fifteen hundred feet above and behind in order to see farther down the road and relay advice to the "shooter" on targets ahead. We would switch positions every ten minutes.

The HVARs we used on those runs were armed with proximity fuses. If we found no targets of opportunity, we had orders to get rid of the rockets before we landed, at any target or in the ocean if we had to, because of problems with the fuses. One day Ted and I had flown back from the north without finding any targets, and when I crossed the lines I swung around and headed into North Korea and fired my rockets at a small bridge that had already been hit.

Ted followed me around and made the same run I did, but his rockets didn't go off. He pulled up and made a level turn until he was headed back toward our lines. I saw him push over into a shallow dive, but just as I started to call to tell him not to fire, the rockets were on their way.

I had a terrible sickening feeling as I watched those rockets hit and explode. Some of our troops had recently taken casualties from friendly fire. Our group commander had put out orders saying that the next time it happened the flight leader was going to be held responsible and court-martialed. I double-checked the impact point against the landmarks I could see and compared it with my flight chart. It looked as if Ted's rockets had hit behind our lines. My first concern was whether any of our troops had been hurt, but I could also see myself sitting at a court-martial table.

We flew back to K-3, and Ted and I went to the operations shack to tell them what had happened. There on the wall was a bigger chart than I had in the plane, and it showed that the lines had shifted more than half a mile south. Ted had fired into an enemy troop pocket. He said he knew it all along, but I kidded him anyway about almost getting me court-martialed.

I never took off on another mission without closely checking the ops office chart against my flight chart. I made that part of the squadron briefing from then on.

We used various strategies for defeating antiaircraft

fire. They could send up a blanket of it, from a few feet off the ground to eighteen thousand or twenty thousand feet. On a dive-bombing run, to keep the gunners from concentrating on one approach corridor, we circled the target and came at it from all points on the dial in what we called an around-the-clock approach.

About three months into my tour, intelligence picked up word of a North Korean troop and equipment buildup in the Haeju peninsula. The Joint Operations Center wanted to obliterate the area to prevent a big attack, and assigned both K-3 squadrons to the mission. It was VMF-311's turn to lead, and as ops, I made the plan. This would be a napalm run. They were particularly vulnerable to antiaircraft fire, because the runs were made low and slow for accuracy, and we expected the AA to be heavy.

The plan involved some elaborate choreography. We came at the North Koreans in three lines, from opposite directions. The two outside lines flew south, the middle line north. Each line took a target path several hundred yards wide. The first plane in each line was assigned a target area at the far end of the line. That way, the smoke and flames from the napalm wouldn't obscure the next plane's target, but would interfere with the vision of the antiaircraft gunners trying to hit us as we jinked out of there. Each drop was a little short of the one ahead of it, so we gradually blanketed the entire area. I went in first in one line, coming in very low. We were passing each other in opposite directions, one line going out with the other coming in. I imagine it looked like some kind of chorus line. It could work only in a large area, but it worked well. We made two attacks on the marshaling area that day. We took some small-arms fire, which usually happened when we came in low with napalm, but the antiaircraft for once was ineffective.

Secondary explosions, when we managed to penetrate a fuel or ammunition bunker with a steel-tipped armor-piercing bomb, could be as dangerous as antiaircraft fire if you were down too low. Ted was flying as my wingman one day on a run over the Haeju peninsula. He was pulling out of his run when he radioed that he'd been hit. I flew around his plane looking for damage as we headed to the rendezvous. I saw a hole in the bottom of his tip tank but no hole in the top. I couldn't figure it out. We got back to the base, where the crew removed the tank and found a rock inside, apparently thrown up by the secondary.

The deeper our missions took us into North Korean territory, the greater the danger from the Soviet-built MiG 15s that flew down out of China to try to stop our attacks on ground targets. The Panthers were slower than the MiGs and less maneuverable. The job of keeping the MiGs off our tails belonged to the Air Force's fighter-interceptor squadrons with their North American F-86 Sabre jets.

I enjoyed the kind of air-to-ground combat we were doing. Flying in support of troops on the ground is what had attracted me to the Marines when I heard about Guadalcanal way back at Corpus Christi. Marines look at themselves as a team, and in Korea we had a front-line visitation program. Two or three of the squadron's pilots would be taken off the flight roster to spend a few days at the front visiting the troops in the trenches. It reinforced the team concept, and the troops I talked with always said they appreciated our air role.

But I also hoped for air-to-air combat. That was the

ultimate in fighter flying, testing yourself against another pilot in the air. Ever since the days of the Lafayette Escadrille during World War I, pilots have viewed air-to-air combat as the ultimate test not only of their machines but of their own personal determination and flying skills. I was no exception. You believe you're the best in the air. If you do, you're not cocky, you're combat-ready. If you don't, you'd better find another line of work.

CHAPTER 10

I had flown sixty-three missions with VMF-311 when I got word in the late spring of 1953 that my application to go on exchange duty with the Air Force had been approved. This meant flying F-86 fighter-interceptors, and maybe fighting MiGs.

I already had some time in F-86s. Even before I had left for Korea, when I was still at Quantico, I knew of the exchange program. It was aimed at giving the Marines a cadre of pilots with air-to-air combat skills. I wanted to get in it if I could.

It happened that Colonel Leon Gray, who had headed the training wing at Williams Air Force Base when I was assigned there from the Naval Training Command at Corpus Christi, had gone on to an assignment as the commanding officer of an F-86 training group at Otis Air Force Base in Massachusetts.

When I received my orders to Korea, I called Colonel Gray. "I'm going to Korea," I'd told him. "I'd like to get into the F-86 program when my Marine missions are done. What are the chances of my taking some leave and coming up there and getting checked out in the F-86?"

He had said it normally took some kind of approval process, but he thought it would be all right if I came up on leave, and he'd make some arrangement. I went through an abbreviated F-86 ground school and had several solo flights in the Sabre over a four-day period. I told the colonel I'd be forever in his debt.

When my Marine flight assignment was drawing to a close, I applied for an Air Force slot. One was coming open, and I was selected to fill it at least in part because of my Otis checkout. I was assigned to the Twenty-fifth Fighter-Interceptor Squadron at K-13, the base at Suwon, south of Seoul, where the Fifty-first Fighter-Interceptor Wing was headquartered.

The Air Force put exchange pilots through an orientation process to familiarize them with the Sabre. It was more detailed than my four days at Otis. When it was over, you practiced air-to-air combat with squadron members. The next step was to fly missions as a wingman, and that continued until you had a feel for the missions and what was likely to happen on a flight. Only then might you lead a two-plane element or a four-plane flight.

Major John Giraudo was commander of the Twenty-fifth FIS. He was tall, with movie-star good looks, and a gung-ho pilot who had been shot down over Germany during World War II and been a prisoner of war. I flew my checkout missions as his wing.

Our job was to set up an aerial screen at the border between China and North Korea, to prevent their MiGs, which were based in Manchuria, from coming down and interfering with the kind of ground attacks we had been doing in VMF-311. We patrolled just south of the Yalu River in long, narrow figure eights that covered fifty to seventy-five miles of the border. Usually eight planes would patrol one sector, four flying high at thirty-five thousand to forty thousand feet, and four low, at fifteen thousand to eighteen thousand. We flew those long figure eights, looking for MiGs, and always made our turns toward the river to keep from being surprised.

Identifying the MiGs was a challenge. Their swept-wing profile was very similar to the Sabre's. The planes were similar in many ways, in fact. They both had six thousand pounds of thrust in their jet engines, and both would dive at supersonic speeds. The MiG was smaller and lighter, and had a higher ceiling and a faster climb rate. The Sabre was faster in level flight and diving, had a greater range, and could turn tighter in a fast dive. The Sabre had six .50-caliber machine guns and 1,600 rounds of ammunition to the MiG's single thirty-seven-millimeter and two twenty-three-millimeter cannons, for which it carried 220 rounds of ammo.

We used binoculars to search for MiGs. Our ground control radar was of some help, but it was too far from the Yalu to give us more than general directions when an enemy plane crossed the border. Instead, we scanned the sky in regular patterns, searching for a glint of sunlight on a wing or any movement that would indicate an airplane. A silk scarf around the neck was doubly necessary now; we turned our heads so much we could look almost directly behind us, like rubbery cartoon characters.

My first checkout missions were uneventful. As I became familiar with the Air Force routine and worked my way into the squadron, one of the squadron's sign painters came around and asked what I wanted to name my plane. We didn't do that much in the Marines, because we swapped planes more than the Air Force squadrons did. But I liked the Air Force custom. You spent so much time with your plane it was like a trusted sidekick. It was a relationship upon which your life depended, so the plane should have a name that accorded a proper mixture of affection and respect.

Soon my F-86 sported the letters *LAD* on the fuselage, behind which the names *Lyn, Annie,* and *Dave* were written out.

As the summer began and rumors of peace talks made me think the war might end before I could experience any air-to-air combat, I started moaning about there not being any MiGs out there. I must have moaned a lot, because one morning I went out on the line and there on the side of my plane under the cockpit, painted as big as a billboard, was a huge M with letters trailing from it so that it read, "MiG Mad Marine."

I thought it sounded a little too aggressive, too swashbuckling or something, and that I should have it taken off. Then I thought, "What the heck. I'll just fly it that way."

Because it was right. I did want to get up there after the MiGs. I flew with a focus on the target, and if it meant flying on the edge, that's what I did. I was doing the best job I could in Korea. The other guys had started this. It was up to us to finish it. I wasn't bloodthirsty, but I had trained for combat, this was it, and it was important to our country. Flying against another pilot in a MiG, when it was him or me, would be the ultimate in combat flying. Maybe I could be the second Marine ace in the Korean

War, after my friend Captain Jack Bolt, if I could get five kills.

I continued to fly wing with John Giraudo as the Sabres deployed to set up a screen against the MiGs. Then our orders changed. The fighter-interceptor Sabres (others flew as fighter-bombers) had been restricted to air-to-air combat. Now we were told that if the MiGs weren't flying on a given day, we should save enough fuel to fly the roads on the way home, and attack any convoys that were headed down toward the front lines.

In the middle of June, John was promoted to lieutenant colonel. A couple of days later, he and I were coming back from the Yalu without having spotted any enemy planes. There was an overcast, so we dropped down to fifteen thousand feet to get under the cloud deck. Just over Sinuiju, we spotted a line of trucks. I knew Sinuiju was a flak trap, but we went on the attack. John started a run and I went in behind him. We were hitting the trucks, and they were pulling off the road on both sides, but the AA was coming from both sides, and straight up from the trucks. He pulled up from the run, and radioed that he'd been hit. I followed him up as he zoomed to ten thousand or twelve thousand feet, then his plane nosed over and went down.

The hit he had taken had knocked out his elevator control, the horizontal stabilizer that on the F-86 was a one-piece "flying slab" rather than a stabilizer with elevators on the rear. It caused him to go through a series of zooms, up four thousand feet, then nosing over and plunging, then up again as the lift increased. John adjusted the power to try to get the plane leveled out so it would cruise. He wanted to get out over the water so he could eject with a better chance of rescue than if he ejected over land.

But each zoom up and down left him at a lower altitude, until finally he radioed that he would have to eject. He was in a dive at five thousand feet and a mile from the water when he ejected. His chute opened at four thousand feet and he came down on a mountainside that was covered with low scrub trees a little distance from a village on the shore. I radioed air-sea rescue at Ch'o Do Island and started circling the area where he went down. I couldn't see him, but I knew he had landed okay, because I saw his parachute being pulled in so he could bury it and hide until the choppers came.

Ch'o Do Island was eighty miles away. Air-sea rescue said the choppers were coming. Two other planes had joined me, but we were getting down toward "bingo" on our fuel—the point where we had just enough to get home. I radioed them to head back, and said I'd stay because I knew exactly where he was.

The choppers didn't show. I did some calculations in my head, and figured the eight hundred pounds of fuel I had left would get me high enough to glide back over the front line. I climbed to forty thousand feet and flamed out with the fuel exhausted. My plan was to get back across the thirty-eighth parallel and eject over South Korea. But I had a good tail wind, and I glided back 108 miles with no power, frost forming in the cockpit and on the windscreen from the cold, and made a dead-stick landing at K-13.

I had radioed the base to have another plane waiting so I could go back up and show the rescue choppers where

John was. A jeep was waiting at the end of the runway. I jumped out of the plane and took a fast ride to operations to show them the spot on a map. Then I got into another plane and took off with three more following. We reached the spot and circled for at least forty-five minutes, but never saw anything at all. The helicopters never showed, and I never found out why.

It was a serious group at the debriefing, having lost the squadron CO. At the end of it somebody came to me and said, "You ought to say something to Father Dan. He was really in your corner today."

Father Dan Campbell was the Roman Catholic chaplain, a priest from Washington University in St. Louis. He was out on the flight line every morning when the squadron took off, in his jeep beside the runway, blessing the planes and calling at the pilots to "give 'em hell." He and I had spent a lot of time in the ready room, talking about religion and the Catholics and the Presbyterians. He had become a close friend.

Father Dan was pensive and distracted, like everybody else at the debriefing. I went over to him and said, "Thanks for what you did today. They tell me you were running that rosary like a monkey on a flagpole."

He smiled. "Thank you, too," he said. "We're going to get him back, I'm sure of it."

After John was lost, I began leading two- and four-plane flights. I much preferred this to flying wing, because as I had written Dave, who was seven then, the wingman is there to protect his leader, and the leader is in the shooting position out in front.

I was flying with Sam Young, a first lieutenant, on my wing when I spotted my first MiG. We were up high at forty-three thousand feet and bounced—started to attack—five planes that turned out to be F-86s. Then, at twenty-three thousand feet, I saw a MiG well out ahead of us, fifty miles inside North Korea and low. I bounced it. I lost the MiG in the clouds but picked up another one behind the first and chased it across the river at about two thousand feet. Our rules of combat were that we could enter Chinese airspace across the Yalu only if we were in hot pursuit. I kept after him at seven hundred miles per hour, gaining very slowly. I finally caught him forty miles inside Manchuria.

Suddenly, the MiG slowed down as if the pilot had cut power. I almost ran up his tailpipe. I realized there was an airfield ahead, and he was trying to get back in.

I opened fire with the .50s when I was within a thousand feet of him, and hit the fuselage and wings. The bullets ripping through the metal seemed to send up sparks. Flames burst out where the right wing joined the fuselage. He hit right on the edge of the field and exploded in a fireball. I was only fifty feet off the ground, and I flew through a cloud of his debris to find myself looking right down the runway of a Chinese air base. The control tower was directly ahead, so I fired and saw the glass explode, then pulled up in a wide left turn to rendezvous with Sam and head back home. I looked back and saw the MiG spread out over a hundred yards of ground. The fire looked like the explosion of a napalm tank.

MiG number one.

I almost ran into another MiG that was coming down out of the clouds as I was climbing, then saw a second, both heading back to base. I didn't have enough fuel to turn on them, and by then the antiaircraft was getting heavy, so I kept climbing up through the clouds and hooked up with Sam for the run home.

The Sabre's nose was blackened, a characteristic of heavy gunfire that signaled the ground crews when I landed at K-3 that there had been a battle. Father Dan was out by the taxiway whooping it up. All the pilots and enlisted men came out, and somebody ran the skull-and-crossbones and red-star flags up the flagpole. The crew chief exulted, because my Sabre was brand-new, and now he could paint its first red star under the cockpit.

That night I wrote a letter home: " 'M' Day, 12 July, 1953. Today, I finally got a MiG cold as can be. Of course, I'm not excited at this point, not much!"

I described how it happened and then wrote, "Funny how the bullets sparkle when they hit a plane like that. Just light up like little lights every time a bullet hits."

A few days later—in fact, it was the day after my thirty-second birthday—my element was flying figure eights along the Yalu about forty miles inland when we saw four MiGs coming into North Korea. We maneuvered to attack. As we closed I caught a glimpse from the side, scanned the sky, and saw twelve more MiGs. We had run into a flight of sixteen that was too spread out to see at first.

We closed on the first four MiGs and I opened fire. Then our odds got better, with four more F-86s diving into the battle. Suddenly planes were swooping everywhere, hunter and hunted, trying to take aim on one another with fixed guns, or maneuvering to avoid getting hit. Pilots had used the same air-to-air dogfighting tactics since World War I, only we were flying six hundred miles an hour instead of barely a hundred. Planes coming at each other closed at twelve hundred miles per hour. You saw a plane flying at you, and then it was gone and almost out of sight. The arena expanded to accommodate the extra speed and distances. The aerial ballet at slower speeds had been almost stately; now it was like fragments moving at atomic speed.

I closed too fast, got a momentary shot, overran, and pulled up in a hard turn. The MiG was behind me; now I was the target. He fired in my direction, but the hard, climbing turn didn't give him enough lead to hit. I was pulling into a diving turn, where I knew the F-86 could outturn the MiG, when my wingman, Jerry Parker, came across from outside and hit the MiG with two bursts.

Then Jerry was in trouble. He radioed that he had only partial power and couldn't fly above a low cruise speed. He headed for the water in case he had to eject, and I peeled off to stay with him and protect him. Quickly we were out of the melee of aircraft and were alone. I kept my speed up, made a couple of circles, and went from one side to the other over him, in case there were still MiGs in the area. There were.

I spotted six of them in single file turning toward us.

The odds weren't good, but earlier in the Korean War, superior numbers of MiGs had broken off attacks when the F-86s turned toward them and "lit up the nose"—fired a burst from the machine guns that, although far out of range, was highly visible. That's what I did, and it worked. The leader turned north toward the Yalu, and the others followed, still flying single file.

Closing fast, with plenty of speed, I pulled within range of the last MiG in line and fired. The bullets hit; he flamed and went down. At that the others turned, made another large circle, and came back at us from different angles, a repeat of the uncoordinated first attack. I flew at them again, firing from way out of range, and again they broke off. Finally, they banked away and flew toward Manchuria. I squeezed the trigger one last time, but my guns were empty.

The MiGs' tactics were so poor I could only imagine it was a training flight, or they were low on fuel, but we were unbelievably lucky.

Jerry had flown out of sight while the MiGs concentrated on me. I radioed him, and he said his engine was holding okay and he was headed back to K-13.

Three days later, on July 22, I got my third MiG in a dogfight in which two other planes in my element shot down MiGs as well. It was the only time in the war a four-plane Sabre flight had shot down three MiGs, and they were the last kills of the war. Bad weather grounded us until an armistice stopped the fighting five days later, after I had flown twenty-seven Sabre missions.

A final tally showed the superiority of Air Force planes and tactics. Sabre pilots had shot down 792 MiGs during the Korean War, to only 76 Sabres lost to MiGs.

The issue that had kept the peace talks dragging for so long had to do with the exchange of prisoners. Our side wanted the Communist prisoners of war to have a choice about returning, while the North Koreans and Chinese wanted all their POWs back. The prisoner exchange that finally was worked out was called Operation Big Switch. It happened in October.

John Giraudo's name wasn't on the list of prisoners who were being sent back to our side. I knew he had been alive after he went down, but all I could think when his name didn't show up was that he had been killed when he was captured.

Senior officers were the last to be exchanged, however. And on one of the last lists to reach our side, toward the end of the Big Switch, John's name appeared.

Tom Miller was still in Korea, and I knew he had been assigned as liaison officer by the First Marine Air Wing to receive any Marine pilots or Marine air crew coming out. I was back with the Marines at K-3, for the last months of my tour, and I called Tom to ask him to look out for John. Tom went one better. He arranged for me to be his assistant, and I went with him up to Freedom Village at Panmunjon, where the exchange was taking place, to be there when John came out.

I know I was grinning when I spotted him. John took one look at me and said, "You son of a bitch. I had it in for you."

He was grinning, too, but it set me back a little. "Why? What's wrong?" I said. "What happened?"

"You caused me all kinds of problems."

"I was trying to get you out of there, for gosh sakes."

"I know you were," he said, laughing. "But let me tell you what happened." John said he had hidden his parachute and was lying low to elude a search party of North Korean soldiers. They were moving away from him when the life raft that was part of his survival equipment suddenly inflated, giving him away. A soldier with a machine gun put a slug through his shoulder. He thought he was a dead man, but the soldiers stood him up and started marching him away. "It hurt enough as it was," John said. "But then you came back and started circling. Every time you came over, they'd throw me down in a ditch and sit on me. It hurt like hell. I was thinking, 'John Glenn, would you please just go home.'"

After the truce, there were still patrols to be flown along the North Korean coasts, but with little or no activity in the north, they were routine. My tour continued through December. I wrote the kids about why I couldn't be home with them for Christmas, and told them to put some snow down Annie's neck. "Sneak up on her and do a good job of it," I wrote.

The holiday spirit didn't go wasted. The brush-covered hills around K-3 were ideal pheasant habitat. Few Koreans had shotguns for hunting, and although they caught some of the birds in baited box traps, the pheasant population was basically thriving. Some of the Marines at the base were enthusiastic bird hunters, and there were shotguns on the base. They formed hunting parties, and brought back several jeeps loaded with pheasant that became Christmas dinner for the base and for the Korean orphanage we supported in P'ohang.

On Christmas afternoon, we brought the kids from the orphanage out to the base for a party. Pilots on R-and-R in Japan had brought toys and games, but the kids had the most fun playing pilot in the cockpits of the F9Fs. They barely came up to the sides of the cockpits, and wearing our hardhat helmets that were far too big for them and goggles that covered half their faces, they were a sight of heartwarming innocence as they peeked down and laughed and giggled at their friends who were waiting their turns. Somehow it made being away from our own families and children at Christmas easier. And you remembered that in all wars, it is the innocent who suffer most.

Annie and the kids and I had a belated Christmas when I got home in January.

PART THREE
FLIGHT

CHAPTER 11

Annie and I had made a career decision before I came back from Korea. We both were in our early thirties. Dave would be eight in December of 1953, Lyn almost seven; our family was starting to grow up, and we had corresponded about how I should pursue my career with the Marines.

Ambitious Marine officers knew they had to get some experience in all aspects of the service if they were to rise through the ranks. The Amphibious Warfare School junior course I had attended as a captain in 1951 was only one of a number of schools at Quantico and throughout the services designed to strengthen an officer's experience and command abilities. It took extensive training—including cross-service training at the Naval and Army War Colleges, and the Air Command and Staff School—as well as seasoning under fire, education, and political and social skills to keep ascending the ladder. I would have had to request a whole series of schools and assignments that would get my ticket punched at all the right stops. Even then, promotion was not a sure thing. The officer ranks were full of colonels who had bumped up against their ceiling and were marking time to retirement. That was a fate I didn't want.

All things being equal, I felt I could compete with the rest of the officer corps. But when I was assigned to the general staff at Quantico, I detected a bias among the ground officers with whom I worked as the only pilot. They respected pilots for their particular abilities but didn't think they were the equal of ground officers in terms of overall Marine Corps leadership. This bias lies at the heart of the Corps' unique role. It alone among the services can operate independently of the others in conflicts involving land, sea, and air operations. But infantry has always been its core component, and the officers commanding the men on the ground have been the most likely to rise to the very top. Every Marine pilot knows that ever since the Marines started flying in 1912, no aviation officer has risen to commandant of the Marine Corps.

The outside chance of heading the Corps' Division of Aviation, or becoming assistant commandant, were worthy long-term career goals, but not ones that appealed to me at the end of my service in Korea. I wanted to keep flying, and I was willing to let the career take care of itself.

In the letters I wrote Annie, I explored the idea of applying for test pilot school. She was cool to the idea at first. There was no getting around the fact that it was not as safe as a job at the bank. But she also knew I loved it and was good at it, and when I pointed out that a whole new stable of U.S. Navy and Marine Corps fighter and attack jets were coming into service, and that my experience could help make them safer and more combat worthy, she allowed herself to be persuaded. And I could think of no better purpose I could serve.

The Naval Air Test Center (NATC) at Patuxent River, where I had spent some of the final months of World War II putting service time on Grumman Bearcats, was a mecca for would-be test pilots from the Navy and Marines. I put in my application from Korea, along with endorsements from my superior officers to backstop my combat record, which now included four Distinguished Flying Crosses and eighteen Air Medals.

If Annie wasn't fully convinced when I wrote to tell her I had been accepted, she did a good job of making me believe she was.

I reported to Pax River at the end of January 1954. Annie stayed in New Concord with Dave and Lyn. We wanted to let the kids finish out the school year, and test pilot school was going to demand all of my attention.

My class was small, and loaded with Naval Academy graduates and pilots who already had advanced degrees. I had had to request a waiver of the degree requirement when I applied; having left Muskingum in my junior year, despite courses from the Armed Forces Institute when I was at El Toro and night courses at the University of Maryland's extension school at the Pentagon when I was at Quantico that gave me a surplus of credits, I still had no bachelor's degree. The test pilot school gave short refresher courses in algebra, trigonometry, calculus, and mechanics, among other subjects. The joke among us was that it was like drinking from a fire hydrant—you got a little in you and a lot all over you. I recalled algebra and trig easily enough, but taking calculus for the first time was something else.

I had to get it largely on my own. I stayed up until two or three o'clock a lot of mornings, poring over calculus primers and working my way up to speed. Walt Hess headed the academic side of test pilot training. Ruddy-faced, friendly, and a willing tutor, he took time to help me out along the way.

The academic work was only part of it. We ran a flight schedule, doing practice tests on familiar in-service airplanes, almost every day. At night we wrote practice flight reports. These linked the theory we were learning in the classroom with the flight-testing procedures and our observations about the planes in flight.

It was an intense six months, and one of the most challenging times of my life. I graduated in July 1954, with Annie and the kids looking on.

Test work was serious business. The new generation of warplanes was faster, more powerful, and more heavily armed. The squadrons who would fly them occupied forward positions in a world increasingly defined by the tensions of the Cold War. The United States and the Soviet Union faced each other over stockpiles of nuclear bombs, but one of the things that kept both sides from using them was parity in conventional weapons. Soviet military technology seemed to be closely matched to ours. It was a test pilot's job to make our airplanes the best possible. We were there to wring them out before they were released to the carrier and field squadrons, to make sure that if the Cold War turned hot, the edge would be ours. It was a big responsibility, and I took pride in being part of it.

Aircraft design is an imperfect science, especially when weapons systems are combined with a supersonic airframe. Designers and engineers, no matter how good they are, cannot foresee every possible interface. Methodical testing is required to determine limits and deficiencies. The Hollywood version of the hell-for-leather test pilot, who goes for broke on every flight, wouldn't last long in the real world of test flying.

The NATC was organized to conduct tests on all aspects of a plane's performance, according to standards, instructions, and directions issued by the Navy's Board of Inspection and Survey. The tests were called BIS trials. Flight tests assessed a plane's aerodynamic characteristics and carrier suitability. Service tests put time on a plane, giving it a lifetime of use in as short a period as possible to find the weak spots. Electronics tests stressed its instruments, electrical circuitry, and communication and navigation elements. Armament tests employed the weapons—guns, rockets, and bombs—in every possible combination of G forces, speed, altitude, and quantity of fire.

Testing in each area began with the so-called design envelope—the chart of speed, altitude, Gs, loading, and weapons-system performance that the designers believed the plane capable of. "Filling out the envelope" meant systematically doing flight after flight under different conditions to fill in blank spots on the charts—the envelope—in each test area to determine what the plane would actually do safely. You started at the lower left with points in the minimum-performance range and gradually worked points to the right and up, out to the design limits. When you reached the upper right corner of the chart you were "pushing the envelope."

The flying had to be methodical and exacting, and done within strict limits on every flight. The planes were fully instrumented, and the instrument readings were analyzed after each flight to spot incipient failure. Both deficiencies and potential improvements resulted in engineering change proposals—ECPs—to the manufacturers. There could be hundreds of these for each plane.

I was assigned to do armament tests, an assignment I welcomed because it took advantage of my recent combat experience.

My first project plane was the Navy's FJ-3. Nicknamed the Fury, the FJ-3 was an F-86 Sabre with a tailhook added for aircraft carrier service. The tailhook and its support structure added weight, which made the plane a little slower, but the main difference was that it used four twenty-millimeter cannons instead of six .50-caliber machine guns. The cannons, like the machine guns, were in the nose of the plane, just aft of the jet air intake.

The FJ-3 gave me a stark example of the value of testing. An airplane that's great aerodynamically, even one with a long service life, isn't much of a combat airplane if it can't incorporate the weapons it's designed to use. Armament tests on the FJ-3 revealed a potentially fatal problem. Extended gunfire weakened the panels on the nose around the cannons, so that they sometimes flexed more than expected and popped rivets that went down the air duct into the engine.

This was a flaw that was not anticipated on the drawing board. The .50-caliber machine guns the Sabres had used to such effect in Korea had never caused a similar problem, but their fire was not as jolting and explosive as the cannons'.

The increased pounding on the airframe created other problems, too, ones that came together in a way that almost killed me.

I was at forty-four thousand feet over the Atlantic, getting ready to run one of the last of our high-altitude firing sequences. I flew into the test area and squeezed the trigger, and as the cannons fired, the canopy seal blew. One second I was in the pressurized equivalent of ten thousand feet of altitude, the next at an unpressurized forty-four thousand feet. At the same moment, the oxygen regulator went. I routinely switched to emergency backup.

I assumed that having weathered the near-explosive decompression, everything was under control. The first clue I had that the emergency backup had also failed was when my eyesight started deteriorating rapidly. Large patches of black started to float in my vision. I recognized the symptoms of oxygen deprivation. The oxygen regulator and its emergency function were virtually fail-safe. They never went out together. But this time they had.

I was eight miles high, with only a few seconds of vision left, and not many more of consciousness.

There was one backup left. A small bottle in my parachute pack, connected to my oxygen mask by a rubber tube, contained ten minutes of oxygen for high-altitude bailouts. I held my breath and put the plane into a steep dive as I groped for the "little green apple"——the green-painted, golf-ball-sized wooden ball at the end of a wire cable that would trigger this last-ditch supply. The black patches painted my vision down to nearly nothing. My lungs were bursting, and my system begged for oxygen. I was on the verge of unconsciousness when my hand closed around the wooden ball. I pulled it, exhaled in a gasp, and drew oxygen into my lungs again.

Seconds later my vision began to return and I pulled out of the dive. I took the little green apple home that night as a souvenir of survival; I still have it.

Annie and Dave and Lyn had moved to Pax River a few weeks before I graduated from test pilot training. We lived in an apartment just off the base, and then moved to a house we had had built in a development a few miles north. It had the fancy name of Town Creek Manor and the even fancier mailing address of Hollywood, Maryland.

Our house backed up onto an inlet of the river. Annie and I had kept our love of the water and water sports, and I planned to keep a little water-ski boat out behind the house. It turned out that our inlet was a mud flat at low tide. But there was a good skiing beach about half a mile away, so I bought the boat anyway. It was a sixteen-foot aluminum runabout with a 35-horsepower outboard, and it was a good, fast ski boat. An eccentric Chesapeake Bay waterman named Stanley looked after it at the Town Creek Marina, and in the summer months, on Saturday afternoons and on Sundays after church, we would take the cooler and some charcoal to the ski beach and, along with a dozen or so other families, cook hot dogs and ski until dark.

Our social circle consisted of other pilots and their wives. Bill and Ruth Payne and Gordon and Alice Otis lived in Town Creek Manor. John and Sibyl Chalbeck, Jim and Sibyl Stockdale, and our next-door neighbors Bill and Witzie Russell all were fighter and attack plane test families. When we gathered around the fire to roast hot dogs and marshmallows on the ski beach, or in the winters when we sometimes dined together at the officers club, the conversation among the pilots would flow to our kids, the cars we liked to drive, who had what new ski boat with what hot engine, and the smorgasbord of planes we got to fly. We often swapped flights or filled in on other projects when our assigned planes were grounded for maintenance or engineering changes. I accumulated time in more than twenty different planes. Pax River was like a candy store for avid pilots.

Rarely did we talk about the dangers we accepted day in and day out. The blown canopy seal and the failure of the oxygen system on the FJ-3 was discussed at project meetings and the ready room but not social gatherings. It wasn't something we consciously avoided. It's just that nothing would be gained. We were all volunteers, loved the flying we were doing, and took the risks with the rewards. None of us would have traded jobs with anyone.

Annie and I didn't talk at home about the dangers, either. We had been through two wars by this time. We weren't so hard-shelled as to ignore the risks, but we were seasoned enough to put them in perspective.

So life at Pax River, outside the daily business of filling in the envelope up in the air, was pretty normal. Annie and

I and the kids were part of a Presbyterian church group that met in a movie theater in Lexington Park outside the base. Annie played the organ, and I taught Sunday school in a class that included Dave and Lyn. The congregation was building a small church near the Town Creek area, and moved there when the work was finished.

Death didn't come up even in my prayers or, as far as I know, in Annie's. As in World War II, I didn't pray to God to keep me alive in the air. I was not among those who believed that I would never die until my "number was up," as many pilots did. Nor did I believe in praying just when I was scared: "Oh, God, if you just save me from this mess, I'll be so good even you will be surprised." All I could ask from the God I knew was the good sense to prepare well and make the right decisions. I prayed for guidance, not deliverance.

The Chance Vought F7U Cutlass was an innovative but flawed plane. It had a delta wing, twin rudders, and no tail, and a pressurized cabin with a long, rounded canopy. On the ground it looked like a praying mantis. This was because of its extended nose gear, which made the nose ride high; it was a requirement in delta-wing planes because of the angle needed to get the necessary lift for takeoff. It was one of the first planes with an all-hydraulic control system. The joke about the Cutlass was that it came with its own supply of buckets, to catch the hydraulic fluid from all the places where it leaked after most flights.

The Cutlass flew well enough in normal use. Its afterburners gave it a higher rate of climb from takeoff to eighteen thousand or twenty thousand feet than any other plane the Navy or the Air Force had. It would really scoot up there. Beyond that, it was only okay. And testing revealed some peculiar aerodynamic characteristics: The company's chief test pilot, John McGuyrt, got into a tight spin and found it impossible to recover. He fought it from twenty thousand feet down to thirty-five hundred feet and just managed to eject before it crashed.

I was the project officer on its armament trials, and we had some strange experiences there, too. It was a twin-engine airplane with the cockpit well forward and the engine inlet ducts on each side, right beside the cockpit. There were four twenty-millimeter cannons, and the gun ports for the cannons were mounted at the top edge of those inlet ducts.

The low-altitude armament tests had all gone well, and I was starting the high-altitude tests. I was at twenty-eight thousand feet and going through a firing sequence when one engine quit. The engine wouldn't restart that high, because the air was too thin. I glided down to ten thousand feet and did an air start with the engine wind-milling in the denser air. Then I climbed back up to try the firing point again. The same thing happened, but in the other engine this time, and my instruments showed a transient overtemp in the first engine, indicating it had almost flamed out again.

Over a few days of altitude gunfire tests, the cannons put the fire out in the engines a high percentage of the time. Twenty thousand feet seemed to be the critical altitude. From there on up, the likelihood of flameout during gunfire increased.

This was not a characteristic you wanted in a combat airplane.

We deduced from the test instrumentation that firing the cannons created a reverberating standing wave in the engine inlet duct. It was like the wave resonating in an organ pipe that makes the sound. Because cannon muzzles were located on the upper lip of the inlet duct, the wave traveled down the duct to the face of the engine, with pulsations that interrupted compressor airflow and operation of the fuel control—flameout.

Colonel George M. Chinn was a career Marine stationed at the Naval Ordnance Depot at nearby Dahlgren, Virginia. He was one of the world's leading armaments experts and had developed a number of weapons, including the twenty-millimeter aircraft cannon. Colonel Chinn thought the answer to the problem was a baffle like the Cutts compensator on a shotgun, which uses louvers or slots to redirect the expanding gas and reduce recoil as the shot emerges from the barrel. He designed a stainless-steel extension to be screwed onto the cannon barrel, with perforations that would deflect the blast effects that formed the standing wave.

We set the fully instrumented F7U up on the ground, aimed into the gun butts on the firing range, and, with the engines running, fired the cannons with their twelve-pound extensions of milled stainless steel. Everything worked fine. The instruments showed greatly reduced standing waves down the ducts.

Then I took the plane up to see if the problem really had been fixed. I fired burst after burst at twenty-eight thousand feet, where the problem had originally occurred. The engines didn't flame out, and there was hardly any fluctuation in the gauges that indicated reverberation up and down the ducts. Next I took the plane up to thirty-nine thousand feet, just below its maximum altitude, and started firing from there. Once again, no problem. We decided that the compensators had done their job, and continued with the armament test points. I fired several hundred rounds out of each cannon in the next two flights.

But on the third flight I was up around maximum altitude, shooting a full load of ammunition, when suddenly there was a godawful racket, with extreme vibration, and it looked as though every emergency light in the cockpit was lit up.

I didn't know what had happened, but every left-engine gauge showed high in the red or zero. I pulled the left throttle to off to stop all fuel flow to that engine, and the racket quieted. I double-checked ejection procedures, headed back toward land, and made some turns to be sure I wasn't trailing smoke. The right engine was operating normally, but I made a higher-than-normal approach on the long Pax River runway just in case there was other damage. I was set to blow the landing gear down with emergency air pressure, but the wheels came down "three in the green," and landing and taxi to the ramp were normal.

Looking the plane over, we discovered that the end of the compensator on one of the cannon muzzles was missing. It had burned off after repeated gunfire had weakened the metal. An eight-pound hunk of stainless steel had gone down the inlet duct, hit the first-stage compressor blades turning at 18,000 or 20,000 rpm, and literally stripped the engine. Every stage of the compressor had some disintegration. Blades had spun off like shrapnel from a bomb. They had pierced the fuselage and the bottom of the wing, bored through the center fuselage into the right engine compartment, and punctured the top of the fuselage behind the cockpit where two fuel tanks were located. It was a miracle that they weren't ruptured, which would have blown the plane to pieces.

Metallurgical analysis of the three remaining compensators showed that they too were on the verge of failure.

Several Cutlass squadrons were put into the Navy inventory and deployed, but within a short time they were limited to land use only, and soon after that were phased out. In many ways, the Cutlass was a design ahead of its time, but there were just too many problems. Besides, a much better successor was coming along.

Chance Vought's F8U, the Crusader, was as good a plane as the Cutlass had been problematic. It was a sleek, swept-wing, low-fuselage fighter with an engine that generated over ten thousand pounds of thrust in normal, or military, power, and sixteen thousand pounds in afterburner, or combat, power. It was the fastest Navy fighter in the world, capable of sustained supersonic flight.

In all modern armed forces, combat aircraft in naval use routinely suffer from the need to meet aircraft carrier design requirements. The planes have to be able to withstand takeoff catapult shots accelerating them to an airspeed of over 145 miles per hour in the length of only 185 feet, and high-sink-rate, tailhook-arrested landings that bring them from 145 miles per hour to a halt in less than two hundred feet. Everything needs strengthening: bomb racks, missile launchers, fuel cells, landing gear, wing-fold mechanisms, catapult attach points, and tailhook assemblies. The load for all these things has to be distributed throughout the airframe as well. The resulting weight increase means that exclusively land-based planes usually are lighter and can therefore climb higher and fly faster.

The Crusader, however, was the equal of any land-based plane. Chance Vought peddled it to the Navy as "freedom's bright new sword." Not since the early days of the F4U Corsair had the Navy and Marines had a plane like it.

The Crusader set an official national speed record of 1,015 miles an hour in 1956, a year after its first flight. Flight tests at Pax River confirmed its speed, altitude, and maneuvering capabilities, while I was running the armament trials on the other side of the base.

The Crusader, like the Cutlass, had among its armament four twenty-millimeter cannons. I had done most of the high-altitude and supersonic gunfire points. They had gone well, without incident. Some of the low-altitude high-Q points remained to be done. "High-Q" refers to the conditions of greatest aerodynamic pressure produced by a combination of speed and air density. One of the high-Q points to be tested was at three thousand feet. The airplane couldn't quite sustain supersonic speed that low, so you'd dive into it, level out, and just below Mach 1 fire the cannons.

I was out over the Atlantic, planning to do two-second bursts. I set up for the point, just subsonic, and fired. Suddenly the airplane did a ninety-degree roll to the right. I caught it before it could go farther and brought it back to normal flight again, but could keep it there only with pressure on the stick. The Crusader had an electric lateral trim pod on the stick that you could use to counteract roll forces. We had had trouble with that trim pod during high-altitude and supersonic gunfire; it had jumped several notches out of calibration, and I thought that was what had happened now. I could retrim it, but, oddly, the neutral-point reading was several numbers displaced.

I went through a commonsense checklist before trying to repeat the test point, even though I thought I knew what the problem was. Control in roll, pitch, and yaw was okay. Trim was okay in all three axes, although off the normal number in aileron roll trim. But when I changed airspeed, things got quirky. I wanted to slow to landing speed and put the gear down to make sure I had landing control. But as I slowed, the plane's tendency to roll right increased. It was controllable, but something was obviously wrong. I decided against doing any more fire points that day, and headed back to the base to talk it over on the ground.

After I landed, faster than normal because of the increased tendency to roll, I taxied to the line. We routinely used the wing fold, both to exercise the mechanism and to save space. I was folding the wings up when I spotted the line crewman pointing up to the right wing. Then I saw other people running out from the hangar to look at it.

I turned around to see what they were looking at, and got one of the shocks of my life. Looking back from the cockpit in flight, I could catch only the leading edge of the wing, but now, as the outer part of the wing folded and came into my view, I could see that the wing structure from the leading edge of the tip, back and inboard almost to the right aileron, was missing. It was simply gone, with a jagged edge of metal that looked as if a giant set of teeth had taken a bite out of it. I understood why the plane had rolled; with two different wing areas, the lift pattern changed as airspeed changed. We later found that nineteen square feet of wing was gone. That's a pretty fair chunk. And we had no idea what had caused it.

There was no question of keeping the plane in service. While we investigated, we grounded the ones we were flying at the Air Test Center and the ones we were getting ready to send out to the fleet.

Chance Vought went into a full-court press. Experts at the plant near Dallas and on both coasts, including those at Edwards Air Force Base in California, where the Air Force does its testing, were enlisted on a high-priority basis.

After a lot of head scratching and number crunching, a team of physicists and engineers figured it out. The fuselage-mounted cannons were free-firing, each seeking its own individual natural fire rate of many hundred rounds per minute. The mathematical probability of all four firing together was extremely slight, but that's what the instruments showed had happened. Sometime during the firing burst, all four cannons had pulsed together for a tiny fraction of a second. All four were inputting vibration into the airframe, and—this was the part that took some analysis—the rate of fire was a resonant frequency of exactly the distance from the guns to the wingtip. The result was that a flexing and reinforcing standing wave had traveled out to the right wing tip like a bullwhip and just flicked it off.

The chance of that happening was so unlikely that nobody had foreseen it or anything like it. The fix was simple: interruptor circuits on the guns that, if they approached synchronized firing, would momentarily cut one of the guns out. I fired thousands of rounds after that without any problems. The order grounding the Crusader was lifted. Its first version, the F8U-1, went out to the fleet and Marine squadrons, where pilots welcomed it as the best fighter they had flown.

At Pax River, we went back to work to make it even better.

In November 1956, after nearly two years at Pax River, I was transferred to Washington as a project officer in the Fighter Design Branch of the Navy Department's Bureau of Aeronautics.

Dave was nearly halfway through the fifth grade, and Lyn was in the fourth grade. They were comfortable and getting along well at Hollywood Elementary. Rather than uproot them, I decided it made more sense for me to commute to Washington. That required a long day. I was up each morning and out of the house before dawn. The drive to Washington on two-lane U.S. 235 took an hour and forty-five minutes, and the same coming home, so I usually arrived after dark. I drove a used Studebaker Champion because it was cheap and got good mileage.

BuAer, as it appeared in the shorthand used in office correspondence, occupied so-called temporary quarters that had been built as the War Munitions Building back in World War I. Years later the Navy had taken it over. The building was near the Lincoln Memorial Reflecting Pool, at the site of today's Vietnam Veterans Memorial, in the heart of scenic Washington. Sometimes Annie would come up on a shopping trip. Often we went to the Capitol and sat in the Senate gallery to observe and to listen to debates. The process of lawmaking had fascinated me ever since my days in the high-school civics class of Harford Steele.

At BuAer, my responsibilities included some work on phasing out the Cutlass. Mainly, however, I was involved with further development of the Crusader. I spent a fair amount of time traveling between Washington and the Chance Vought plant. There were still a lot of change proposals coming out of the continuing test program. My job was to help sort out which changes were mandatory for safety reasons or combat capability and which would just be nice to have. A third factor, which I faced for the first time now that I was a bureaucrat, was to determine which ones the Navy and Marine Corps could afford. The ECPs I had made without regard to cost as a test pilot, I now learned, could be amazingly expensive. I liked to tell the Chance Vought engineers they should give us the plane for free and live off the change proposals.

Pratt & Whitney manufactured the Crusader's J-57 engine. It had an excellent record, but at Pax River I had sat in on discussions about whether it had logged enough hours at combat power at high altitude to be sure no bugs remained. I started thinking about how to kill two birds with one stone. We could do the test and also call attention to the fine plane the Navy had purchased as its frontline fighter. An Air Force Republic F-84F held the current transcontinental speed record, but I thought the Crusader could top it and, for the first time, cross the country at supersonic speed. And when I realized that the Crusader flew faster than the muzzle velocity of a bullet shot from a .45-caliber pistol —586 miles an hour—I had a name: Project Bullet.

Project Bullet taught me indelible lessons in bureaucratic persuasion. In addition to my regular job, I worked

through the end of 1956 and into the spring of 1957 trying to convince the Navy and the Pentagon not only that
the F8U could cross the country faster than the three-
hour-and-forty-five-minute record, but that it would be a
perfect capstone for the testing program: a sustained max-
power, max-altitude, max-Mach-number long-distance
flight. My theme was, "We'll never have a better time than
now." I pointed out that it would also convince the public
that the Navy was spending its money wisely, on a good
machine. The admirals and generals pored over the proposal in several meetings, and finally gave it their okay.
Some of them were a little envious. It was a pilot's dream.

They threw me a curve when a second pilot and plane
were assigned to Project Bullet. I hadn't expected that,
but I had to admit it made sense. The cost of the planning
and arrangements dictated a backup in case one pilot got
sick or one of the planes had a problem. Navy Lieutenant
Commander Charles F. Demmler joined me on the
project.

The speed run was set for July 16, 1957, two days before my thirty-sixth birthday. As July approached, I went
over the logistics of the run again and again. With our engines running in afterburner at full combat power for
most of the flight, we would need to slow down three
times for refueling during the 2,445-mile flight. The refueling operations would have to be precise to avoid losing
time. The Air Force had the only jet tankers. We made a
request, but it was turned down in language that smacked
of bureaucratic hokum. It went something like this: "We
have carefully considered your request and would normally welcome the opportunity to take part in Project
Bullet. However, our existing operational commitments,

now and in the near future, are such that participation is impossible."

Reading between the lines, it was fine with the Air Force if Project Bullet failed and the Air Force kept the record.

That left us with the Navy's twin-prop AJ refuelers. They could get up only to twenty-five thousand feet with a load of jet transfer fuel, and they poked along at just three hundred miles per hour maximum. The Crusaders would be flying at over a thousand miles an hour and fifty thousand feet and would have to drop to the tankers' altitude and speed. Using the AJs to refuel would add at least twenty minutes to the flight.

We practiced the rendezvous and in-flight refueling over and over. It took precise maneuvering. The tanker trailed a fuel line with a funnel-like drogue attached. The Crusader had a probe that extended from below the cockpit out to one side. You had to mate the probe with the drogue, fly for seven minutes during the fuel transfer, then back off carefully enough to break the connection without hurting anything, retract the probe into the Crusader's fuselage, and speed up again.

On one practice run north of Dallas, I had just plugged into the tanker when the AJ started belching black smoke from both engines. I pulled out immediately and flew alongside as the smoking AJ started a slow descent. I talked to the pilot on the radio. He tried everything, but nothing worked. Finally at thirty-five hundred feet, the three crew members bailed out and the pilot trimmed the controls to guide the limping tanker toward an uninhabited area before bailing out himself. The explosion when the AJ hit looked like an atomic bomb. A refueling error had filled some of the AJ's tanks with jet fuel. Basically high-grade kerosene, it didn't give the AJ's gasoline-burning engines enough power to stay airborne.

Charlie and I knew that to qualify for an international record, we had to have "sporting licenses" from the international sanctioning body for official aviation records, the Fédération Aéronautique Internationale in France. We got them a few weeks before the flight and bantered good-naturedly about who might actually win the record. He suggested I might come down with a mild disease or break a leg. I told him I had ordered the ground crews not to put any ammo in his guns so that he wouldn't shoot me down over Pittsburgh if I was in the lead.

Under a six-hundred-foot thin overcast, I took off from Los Alamitos Naval Air Station in California at 6:04 Pacific time on the morning of the flight, in a camera-equipped reconnaissance version of the Crusader, the F8U-1P. I had started the several cameras before taking off, hoping to produce the first coast-to-coast strip picture of the United States ever taken. Charlie was following in half an hour, in a regular fighter version.

I popped up through the overcast in seconds, climbed to thirty thousand feet, and accelerated to just under a thousand miles an hour. The plane got lighter as it burned fuel, and I held a 1.25 Mach number, letting it slowly accelerate to Mach 1.48 during a gradual climb to fifty-one thousand feet.

The plane flew beautifully, with no strain at all. With no clouds and the Earth far below, the sensation of speed was less than in a car at thirty miles an hour. I was doing my own navigation, hitting checkpoints every three and a half minutes. Staying on course was no problem, but contact with the tankers up ahead was crucial. Two of the AJs were circling at each refueling point.

I approached my first refueling rendezvous over Grants, New Mexico, west of Albuquerque. It took forty miles just to slow down. I mated the probe with the drogue on the second pass.

"Check under the hood?" the tanker pilot joked as the fuel was flowing down the long line into the Crusader's tanks.

"Don't bother. I'm in sort of a hurry this morning."

"Come and see us again. I'll mail your Green Stamps in the morning," the tanker pilot said as I backed the probe out and hit the afterburner to resume full power. The pit stop had taken only twenty seconds longer than I had planned.

The second, over Emporia, Kansas, went perfectly. Meanwhile, back in Albuquerque, Charlie hit the drogue a little too hard. He was unable to take on fuel and had to land. The message I heard over the radio was cryptic: "Charlie went in at Albuquerque." It could have meant that he had crashed, but another call verified that he had landed okay.

My third refueling was scheduled to take place over Indianapolis. Visibility was a little reduced at twenty-five thousand feet, and I kept trying to pick the tankers up on the bearing my direction finder showed. My fuel was almost gone when one of them spotted me and radioed. Seconds were precious. It was the backup AJ, and I used it even though it could pump fuel only half as fast as the primary. The flow stopped before I had a full load. I broke the connection without waiting to find out why, hit the after-burner, and climbed again with a thousand pounds less fuel than I had planned.

My route east took me within ten miles of New Concord. I was climbing at Mach 1.4 and, without know-ing it, sending sonic booms reverberating forty-five thou-sand feet down to the ground. My parents and their friends and neighbors knew about the run. They were out in the yard to see if they could catch sight of the Crusader or its vapor trail when—they told me later—two sonic booms rattled New Concord and one of the neighbors came run-ning down the street calling to Mother, "Mrs. Glenn, Johnny dropped a bomb! Johnny dropped a bomb!"

The sonic booms, created by perfect atmospheric conditions over the eastern third of the United States, hit the ground behind me as I flew. A ceiling collapsed in Indiana, windows broke in Pittsburgh, and people grum-bled. Somebody called the Pennsylvania National Guard about an unexplained explosion, and the Philadelphia Daily News reported that "the flight and the resulting blast caused quite a stir here."

I had no idea that was going on below my flight path. I was just worried about maintaining supersonic speed and get-ting to New York with enough fuel to land. I played the power on descent to stay close to my fuel consumption charts.

Floyd Bennett Field on Long Island was my destina-tion. I had less than three hundred pounds of fuel left when the official timers clocked me past the tower, which was the official finish line. I went to idle throttle, pulled up sharply into a high climbing turn, and glided around to a landing. My final worry was having enough fuel to taxi back to the ramp and the welcoming party.

The trip across the country had taken 3 hours, 23 minutes and 8.4 seconds, 21 minutes faster than the old record and, at an average speed of 723 miles per hour, 63 miles per hour faster than the speed of sound at thirty-five thousand feet. (Sound travels faster at sea level, about 760 miles per hour.) I had crossed the country in considerably less time than the project's namesake bullet would have taken, and the record now belonged to the Navy and Marines. The engine had performed flawlessly. So had the cameras, and we had a coast-to-coast strip shot of the country. It was, altogether, some birthday present.

Rear Admiral Thurston B. Clark, commander of the Naval Air Test Center, was waiting on the ground. He and the rest of the NATC staff looked proud and relieved. There was also a cadre of reporters, and television crews lugging cameras with huge film magazines that looked like Mickey Mouse ears.

The sight I liked best, though, was of Annie, Lyn, and Dave waiting in the roped-off area that held the welcoming party and a military band. The sun shone and the day was beautiful. The band was playing "Anchors Aweigh" and the "Marine Hymn." Annie looked gorgeous in a white hat, a choker necklace, and a print dress that showed off her shoulders. Lyn wore a striped dress and white gloves, and Dave a white shirt and pleated pants. They were a family to be proud of. I had carried Dave's Boy Scout knife and a brooch of Lyn's in the shape of a cat, and I pulled them from my flight suit and presented my children with their "supersonic souvenirs."

I was proudest of Annie, because I knew how difficult

it was for her to be out before the cameras. Inevitably, I knew, reporters were going to want to know what she thought of all this, and Annie, with her stutter, would have trouble telling them. I had a pang of conscience that stepping into the spotlight like this might not be fair to her.

None of it seemed to faze her, though. We posed for pictures for a profile *The New York Times* was running the next day. She went through the photo session with no problems, and the article appeared with a picture of all of us hugging. Lyn and Dave handled all the public exposure well, too. I came away confident that we would all survive our little brush with fame. It surely wouldn't last. After all, as the *Times* profile put it, "At 36, Major Glenn is reaching the practical age limit for piloting complicated pieces of machinery through the air."

CHAPTER 12

We stayed in New York for a flurry of news sessions and appearances. Thomas S. Gates, the secretary of the Navy, announced that I would receive a fifth Distinguished Flying Cross. The Navy wanted to get all the publicity it could out of the successful speed run and was even planning to send me on a tour. I looked forward to this. I was proud of what Project Bullet had achieved, and I had some things I wanted to say about the spirit of teamwork that the run represented. I thought people ought to share the same enthusiasm I felt for what was essentially a national accomplishment.

Between sessions, we spent some time shopping. Dave and I were at Macy's when I noticed a woman who kept looking at me. I thought it was because she had seen the profile and photo in the *Times*. Instead, she came up to me and asked if I had ever seen the television program *Name That Tune*.

I knew the program, although usually I was still on the

road back from Washington at seven-thirty in the evening, when it aired. She asked me if I'd be interested in learning more about the show and possibly appearing on it.

I had played the trumpet, sung in the glee club, collected big-band records, especially Tommy Dorsey's, and always listened to music of all kinds. My tastes were far-flung, from barbershop quartet harmony and the popular music I had grown up with to the standard works of classical music. It was a love that I shared with Annie. Her big regret was that despite my pleasure in music, I had two left feet when it came to dancing. But I had spent hours singing with Monty Goodman and whoever else happened to be around, usually at officers-club parties, as far back as World War II. In China after the war, someone dubbed the singers in the squadron the Little Madrigal Society, a name loftier than our singing warranted. We harmonized and sang old favorites that most nights got embellished with raunchy "overseas lyrics" by the end of the evening. That continued in Korea, where we had a recalled reserve pilot in the squadron next door named Woody Woodbury, who was a professional entertainer. We all chipped in to buy a piano and have it shipped over from Japan, and after he finished his missions, Woody would play in the Quonset hut that served as the officers club, and we'd gather around and sing. Our generosity bit us back when the wing commander heard Woody and sent him out on a tour of Far East bases. But we had a record player and plenty of good records, and I played them all the time.

So I said, "Yes, I know a little about music, and I'd like to learn more about the show." I had to get clearance from my bosses at Pax River. They said it was okay, and I went back to New York a few weeks later for an interview with the producers. They wanted to find out how many songs I knew. A woman would hum a tune, and I'd try to name it while the producers scribbled notes. She must have hummed hundreds of tunes. It surprised me that I knew so many of them. At the end of the session they asked me if I wanted to be on the show. I had to check again with the Navy at Patuxent and the Marine Corps, and they gave me the go-ahead.

Before I appeared on *Name That Tune,* the cross-country flight brought a moment of pride to my hometown. New Concord invited me to share the spotlight with the annual potato and flower show during its traditional Labor Day weekend festivities. The local Board of Trade put on a parade, and Annie, Dave, Lyn, and I rode in a new Chevy convertible that was a far cry from the old cruiser I once had piloted around the streets of New Concord. New Concord High School joined the fun by announcing the Major Glenn Trophy, to be engraved with the name of each year's highest-scoring math student. Apparently my alma mater had not heard of my struggles with calculus.

Name That Tune aired live on CBS on Tuesday nights. The host was a jovial fellow named George DeWitt. He had a million-dollar smile and wavy black hair. I envied his hair, because mine seemed thinner each time I looked in the mirror. I was becoming, as they say, "follicly challenged." I learned just before the first show in September that my partner was to be a ten-year-old from Hattiesburg, Mississippi. His name was Eddie Hodges, he had bright red hair and freckles, and he wanted to be a Baptist preacher like his granddad.

Eddie and I made it through the first two rounds and won $10,000, pushing a button ahead of the clock when we recognized the tune. We won the next week, and the next, adding $5,000 a week to our winnings. A young singer named Leslie Uggams was one of the other contestants. If Eddie and I made it through the fifth and final round of the Golden Medley Marathon, we would split the grand prize of $25,000. Even half of $25,000 was an astounding amount of money to a Marine pilot.

The host wanted to know what we would do with the money if we won. I said it would help pay for Dave and Lyn's education, and if there was any left over, it would go toward Lyn's wedding. She was in the audience with Annie, and I saw her blush and squirm and roll her eyes. DeWitt joked with Eddie about some girl trouble he said he was having, then asked me if I'd ever had girl trouble when I was ten.

I said I had. There was a girl in school who wouldn't talk to me because all the other boys liked her, too.

"Did you ever get to talk to her?" he asked.

"I did. I got to marry her," I said.

Now it was Annie's turn to blush while the rest of the audience applauded.

The Golden Medley consisted of snatches of five songs. We had thirty seconds to identify all five. The only clue we had was the identification before each tune of the dance you did to it. The first was the Peabody, and the music was a Gershwin tune, "Liza." Then "La Paloma," a tango. The third song was a square dance, "The Chicken Reel." Next was a waltz, "Don't Ask Me Why."

We had nine seconds left at the beginning of the fifth and final song. You could hear a pin drop in the audience after DeWitt said the dance was the gavotte. I had never heard of it. But then the sound track broke into a lilting song I recognized. Eddie looked up at me, and I pressed the buzzer with two seconds left and whispered in his ear.

"'Amaryllis'?" he said to DeWitt.

"'Amaryllis' . . . is right!" DeWitt said, and Eddie and I hugged each other. Eddie sang "Gospel Train," and DeWitt and I joined in.

Eddie and I divided up our jackpot. He went on to play a juvenile lead in the original Broadway cast of *The Music Man*. Annie and I were there on opening night to hear Eddie perform, including the wonderful solo "Gary, Indiana." My winnings, after taxes, went to start a college fund for Dave and Lyn, and to buy Annie an electric organ.

The strains of the Golden Medley had barely faded when, on October 4, 1957, the Soviet Union shocked America by launching the first Earth satellite, a 184-pound ball of steel and wire called *Sputnik*. The United States had announced that it planned to put a satellite in orbit in 1958. To be beaten to the punch dealt a devastating blow to Americans' self-image of technological superiority. The Soviets had already successfully tested their first intercontinental ballistic missile. Now, with *Sputnik* sending back radio signals as it orbited, it was as if they had an eye in the sky looking down on us. We heard Soviet Communist Party leader Nikita Khrushchev taunt, "The United States now sleeps under a Soviet moon," and announce a program to bring more third-world students to the Soviet Union to study science and technology, as a step to promote Communism around the world.

Any euphoria that remained from my transcontinental

speed run faded. Supersonic flight had been outdone as a yardstick for measuring military superiority. The American-made television sets, transistor radios, and cars with tail fins that meant technology to U.S. consumers in the rich post-war years seemed frivolous next to the evidence of Soviet scientific achievement beeping overhead. Suddenly the Cold War converged around the next frontier—space—and the Soviets had gotten there first.

The Soviets hoisted an even larger satellite into orbit a month later, on November 3. This one weighed over a thousand pounds, and carried a female dog, Laika, whose pulse, respiration, blood pressure, and heartbeat were measured by instruments in the space capsule and relayed to Earth.

While the Soviets had proved they could get a half-ton payload into orbit with a living creature on board—*Sputnik II* wasn't designed to bring Laika back again, an aspect of the mission the American public would not have tolerated—America's first effort to launch a satellite was a colossal and widely viewed failure. In December a Vanguard rocket carrying a three-pound satellite struggled to lift off from Cape Canaveral, only to settle back and explode in a ball of orange flame. Nobody knew how many failures the Soviets had had before they got a rocket off the ground. They weren't sharing that kind of information with the world. But President Eisenhower's commitment to openness meant ours was broadcast on national television, and produced headlines like "KAPUTNIK!"

So the year that had seen a moment of great triumph for me and for America's military ended with us playing catch-up in a game in which, it seemed, we were desperately behind.

The commuting to Washington was getting old. I was on my third used Studebaker. One I had worn out. The second was totaled one rainy morning as I was on my way to work, when a big Oldsmobile driven by a sailor rushing to Pax River to make Monday morning muster slid over and knocked me off the road and down an embankment. I dreamed after that of commuting in a Marine six-by-six loaded with ten tons of sand, but it wouldn't have gotten the Studebaker's mileage. Annie and I decided we would move to the Washington area at the end of the 1958 school year, when Dave finished elementary school. Tom Miller had been assigned to the Bureau of Aeronautics, and since my old friend was also based in Washington, we thought we might find houses close to each other.

Tom and Ida Mai had had three children, a son and two daughters, since the tragic death of their first baby during World War II. They, too, were interested in settling in a good school district. Annie and Ida Mai studied the possibilities, and we decided on Arlington, Virginia. Prices were high, but we found two lots side by side across the street from one of the best schools, Williamsburg Junior High. Tom knew a builder who had some stock plans, and we started planning a pair of ranch houses that would look from the front as if they had one story, but opened up on the lower level onto a beautiful tree-shaded hillside.

While the house plans were being drawn, I continued to commute from Pax River. Rumors wafted through

BuAer that the government was laying plans to put a man in space. The Air Force was advocating its program, called MISS, for Man in Space Soonest. The National Advisory Committee on Aeronautics—NACA—was also in the game, testing a winged rocket plane, the X-15, and exploring other possibilities.

In late January or early February 1958, word reached BuAer that the NACA Research Center at Langley Air Force Base in Virginia was looking for a few test pilots with extensive flight experience to do some part-time studies together with NACA's regular group of test pilots. NACA scientists were trying to come up with parameters for a piloted spacecraft. They wanted people to make some runs on a space flight simulator for an investigation of reentry characteristics of various spacecraft shapes.

Navy Captain Joe Smith was my boss in the Fighter Design Branch. He was an experienced pilot, a great guy, and a good friend who had approved my time away from BuAer to do Project Bullet. I went to him and asked for a couple of days here and there for the Langley assignment, and he agreed.

The NACA-Langley job was my introduction to orbital mechanics—the forces needed to get into orbit, how to adjust an orbit from oval to round or vice versa, where to apply the power and how much, what is involved in making a reentry, and how to control it all by hand from inside a spacecraft. I spent several days at Langley, after which NACA asked me to go to the Naval Air Development Center at Johnsville, Pennsylvania, to make some runs on the centrifuge there.

I had experienced reasonably heavy G forces, as had most jet fighter pilots. Four to six Gs—multiples of the force of gravity—were common during a pullout from a steep bombing run or maneuvering in air-to-air combat. Sometimes you'd hit eight and even approach nine, which was the design limit for most fighters and attack planes. During this, the pilot was sitting upright in the cockpit, usually wearing a G suit to help prevent blood flow away from the brain toward the lower extremities.

But the centrifuge at Johnsville was like nothing I had experienced at the controls of an airplane. It had a fifty-foot arm and, spinning at high speeds, could produce literally crushing G forces of twenty-five times the weight of gravity in the pod at the end of the arm. No human could take that many Gs, but I rode it at speeds simulating the eight Gs expected in a space launch, from positions ranging from upright to supine. The person going into space most likely would be lying back, taking the Gs in the chest as if he were lying in a bed that was being accelerated straight up. Even at high G forces, I found it possible to keep track of the instruments and indicators and use a three-axis hand controller to make the kinds of adjustments needed in roll, pitch, and yaw.

As I rode I was wired up with leads and monitors that kept track of my pulse, blood pressure, heartbeat, and respiration. The similarity to the dog the Soviets had sent into space was all too obvious. The NACA scientists wanted to compare data from the centrifuge runs to data from the simulator runs at Langley. Whirling around at the end of that long arm, I was acting as a guinea pig for what a human being might encounter being launched into space or reentering the atmosphere.

When the centrifuge runs were finished, another temporary duty assignment took me to the McDonnell

Aircraft plant in St. Louis, where a team was working on space capsule design. Years earlier, an NACA engineer named Harry Julian Allen had figured out that a blunt reentry shape would disperse heat where a needle nose would not. Maxime A. Faget and his engineering group at Langley had refined the concept into a capsule that would sit on top of a rocket, be lifted skyward, and re-enter blunt end first. McDonnell was perfecting a manned version. Since I now knew something about what the capsule and its pilot would go through, I was the Navy's representative in talks about its configuration. The capsule was designed for fully automatic flight, but had manual controls. Scientists didn't know if a human could control an orbiting spacecraft, and this was part of the research. But there was never a question in my mind, even at that early stage, that the person in that capsule was going to have to be able to control it, as well as do the onboard research projects that were also in the process of being designed.

Without really knowing it, I was seeing the very beginnings of the U.S. manned space flight program. At that stage, however, it consisted of research and rumors. If there had been a call for volunteers, I would have been at the front of the line. But the Eisenhower administration had not publicly committed itself to sending a person into space, nor said whether a program to do so would be military- or civilian-run.

That made it, for me, a somewhat frustrating time. The research going on at Langley and Johnsville was fascinating, as was the space capsule design. But I found it impossible to connect that research with my own future. I was almost thirty-seven years old, a United States Marine Corps major stuck in a desk job. I had continued to take courses at the University of Maryland's Pentagon extension and felt sure that with my previous correspondence and extension courses and my course work at test pilot school I had the equivalent of a master's degree, but in fact I still had not been awarded even a bachelor's. Without a degree, my military advancement was uncertain. I could feel the time ticking away.

Work on the house in Arlington started in the spring, and we moved in at the beginning of June. My family life, at least, was settled. Our house, and the Millers' next door, was made for comfortable, casual living. Annie had given the house an early-American motif. It had plenty of space for the kind of family gatherings we liked, around the fireplace and dining table.

Annie and I had always stressed, and more so since Dave and Lyn came along, the need to talk to each other, to communicate, to share views, feelings, and thoughts. Annie started a family tradition of candlelight dinners. When we were all together, we usually would eat by candlelight and talk about what was on our minds and what we were doing. This replicated the talks my parents and Jean and I had shared around the table while I was growing up. I thought such family conversations were vital to Annie and me, and Dave and Lyn. I was gone a lot, so we needed to carve out moments for communication, to make up for lost time. But I always had to remember not to let my interest in what the kids were doing turn into an inquisition. I let them come to me and open up gradually.

That summer, near the end of July, Eisenhower signed the National Aeronautics and Space Act into law. At last it was official. The country was going to put a man in space, in a program that would be open to the world's scrutiny and, being civilian-run, would emphasize scientific rather than military goals.

In October, after Dave had started the seventh grade at Williamsburg Junior High and Lyn was in the sixth grade at nearby Jamestown Elementary, the National Aeronautics and Space Administration (NASA) came into being, succeeding NACA. The new agency inherited a $100 million budget; three laboratories, including the one at Langley; the sections of Edwards Air Force Base dedicated to high-speed flight probing the limits of the atmosphere; a rocket-sounding facility at Wallops Island, Virginia; and 8,000 employees, including 177 in the agency's decidedly unspacelike headquarters in the old yellow Dolley Madison House at the northeast corner of Lafayette Park, across from the White House. The Air Force MISS program notwithstanding, NASA was to have the prime manned space role. The goal of its Langley-based Space Task Group was to launch an orbital flight, see how man fared in the new environment of space, and bring him and the capsule in which he flew back safely to Earth.

The existence of an actual program was a tonic that eased my earlier frustration. A few blocks away at BuAer, when word spread quickly that NASA was combing the services for men who fit a profile for potential space flight, I began to see that I might fit that profile.

Nobody knew exactly what the criteria would be. In fact, NASA's initial plan was to put out a public call for applicants, including not only pilots but people who had undertaken other demanding and hazardous assignments such as research submarine duty, polar exploration, mountain climbing, and deep sea diving. The only word that reached me, however, had to do with age, weight, height, and education requirements.

I determined that when the call came, I would be ready. Here was something that was both scientifically interesting and a logical extension of what I had been doing, and doing well, all my adult life. It would be a welcome far cry from my desk job at BuAer. Maybe it would even advance my Marine career, adding a couple of ticket punches to replace the senior-level staff schools I had neglected to apply for.

The cutoff age was supposed to be thirty-nine. There was nothing I could do to change my birthday, but I had a couple of years to go. I knew from my talks with the McDonnell engineers that the height and weight of the people chosen would be limited by the size of a capsule that would fit on top of a relatively small rocket. My father's genetic legacy meant I was prone to overweight. In fact, thanks to the desk job, snacking on the drive to and from D.C., and lack of exercise, by the time I got a hint of the requirements I had ballooned to a slack-muscled 208 pounds on a frame just under five-eleven. That was way too much. I started serious dieting and working out at the Pentagon gym. I ran, swam, jumped rope, pogoed on a trampoline, and worked out with barbells, aiming to get down to 178 pounds and maybe less. Tom Miller worked out along with me. He said I'd never make it, but I did.

My old friend Tom and I are in dispute over the next part of the story. The height limit for the future space travelers was supposed to be five-eleven, and I made that by a fraction. Tom for years has claimed that I was six feet even

and that I spent two hours every night with a stack of encyclopedias strapped on top of my head to try to get shorter.
He has even claimed I slouched to make the height limit.
Where he dreamed that up, I'll never know. If Tom was
trying to say I wanted it badly enough to do something as
ridiculous as that, he wouldn't be wrong. But I didn't have
to. As far back as my application for Civilian Pilot Training
when I was a sophomore at Muskingum College, and all
through my Marine years, my height was recorded as sev-
enty-one inches, or five-eleven. Encyclopedias on the
head? I think Tom was just having fun with some reporters.

Educational requirements were another matter. My
lack of a degree was the factor most likely to stymie my
ambitions.

By the end of 1958 NASA's plan to draw its spacemen
from a roster of daredevils and adventurers was scrapped
without becoming public. NASA thought a fair number of
crackpots might apply, and the country didn't have time to
screen them out. The field shrank to pilots, and then to
military test pilots on active duty, with added points for
combat experience. The agency believed people who had
worked under the most arduous and adverse high-speed
flight conditions provided the best point of departure into
what would be an entirely new realm of flight.

NASA screened the records of 508 test pilots and
eliminated four-fifths of the prospects. Some were too
tall, some didn't have enough experience, others had
some glitch in their military records. At some point in the
process, NASA discovered there weren't any Marines on
the list because the Navy hadn't specifically been asked to
include them, even though Marines went to the Navy's
test pilot school. But Colonel Jake Dill, who had been my

commanding officer at the Marine detachment at Pax River and was now second in command in personnel at Marine headquarters in Washington, made sure Marines were on the pared-down list of 110 candidates.

On December 17, 1958, the fifty-fifth anniversary of the Wright Brothers' first flight at Kitty Hawk, NASA announced in Washington that the name of its manned space program would be Project Mercury.

Early in 1959 I received orders stamped "top secret," directing me to report for a briefing at the Pentagon, where I found myself in a room full of test pilots. A pair of NASA officials confirmed the rumors that had been circulating at BuAer. Abe Silverstein and George Low said NASA needed volunteers for its program to send men into suborbital and orbital flight. These men would be called astronauts. It would require a new kind of test flying that had never been done, but it was a necessary step in the space race with the Soviets. Anybody who volunteered could reconsider his choice at any time, they said.

I volunteered without hesitation. From thinking that I might fit the NASA profile, I now began to think for the first time about the reality of going into space.

Annie had always accepted that uncertainty was a part of the life we had chosen. We both had understood, first when I enlisted and went away to World War II and later when I volunteered for service in Korea, that flying in combat carried risks. The same was true of flying as a test pilot. But I was convinced beyond any doubt that these were things I should do.

I am not a believer in irresistible fate or complete

predestination. But I believe strongly that each of us has a unique set of God-given characteristics, talents, and abilities. Our end of the partnership is to use those capabilities to the maximum and for a good purpose as we pass through this existence. The space program was one of those things that was going to require all I could give it.

Patriotism was involved as well. It was the same unhesitating willingness to do something for the country that I had learned from my earliest days in New Concord, the motivation that had driven my father to volunteer for World War I, that had taken us to the cemetery on Decoration Day to play echo taps together, that had required me to enlist after Pearl Harbor and to request assignment to Korea. Now the country needed astronauts.

All this was a prelude to telling Annie about the program and what I wanted to do. It didn't come as a complete surprise. After each trip to Langley, Johnsville, and St. Louis, I had sat and talked to her and the kids for hours about this new program. They had sensed my interest and enthusiasm.

"Fly into space?" she said.

"Not fly, exactly. Be launched," I said.

Her dark eyes grew troubled. "It sounds dangerous," she said.

"Nobody is going to go anywhere unless everybody believes it is safe," I said. "NASA is going to have a complete testing program for equipment, and training for the astronauts. But going into space . . . it will be something nobody has done before in all history. Imagine being able to look down on the Earth from a hundred or a hundred and fifty miles. Annie, the scientists I've been talking to say we have the know-how to do this. And if we can do it, we ought to. We ought to get up there before the Soviets do, if we can. Around the world, a lot of people are looking at us and them, and figuring out whom to put their money on. We're telling people everything about our program. There won't be any secrets. The world has its eye on us, and if we're going to send astronauts into space, I want to be one of them."

I told her what I knew about the training that would be involved, the exhaustive testing that anybody selected for the program would undergo. I told her it would mean more time away from home, but that once again, it was an important job that needed to be done, and I felt lucky just to be in a position to be considered.

"I don't know that I'll even make it," I said.

"If you want it, you'll make it," Annie said.

Annie understood, as she always had, and in her turn set about doing what she had to do to make it work. Once more it meant that we would face uncertainty, and she would be at home waiting for the outcome. I knew, too, that her speech continued to make life more of a burden for her than it was for many other people, and that I could have alleviated that for her if I had been around more. Lyn, from the time she was a little girl, had been Annie's telephone surrogate, answering the phone or calling to make appointments for Annie at doctors' offices and beauty parlors. I don't know what I would have done if Annie had said no. But she didn't.

"Astronaut," she said. "How do you spell that?" And then, after turning the word around and giving it some thought, she laughed and added, "Are you sure we shouldn't take out the second *a*?"

There was reluctance in some quarters of the Air Force to join NASA's program. A number of Air Force test pilots who met the requirements eliminated themselves from consideration at the start. They had assumed—prematurely—that the MISS program would lead America's manned space efforts. When NASA was given that role, some of the Air Force's finest were livid, and reacted with intemperate and caustically critical public comments about NASA's plans and the flight-testing program leading up to manned flight. Out at Edwards Air Force Base, the word was that the astronauts were going to be "Spam in a can." That meant that the men the Mercury program lifted into space would be passengers. They would "ride" a rocket, not "fly" it. Further, they would follow test monkeys into space, since NASA was expected to send monkeys up before launching a manned flight, and they were much too proud to fly where monkeys had preceded them. They didn't like the idea that NASA was civilian-run, even though a military-based program would have caused suspicion around the world about our intentions in space. They thought that signing up for a program about which so much was still unknown would sidetrack their careers.

Nevertheless, Project Mercury wound up with over eighty candidates.

The selection process continued through several more screenings. I went to meet Charles Donlan, the deputy director of NASA who was in charge of astronaut selection, carrying the results of my centrifuge runs at Johnsville because I wanted him to know I had already withstood the G forces of reentry. He had drawings of the proposed Mercury capsule, and I asked if I could look them over after the meeting, since I had been involved in some of the talks about its configuration.

As the field narrowed, my lack of a degree loomed larger. Jake Dill, unbeknownst to me, learned that I had been deselected on the basis of the degree requirement, since the astronauts were all to be assigned specific, and highly technical, tasks within the program in addition to riding rockets into space. Jake knew my background and thought this was unfair. He went to NASA with all my combat and academic records, and my technical flight test reports from Patuxent. I later learned he had met with the selection board and convinced them that I had more than the equivalent of a degree.

Unaware that Jake had done all this, I tried to deal with the problem by having all my extension and correspondence credits transferred to Muskingum. The college didn't argue with my credits, but the faculty committee that had to approve the degree declined my request on the basis that I didn't meet the college's residency requirement. I had lived in New Concord for nineteen years but had no recent time on campus.

Nevertheless, when the list got down to thirty-two, my name was still on it.

A letter arrived at home saying the selection committee had been impressed by my record and interview. If I was still interested, I was to report for medical tests at the

Lovelace Clinic in Albuquerque, New Mexico. No one outside of key people in Washington and at the clinic knew what the tests were for. To keep it that way, I was to travel in civilian clothes and not discuss Project Mercury unless authorized. We were given numbers so we wouldn't have to use our names.

The tests I and the other candidates went through in small groups at Lovelace were the most extensive medical examinations known to humankind. That's the way it seemed at the time, and it turned out to be true.

Lovelace was a diagnostic hospital specializing in aerospace medicine. It had been founded by Dr. W. Randolph Lovelace II, a prominent space scientist and chairman of the NASA life sciences committee, who had conducted high-altitude and pressure suit experimental work at Wright-Patterson Air Force Base. The clinic was private, but there was a strong military flavor to its administration, which was directed by Dr. A. H. Schwichtenberg, a retired Air Force general. The doctors, led by Lovelace, were a hard-nosed group, or so it seemed to those of us they were poking, probing, and evaluating.

For over a week they made every kind of measurement and did every kind of test on the human body, inside and out, that medical science knew of or could imagine. Nobody really knew what that body would go through in space, so Lovelace and his team tried everything. They drew blood, took urine and stool samples, scraped our throats, measured the contents of our stomachs, gave us barium enemas, and submerged us in water tanks to record our total body volume. They shone lights into our eyes, ears, noses, and everywhere else. They measured our heart and pulse rates, blood pressure, brain waves, and muscular reactions to electric current. Their examination of the lower bowel was the most uncomfortable procedure I had ever experienced, a sigmoidal probe with a device those of us who were tested nicknamed the "Steel Eel." Wires and tubes dangled from us like tentacles from jellyfish. Nobody wanted to tell us what some of the stranger tests were for.

Doctors are the natural enemies of pilots. Pilots like to fly, and doctors frequently turn up reasons why they can't. I didn't find the tests as humiliating or infuriating as some of the other candidates did. Pete Conrad was so incensed by having to rush through the hospital's public hallways "in distress" that he told General Schwichtenberg he wasn't giving himself any more enemas, and deposited his enema bag on the general's desk for emphasis. He didn't get chosen for the space program until later. But I thought the tests, obnoxious as they were, were fascinating for the most part. It was all in the interests of science, and going into space was going to be one of the greatest scientific adventures of all time.

After eight days at Lovelace, one candidate washed out for medical reasons and the rest of us, again in small groups, received orders sending us to Wright-Patterson and the Wright Air Development Center's Aeromedical Laboratories. We traveled separately, as we had to Lovelace, to preserve secrecy.

The tests at Wright-Patterson were more familiar. They subjected us to the kinds of stresses test pilots could be expected to endure, heightening some of them in an attempt to simulate the thin reaches of space. Again, the doctors were guessing. They injected cold water into our ears as a way to create a condition called nystagmus, in which you can't keep your eyes focused on one spot, then measured how long it took us to recover. They measured our body fat content and rated our body types as endomorphic, ectomorphic, or mesomorphic. They inserted a rectal thermometer, sat us in heat chambers, ran the temperature up to 130 degrees Fahrenheit, and clocked the rise in our body temperature and heart rate. We walked on treadmills, stepped repeatedly on and off a twenty-inch step, and rode stationary bicycles. We blew into tubes that measured our lung capacity and held our breath as long as we could. We plunged our feet into buckets of ice water while the doctors took blood pressure and pulse measurements. We sat strapped into chairs that shook us like rag dolls. We were assaulted with sound of shifting amplitudes and frequencies that made our flight suits quiver and produced sensations in our bones. We endured blinking strobe lights at frequencies designed to irritate the nervous system. We entered an altitude chamber that simulated sixty-five thousand feet of altitude, with only partial pressure suits and oxygen. We lay on a table that tilted like a slow-motion carnival ride. We pushed buttons and pulled levers in response to flashing lights to test our reaction times. We sat in an anechoic isolation chamber to see how we might endure the vast blackness and silence of space. All the while sensors plastered on our heads and bodies recorded our reactions.

The isolation chamber was simply a dark, soundproof room. A technician led me in, seated me at a desk, and turned out the lights when he left. I had no idea if I would be there for fifteen minutes or fifteen hours. I knew the easiest way to make the time pass would be to put my head down and go to sleep. But I suspected that the doctors wanted mental alertness. I opened the desk drawer after a while and found a writing tablet. I had a pencil in my pocket. "Will attempt to keep record of the run," I wrote on the first page.

Before I was through I had scrawled eighteen pages in utter blackness, keeping track of each line by moving my finger on the page. I wrote down the ways I could think of to pass time in the dark, including exercising, changing clothes, and making a list of as many things as possible before I forgot what I had listed first. I stood up and located a hum in a corner of the room; thought about Helen Keller's dark, silent world, permanent where mine was temporary; and discovered I could make a spark by rapidly tearing off pages of the tablet. Finally I decided the best exercise was writing poetry, because I would have to memorize each line before continuing. Here are the first and last of seven verses on one of my favorite themes:

> To mankind's every broadening store
> Of knowledge, each must give
> His own peculiar talents, so that all
> May better live.

· · ·

> Then use all your inborn talents,
> Use them each and every day.
> Add to mankind's store of knowledge,
> Make them glad you passed this way.

I had moved on to summarizing my thoughts on the isolation experience when the door opened after three hours and the lights came on.

The point of all these exercises, I understood, was not only to measure the ability of our bodies to adapt to the unknowns of space, but to test an array of mental processes, including our motivation, detachment, and resistance to the impulse to panic.

The direct psychological testing was just as intricate. Psychologists and psychiatrists trailed us around, asking questions and making notes. We answered 566 questions on the personality inventory test. We described ourselves by completing the phrase "I am . . ." twenty times. This was easy at first: "I am a man. I am a Marine. I am a flyer. I am a husband. I am a father. I am an officer." Closer to the end, you were describing yourself in ways you hadn't thought of very much.

We took Rorschach tests, which I always thought looked like butterflies. It reminded me of the story about the guy who thought the ink blots looked like pornography. Finally the doctor asked why he always saw erotic images. "How should I know, Doc?" the man replied. "You're the one showing me the dirty pictures."

Of course, by the time you got to the psychological tests, you could pretty much figure out the answers. It was clear NASA didn't want abrasive, jumpy, or edgy men going up in its space capsules, or men who seemed to want to take unnecessary risks. Levelheadedness was the order of the day. The psych testing was as much about what you wouldn't do in a certain situation as what you would do.

The Wright-Patterson tests, like those at Lovelace, were more of a challenge than an irritation. By now I wanted very badly to be selected, but even the testing process itself had piqued my curiosity. Clearly, a vast and epic enterprise was under way. We were on the verge of doing something that had never been done before, and each step of the way was going to be interesting.

Mainly, I thought that no matter which of us were chosen, the astronauts would have a job to do.

I had been back in Washington two weeks when the phone rang at my desk at BuAer. I answered it and heard Charles Donlan say, "Major Glenn, you've been through all the tests. Are you still interested in the program?"

"Yes, I am. Very much," I said, and held my breath.

"Well, congratulations. You've made it."

I don't remember my response. I know I felt a swell of pride—that I couldn't help—but I also felt humble at being a small part of a program that was so full of scientific talent and of such importance to the nation.

Hanging up the phone, I was struck by the fact that the call had come on the day it did. It was April 6, my wedding anniversary. Annie and I had been married sixteen years, and that night we had planned to go to dinner at Evans Farm Inn in McLean and a play in downtown Washington in celebration. There was no greater celebration than sharing the news with her. I told her I had no idea where all this would lead, but wherever it was, we were in it together.

PART FOUR

PROJECT MERCURY

CHAPTER 13

The Mercury Astronauts met for the first time at Langley Air Force Base on April 8, 1959. We were seven pilots, three from the Air Force, three from the Navy, and one from the Marines, but none of us were in uniform, and at that time we were still anonymous. I had just received a routine promotion from major to lieutenant colonel, but as astronauts, we all ranked equally. We wore suits, the uniform of our new service, as we milled about with NASA officials and discreetly tried to check out the other men who had been chosen for this new assignment.

Robert Gilruth was the head of NASA's Space Task Group, which included Project Mercury. Bob had a fringe of thinning white hair and prominent black eyebrows, which gave him a look of Buddha-like wisdom. He was an old-line aerodynamics investigator with a quiet, congenial manner, and he stepped to a podium in the room where we had gathered to brief us on how NASA planned to tie us into the project.

"You're not just short-term hired guns," he said. "NASA wants the benefit of your experience as test pilots and engineers. Project Mercury is a team, and you're part of it. This isn't the military, where direction comes from the top down. We want your direct input. Any problem you have with design, or anything we're doing, you let me know.

"But let me warn you. Project Mercury isn't a continuation of anything. Nobody's ever gone into space before. It's completely new, it's untried, there are many uncertainties ahead. If for any reason whatsoever you decide it's not for you, you can go back to your respective services with no questions asked."

None of us were looking back, however. After Bob's brief introduction, we tried to get to know the men we were going to be working with.

I had met some of the others who'd been chosen, but didn't know them well. They may have known a little more about me, as a result of the cross-country speed run and *Name That Tune*, than I knew about them.

Al Shepard had worked on the Crusader. We had attended meetings together, and comments he had made revealed a sharp, analytical mind. A couple of times Annie and I had been in groups that included him and his wife, Louise, but we didn't know them well. I knew Scott Carpenter and Wally Schirra, because they, like Al, were Navy. I didn't know Wally's work or personality, but Scott had been in my group during the testing at Wright-Patterson. We shared an open-minded curiosity that had made us like each other right away.

The Air Force guys, Gordo Cooper, Gus Grissom, and Deke Slayton, I didn't know at all. I had met them for the first time when we were going through some of the testing.

One thing we did know, from our own histories and what we had gone through at Lovelace and Wright-Patterson, was that we were all extremely competent. The Langley meeting bolstered that impression. The way each man walked, stood, and shook hands exuded confidence, and maybe just a little arrogance. The fact that we had been selected meant we stood on a high step on the test pilot ladder. We were part of an elite group, an exclusive fraternity. Talking to each other, we didn't need preliminaries.

I learned quickly that several of the others had flown in Korea. Gus had about a hundred F-86 missions under his belt. Wally had served as an exchange pilot with the Air Force, as I had, and had shot down two MiGs. Scott had flown P2Vs, a long-range patrol plane; he had only about two hundred hours of jet time, which made his selection a little surprising. Deke and I were the only two who also had flown in combat during World War II; he had done bombing runs in B-25s over Europe. More recently, he and Gordo had been test pilots at Edwards. They had been flying the hottest of the Air Force jets, the Century series, although Deke was in fighter ops and Gordo was in engineering. Gus had been doing electronics testing at Wright-Patterson.

So flying was a big part of what we talked about. We wondered what kind of planes we would be flying at NASA to keep up our proficiency, and if flying was even going to be part of our training.

The rest of our conversation was pretty mundane. We didn't talk about deep space, we talked practicalities. The initial phases of our training would be held at Langley, and we discussed who planned to move there with their families, how much houses cost, and how the schools were. Scott, for example, had been one day away from leaving on a six-month tour of the South Pacific aboard the USS *Hornet* when he got a telegram telling him new orders were on their way. Annie and I, in the few days we'd had to talk about it, hadn't decided whether to move from Arlington.

So many details were unclear. Despite Bob Gilruth's brief assurances, none of us knew if Project Mercury was going to be an ongoing program or just a short look at the feasibility of space flight that would have us back at our respective services in a year or two. It was certainly obvious to me from the levels of preselection testing we had gone through that something unusual lay ahead. But at that point, even NASA was still feeling its way.

NASA was sure about one thing. Walter Bonney, NASA's director of public affairs, told us at the Langley briefing that he had never seen such a frenzy of media interest. A tall, stoop-shouldered veteran of NACA who had gray hair and a mustache, Walter said news organizations were coming out of the woodwork to cover the news conference NASA had called for the next day in Washington to announce the astronauts. "I don't know how we're going to handle it," he said.

Even with Walt's warning, not one of us knew what he was in for. We flew up from Langley in the morning and arrived at NASA headquarters in the Dolley Madison House early in the afternoon on April 9.

As the hour of the news conference approached, we could hear a buzz of expectation on the other side of the curtain. At two P.M. we walked onstage into a glare of television lights. When my eyes adjusted, I saw an auditorium overflowing with reporters, technicians, and photographers. We sat at a long table. A NASA logo hung behind us. Two models, one of the Atlas rocket that was going to put our first spacecraft into orbit, and the other of the spacecraft itself, the Mercury capsule with its escape tower on top, leaned against the front of the table. Each place had a microphone.

Walt shambled to the podium while people handed out press kits with our names and information about Project Mercury. Then we waited for a few minutes with flashbulbs going off in our faces while the reporters with afternoon deadlines scrambled for the phones to alert their offices. They came back, and T. Keith Glennan, NASA's director, who had served in the same capacity at NACA, stood up and said, "It is my pleasure to introduce to you—and I consider it a very real honor, gentlemen—Malcolm S. Carpenter, Leroy G. Cooper, John H. Glenn Jr., Virgil I. Grissom, Walter M. Schirra Jr., Alan B. Shepard Jr., and Donald K. Slayton . . . the nation's Mercury astronauts."

Everybody in the auditorium stood up and applauded. That was a strong hint. If any of us had thought he was going to remain anonymous, the standing ovation caused him to think again. I had had enough exposure to the press to know this was unusual behavior. In my experience, the press could be a little snippy, as *The New York Times* had been in the profile that followed Project Bullet, where the writer commented that my favorite adjective was the word *real*. But this was different. Reporters didn't tend to give standing ovations to people they were covering. We looked up and down the table at one another, and a few of us raised an eyebrow.

Then came the questions. They weren't questions concerning flying, which was what we were best equipped to talk about. They were personal questions. I'd fielded similar questions after Project Bullet, so when they came to me, I had a little more to say. It was important to me that Annie and the kids were behind me in this venture; I couldn't have done it without them, and I said so. My religion had always been important to me, and I said that, too. Expanding on my beliefs, I stated my conviction that we come into this life with a set of talents and abilities and that if we use them as best we can, opportunities arise. I didn't have to make any of it up, and I wasn't playing to the audience. It was just the way I felt. The questions opened a tap, and vintage New Concord, Ohio, came pouring out.

I had thought a lot about what we were going to be doing in the space program, and even then I saw it in big terms. I knew we hadn't been chosen just to be test jocks on more flights out of Edwards or Patuxent River. Obviously, there were going to be a lot of military implications, but this was something else. The world was at the door of a new age, and we were the people who had been chosen to take the first steps across the threshold.

So when it came to the historical significance of what we were doing, I said, "This whole project with regard to space sort of stands with us now like the Wright brothers stood at Kitty Hawk with Orville and Wilbur pitching a coin to see who was going to shove the other one off the hill down there. I think we stand on the verge of something as big and expansive as that was fifty years ago."

Somehow, and without intending to, I found myself speaking for the group. Asked why we had volunteered, when it was my turn I said, "I think we are very fortunate that we have been blessed with the talents that have been picked for something like this. I think we would be almost remiss in our duty if we didn't make full use of our talents. Every one of us would feel guilty, I think, if we didn't make the fullest use of our talents in volunteering for something like this—that is as important as this is to our country and the world in general right now."

These opinions seemed to resonate with the reporters.

Part of the NASA press kit was a rundown of the physical testing we'd been through. Dr. Lovelace, tall and thickset, with a receding hairline, gave the merest hint of it after he was introduced, saying, "I just hope they never give me an examination." So of course somebody asked which of the tests we'd liked the least. I backed into an answer on that one, telling the reporter, "They went into every opening on the human body just as far as they could go. Which one do you think you would enjoy the least?"

By that time I'd probably said too much. I had marked myself among my new colleagues by not being laid back and cool, the way test pilots were supposed to be. But when it came to the Steel Eel, nobody disagreed with me.

And for the record, I wasn't the only one who raised both hands when somebody asked if we thought we'd come back from space alive. Wally Schirra did, too. I wasn't the only one wearing a bow tie, either. There was one on Deke Slayton.

Even when it was over, none of us fully compre-
hended what would happen. It was inevitable that we
would be viewed not simply as pilot-technicians going
about our business. The mission alone dictated that. But
even my idea of the program's pioneering aspects didn't
account for the overwhelming level of attention. The average person could Walter Mitty him- or herself into winning the Indianapolis 500, since everybody drove a car; all
you had to do was imagine yourself going faster and making nothing but left turns. But space was so new nobody
had a way to relate to it realistically. And all most of the
press could do was project an epic vision. People looked at
us as if we had stepped out of the pages of science fiction
or descended from another planet.

And then there were the Soviets. Their strides in
space, combined with our fear of their intentions, placed
the astronauts in the front line of the war for not only
space supremacy but—in many minds—national survival.
The Soviets seemed so joyless and ideologically grim, and
we didn't want to be like them. Soviet premier Nikita
Khrushchev had said, "We will bury you." Americans knew
a threat when they heard one.

So by the time the television news signed off that
night and the papers appeared the next morning, seven ordinary, if courageous and determined, men were carrying
America's hopes and dreams for flying into space, and
were the objects of an insatiable curiosity.

At home in Arlington, I had tried to tell Dave and Lyn
what was going to be on the television news and in the papers when we were announced. Telling the kids that their

father was going to be an astronaut called for some imagination.

I had briefed them after my trips to Langley, Johnsville, and St. Louis, so the idea of space flight wasn't completely new to them. They knew a lot about airplanes, more than most kids, having lived at Patuxent River. Dave had a roomful of models, just as I had had as a boy. We talked downstairs in the family room, where we spent a lot of time when I was home. It had a fireplace with a raised hearth, and sliding doors that opened onto a stone patio. There were bookshelves on either side of the fireplace that contained quite a few books about flying.

I told them this was a step beyond airplanes. It was a first step toward people traveling and doing research up above the air, in space. I said we didn't know where it would lead, but that it would be very important to our country, and that we were lucky to be here at a time when it was all starting. And as we went along and learned more about it, they were going to learn more about it, too.

"It's going to be a new experience. It's exploration," I continued. "We're all going to be a part of it, Mom and you guys, too. I'll be away some more, and that won't be so much fun. But it's going to give us a chance to do things we'd never get to do, a lot of opportunities."

I also told them that they might receive a lot of attention. They were used to that after Project Bullet, but I said I thought this might be bigger, and to try to be prepared. "Just be yourselves," I said.

Annie had had questions, too, that I didn't learn about until later. The idea of going into space was still quite new in religious terms, and she had gone to our pastor, the Reverend Frank Irwin at the Little Falls United

Presbyterian Church, to ask his opinion of man's venturing into space. She didn't think God was waiting on a golden throne just above the atmosphere, but she wanted to explore the religious implications, if there were any.

I had prepared them all as best I could. But nothing that I said anticipated the storm of curiosity that broke with the announcement.

The media attention didn't go away. It continued at a fierce and distracting pace. All the astronauts had similar experiences, but because we were in Arlington and closer to regular media outlets than the other families, we probably got the brunt of it. Television crews virtually camped in our living room; we kept having to move furniture so that they could place their meltingly hot lights and run their cables. Fuses blew. Footprints marred the floors. Soft drinks spilled onto carpets. We were daubed with makeup. At the end of one take, we'd have to do another. The reporters asked the same questions, focusing not on the great new program of space exploration with its worldwide implications, but on us as individuals: Why do you want to do this? Are you a daredevil? Are you suicidal? Tell us about your background. What kind of home life did you have as a child? What does your wife think? Your children? Explain your religious beliefs. What are your favorite foods? Can we have the recipes? And so on. Press photographers—and tourists, too—lurked outside and took pictures of the house and of the kids as they walked to school across the street.

NASA had thought that if we granted a few of the interview requests, the number would eventually diminish.

But it didn't; in fact, the requests increased. Eventually we started saying no when reporters called. I told Walt Bonney it was all too much. "We might be military and government employees," I said, "but our homes are our castles just like anybody else's. There's got to be a better way."

Walt understood the problem. He said NASA had begun to realize that we were getting overwhelmed. "Sounds like you're being nibbled to death by ducks," he said. But he added that there was a legitimate story as well as historical interest in the families. He had thought about it, and he decided we should tell our family stories exclusively through one news organization. That would put the information in the public eye, solve the problem of multiple interviews, and compensate us for opening our homes and giving up our private time.

He had put out feelers, and there was considerable interest, particularly from *Life* magazine, in such a deal. It sounded like a good idea, and we endorsed it pending approval of the higher-ups at NASA and the White House. We wanted to get the media off our doorsteps so we could get on with our training. Another reason I liked it was because of Annie's speech problem. It would be a much better arrangement for her, and that alone would make it worthwhile whether money was involved or not.

Walt said that with all the proposals coming in, we should have an attorney to represent our group, and he suggested a name to consider unless we had other ideas. Leo DeOrsey, he said, was a top tax attorney who represented some of the biggest names in Washington.

Soon DeOrsey called wanting to arrange a dinner with us and our wives "to discuss the terms of our relationship." We met him and his wife at his country club outside Washington. He was a pudgy, jowly little man, affable and easygoing.

We ate a lavish meal in a private room. Over dessert and coffee, Leo for the first time assumed a serious demeanor. He tapped on his glass, cleared his throat, and said he had given the matter of representing us a great deal of thought. He had estimated how much of his time would be required, and decided that if we wanted to work with him, he would do it on two conditions.

"Uh-oh, here it comes," I thought. Leo wore an expensive suit, and the Columbia Country Club with all its dark wood paneling and hunt prints couldn't have been cheap. His conditions were sure to cost a lot of money.

"One," he said, "I will accept no fee. Two, I will not be reimbursed for any expenses I incur in representing you."

Leo sat back with a deadpan expression and awaited our response. I thought I had missed something. Then we all looked at one another and, one by one, began to grin. Leo broke into a broad smile, and we all started laughing together. Leo was another person who saw the astronauts as representing something out of the ordinary in American life. He liked the idea of working with us, thought we needed the benefit of good financial advice, and saw providing us with it as his contribution to the space race. We accepted his offer.

The proposal to provide one news organization exclusive access to the homes and families of the astronauts was approved by the White House over objections from the press. Leo considered all of the offers and settled on *Life*. The deal would pay $500,000, divided among the seven of us, spread out over the three years Project Mercury was expected to run. Before taxes, that would mean almost $24,000 a year for each family, which was a lot more than our military pay. For the Glenn family, even after taxes, it would add a good bit to the *Name That Tune* money in Dave and Lyn's education fund.

Annie and I decided not to move to southern Virginia. We had spent so much time and effort in deciding on Arlington for the schools, and then building a house we liked very much, that it seemed foolish to discard all that after less than a year. Annie felt comfortable with Tom and Ida Mai Miller next door. Our children and the Miller kids spent a lot of time together. It was almost like one family, kids running in and out of both houses all the time. Dave was thirteen and Lyn twelve, ages at which it's difficult for kids to move even though they certainly were used to it. Lyn had taken on an even greater role in running interference for her mother, answering the phone and making calls and appointments for her, but the Millers were another backstop for Annie when she encountered situations that required her to speak.

The level of training Project Mercury would demand was not going to leave much time for the kind of home life I wanted to have, anyway. No matter where we lived, I would be gone much of the time. So we agreed that I would live in bachelor officers quarters (BOQ) at Langley during the week, and commute home to Arlington on weekends when we astronauts weren't traveling.

I didn't want to spend a lot of money getting back and forth. The deal with *Life* was still just a promise. I saw an ad for a new, very small, and inexpensive import that a

Washington Pontiac dealer was selling. The NSU Prinz had a tiny four-cylinder engine, would do sixty wide open on a flat road, and had a list price around $1,400. Two adults could fit in front, but the backseat wouldn't hold much more than a small dog or a shoe box. It got forty-five or fifty miles to a gallon of regular. I got $368 for my 1951 Studebaker Champion on a trade, and with the Prinz could drive the 180 miles from Arlington to Langley for not much over a dollar. It was the second new car we'd ever owned. Our other car, the Chevy station wagon Annie drove, was the first.

Al had a Corvette, and Wally drove an Austin-Healey 3000. My transportation was kindergarten level by comparison, and I took a lot of ribbing about it.

The astronauts started working at Langley toward the end of April, after each of us had a chance to wind up his duties with his service and spend some time with his family. NASA's Langley Research Center occupied one side of the air base. We had seven office-size desks in a single large room, one secretary, Nancy Lowe, for the seven of us, and twenty-five or thirty engineers scattered among several buildings. That was all there was to NASA's Space Task Group and its Project Mercury in the spring of 1959.

We began training in earnest in May, following a program designed mainly by a Navy lieutenant and clinical psychologist named Bob Voas. Bob was studious, analytical, and friendly; he had designed the criteria by which the Navy chose men for hazardous duty, and had put together much of the testing regimen by which we were selected.

He and I had similar ideas about the breadth of the astronauts' duties beyond being just test pilots, so we hit it off.

Surprisingly to some, a specified exercise regimen wasn't a part of the program. Obviously, we were supposed to stay healthy, but NASA wasn't going to roust us out and run us down the road at six in the morning, boot camp style. The agency suggested that we each do at least four hours of unsupervised exercise each week. But the suggestion was hardly necessary; we had been motivated enough to get into the program, and we took it on ourselves to stay in shape.

We took different approaches to the challenge. Wally, for example, had played lacrosse at the Naval Academy and preferred exercise that was competitive and fun. But ever since I had shed my excess weight in the days leading up to the selection process, I had had to fight to keep it off, and I wanted to be in top condition. Running became my recreation; it was the only thing that worked for me, and I did it religiously each morning. Running didn't need a gym, court, pool, ball, racquet, or another player; I could keep it up wherever our travels took us, and on my own schedule. My early morning runs at Langley took me from my BOQ around a parking area, among some other BOQs and administrative buildings, around a small park, and back—a route I had driven in the Prinz and measured at two miles. When we were on the road I ran in the mornings for about the same length of time. At one of our many news conferences, when somebody asked about exercise and I said I ran two miles every day, Gordo dryly observed that he wasn't going to have to run to space.

Scott was the best physical specimen in the group. He had been on the ski team at the University of Colorado, and a gymnast, and he spent his time working out on the trampoline. Deke, who had been a boxer in college, was accustomed to road work and was in good shape, too, although he and Al had both smoked cigarettes prior to selection. I had smoked a pipe occasionally since college, and also stopped during the preselection screening. Wally and Gus played handball avidly, and we all joined in from time to time. All seven of us liked to water-ski.

We all had to fit about the same mold physically, because of limitations on the size of the capsule. While there wasn't a lot of difference between us in terms of size and stature—Gus was the shortest of the group—once we started working and spending time together, the various facets of our personalities came out.

Wally was the most outgoing and gregarious. He could be counted on to lighten up a dull meeting with a joke or two, no matter what the subject.

Gus would have a jovial, joking manner at times, but for the most part was very serious about what we were getting into.

Deke was quieter and more reserved, but gave every impression of being someone to listen to when he had an observation to make. Gordo probably said less than any of the others, but he, too, gave the impression of being a capable person who could be counted on when things got tight.

Scott was outgoing and thoughtful at the same time. He and I had shared the belief that the knowledge to be gained from every aspect of our testing outweighed the inconvenience and annoyance of being stuck and prodded and questioned by doctors, and by having to analyze ink blots and blank pieces of paper. And Scott's relationship with his wife, Rene, seemed much like the one that Annie and I shared.

Al was more of an enigma. One side of him was cool, competent, and utterly dedicated, the other ready to cut up, joke, and have fun. He could defuse a tense situation in an instant with a wisecrack, and he had a way of being able to relax everyone around him and make them perform better. There was a part of him, however, that didn't like the restrictions that came with being a public figure.

We were all different, and yet we all quickly realized that we had a collective strength. It came in part from the enormous attention that was focused on us, and also from the fact that we, and nobody else, were going to be riding fire into the heavens and flying the machines we rode in. Even though I saw the astronaut corps as a new service, Bob Gilruth had been right: The military model didn't apply. The astronauts had a role beyond our rank and salary in the NASA hierarchy.

Differences among us were nevertheless inevitable, and we would have them as time went on. When this happened, we would go behind the closed doors of our office at Langley or wherever we happened to be. We called these sessions "séances," and most of the time we left them with our differences resolved.

CHAPTER 14

We had been in training for about three weeks when NASA took us to Cape Canaveral for our first missile launch. We gathered on the night of May 18 on a camera pad about half a mile from one of the big gantries where an Atlas D waited to lift off.

The Atlas was the United States' first intercontinental ballistic missile (ICBM), built by the Convair division of General Dynamics. Its two Rocketdyne booster engines each produced 154,000 pounds of thrust, and its central sustainer engine generated 57,000 pounds. Two small vernier engines used for guidance added another 1,000 pounds of thrust each. The Atlas was thin-skinned, basically a steel balloon. Its fuel combination of liquid oxygen and RP-1 kerosene had to be kept under pressure or the missile would crumple like a crushed soft-drink can. It was a very fragile vehicle that had been under development since the mid-1950s and had yet to produce a record of sustained success.

The sight of the Atlas on the launch pad was dramatic enough to have been designed by Disney. Searchlights played on the silver rocket, and clouds of water vapor came off it from the liquid oxygen, or lox, cooled to 293 degrees below zero Fahrenheit.

The count went down. Suddenly the engines lit, and the rocket lifted off slowly in a blast of orange flame and billowing smoke. The powerful racket of the engines rolled across the palmettos like a wind. We watched it gain speed, a brilliant phoenix rising into the night sky.

A minute after lifting off, it blew. The explosion looked like a hydrogen bomb going off right over our heads, so close we ducked before we realized the flight path would carry the debris out over the Atlantic. We stood in stunned silence after the roar of the explosion faded. Even Wally didn't have one of his usual wisecracks. Then Al said, "Well, I'm glad they got that out of the way."

We were a sober group of astronauts when we sat down the next morning with B. G. McNabb and his engineers in the Convair launch team. They had little to tell us, except that they were analyzing photo and telemetry data from the launch.

Ten days later, on May 28, NASA launched an Army Jupiter missile from Cape Canaveral. In the nose cone were Able, a rhesus monkey, and Baker, a South American squirrel monkey, both festooned with electrodes to register the effects of weightlessness. The missile achieved an altitude of three hundred miles, and the nose cone was recovered from the Atlantic Ocean with both monkeys alive.

Able died four days later from the effects of anesthesia given for removal of the electrodes. Nevertheless, the heart stoppage and brain alterations that some doctors had

thought might result from weightlessness apparently had
not occurred. At the very least, it was clear living beings
could reach space and return alive—if the rockets that
carried them didn't blow up.

A few days after the Able and Baker flight, we reported to
Wright-Patterson for our first round of centrifuge training. Our physical training, including centrifuge runs,
would simulate the supposed stresses of space flight, such
as high G forces from acceleration and deceleration, the
effects of zero G, or weightlessness, and various bogey-
men the doctors didn't know whether we'd encounter but
thought we should be ready for.

We had all experienced the Wright-Patterson centrifuge in pre-selection testing. Its twenty-five-foot arm
could rotate at speeds that would produce more than
our expected mission profile of eight times the force of
gravity.

The centrifuge at Johnsville, where I had done experimental runs for NACA, was new to the other astronauts.
As time went on, it became a big, and often dreaded, part
of our training. It was run by a computer, and in those
days computers were still rare enough that big ones were
given names. The one at Johnsville was Computer
Typhoon. It filled up a space about half the size of a basketball court, with rack after rack of vacuum tubes, and
every time something went wrong, which was often,
technicians had to crawl into the works searching for the
blown tube.

Our initial runs on the Johnsville centrifuge took us
through the G loadings of the normal launch profile. We

rode strapped into contour seats, each molded to our individual body shapes, with our upper arms and calves supported in half-round troughs. There were two active control handles—the abort handle, which you grasped with your left hand, and the attitude control handle, on the right. The runs began with light acceleration off the launch pad while the booster was still full of fuel, then went to the heavier Gs we would experience as the fuel burned out up to staging, where two of the three engines dropped off, and finally a steady buildup to almost eight Gs at insertion into orbit. From liftoff to orbit was expected to take about five and a half minutes, so we repeated that profile, run after run, until we were familiar with what we could—and could not—do at each stage. The challenge was to raise our arms out of the support troughs at different G levels to operate switches or buttons, to report dial readings to make sure our vision was okay, and to operate the abort handle at different stages.

That was an important check. In a prelaunch emergency, the astronaut would have to release a thumb button safety on the top of the abort handle and rotate the handle forty-five degrees counterclockwise to set off an escape sequence: Explosive bolts would detach the capsule from the booster; the escape tower, propelled by small rockets, would lift the capsule fifteen hundred feet into the air, socking the astronaut with twenty-one Gs for just a moment; more explosive bolts would detach the tower; then the chute would come out and the capsule would land. The abort sequence could also be triggered during the early part of the launch in case of a malfunctioning booster, so we had to be able to operate it under any G load. The abort handle was known as the "chicken switch."

Later in our training, NASA's doctor-scientists wondered what might happen if we had unexpectedly high G loads for whatever reason. Humans had made high-G runs at Johnsville, but none in the Mercury launch and reentry positions, so nobody knew. Al, Gordo, and I were assigned to find out. As in all of our physical training and monitoring, the potentially dangerous high-G runs were overseen by a team of flight surgeons headed by Dr. William K. Douglas, an Air Force major who was an expert in aeromedicine.

The three of us spent a couple of weeks at Johnsville making run after run. We worked up to forces of sixteen Gs—sixteen times the force of gravity. It's something I never want to do again.

The cab at the end of the centrifuge arm was a spheroid. Imagine squashing an eight-foot ball on opposite sides so it's only four feet across when viewed head-on. Inside, it resembled the Mercury capsule.

You got into the cab in its vertical position and sat down as if behind the wheel of an automobile. Then it tilted backward until you were lying down. The arm started spinning, and the faster it spun the more the cab rolled to horizontal, so that you always had the pressure from the same direction rather than having the centrifugal force try to throw you sideways.

At up to five or six Gs, as I had learned in my first Johnsville runs in 1958, all of us found it possible to raise our arms to the control panel to move switches. But as the arm spun faster and the forces built to seven or eight Gs, you could no longer raise your arms. As the Gs ratcheted up, there was the blackout problem. Even when you were lying down, the blood tried to drain from your head.

Tensing your muscles to keep the blood up there was the only antidote. We didn't know what the limits would be. At sixteen Gs, you were straining every muscle of your body to the maximum, just to stay conscious for a few more seconds. If you even thought of easing up, your vision would narrow like a set of blinders and you'd start to black out. Although a Navy lieutenant named Carter Collins pulled over twenty Gs for a few seconds in response to a challenge that Max Faget had issued, the astronauts decided the G limit had been established at sixteen. Without a G suit, that was as much as you could endure and still stay conscious for any length of time.

After those runs, we found we were all splotched on the back with small red marks that looked like measles. These petechiae were ruptured small blood vessels, which had broken under the pressure of the Gs. Sixteen Gs pushed a 170-pound astronaut into the contour couch with a force of twenty-eight hundred pounds. Without the couches that followed the exact contours of our bodies, worse ruptures might have occurred.

There was always another what-if. To explore what would happen if the capsule came down on land, McDonnell engineers dropped pigs in their own molded contour couches to establish impact limits. Pigs were used because they have a torso weight distribution similar to that of humans. The results led to the addition of a cushioning system—a rubber landing bag—between the capsule and the heat shield that would deploy and fill with air after the main chute had opened.

This in turn made us and the engineers wonder what

would happen if the capsule didn't land straight on. If the capsule hit, skipped, tumbled, and hit again on the small end, it would toss the astronaut out of the couch into his lap belt and two shoulder restraint straps. Were they adequate, or did we need a new torso restraint system covering most of the upper body to spread the load more evenly? Losing an astronaut at that stage would be unfortunate.

So still later in our training, we went back to Johnsville to try "tumble runs." EI to EO, we called them—"eyeballs in" referred to the forces of normal launch acceleration and reentry, while "eyeballs out" reversed those forces.

A particularly sadistic feature of the centrifuge allowed 180-degree rotation of the cockpit in the space of less than three seconds. You could go from EI to EO just like that. On these runs, our helmets were strapped down to keep our heads from snapping and doing serious damage to our necks.

This was an interesting exercise when we shifted from two Gs going to two Gs coming. But we were filling out a brand-new envelope, probing for limits that had never been set. We worked up slowly through higher and higher deltas until we had completed nine to nine, a delta of eighteen Gs. Those were rough runs. They produced petechiae that were even more pronounced. You could see the pattern of our shoulder straps across our chests marked in ruptured blood vessels. We also experienced some shortness of breath from the compaction of the oxygen-absorbing tissues in the lungs.

The doctors didn't normally go through the same runs we did, but Bill Augerson, one of the flight surgeons, was participating in this particular experiment and was

taking the runs ahead of us. Bill went on a run that took
him from ten Gs EI to ten EO. When he got out of the cab,
he was coughing, and he couldn't stop. We stopped the
test. Once he had recovered, we set up an anthropomor-
phic dummy and rotated it through the same movements
Bill had experienced. Its assemblage of body parts and or-
gans allowed us to see what had probably happened. The
doctors finally decided that the changing force vectors had
literally thrown Bill's heart into the back of his left lung,
knocking the air out of that lung. That was why he was
hacking away, unable to catch his breath.

We decided to end that test.

In September 1959 the Soviets reasserted their superiority
in space by launching a rocket to the moon. *Luna 2* crash-
landed on the moon's surface, but as the headlines pointed
out, it was the first man-made object ever to reach the
moon.

A month later *Luna 3* sent back the first pictures of
the dark side of the moon. Khrushchev's boast that
America slept under a Soviet moon sounded more threat-
ening than ever.

On December 4 NASA launched a monkey named
Sam to an altitude of fifty-five miles and safely recovered
him. This was more fuel on the fire for the test pilots at
Edwards. By now, of course, they were smarting over the
public attention we were getting, courtesy of *Life,* and
over the failure of the Air Force to win support for its
Dyna-Soar program——a plan to put a single-orbit, piloted
military spacecraft into the nation's arsenal. *Dyna-Soar* was
short for "dynamic soaring," and the program's mission

was to see what Soviet satellites were up to. We were at Edwards often enough to get wind of the continued derision some of the test pilots had expressed for the astronaut program. The monkey flights, first Able and Baker, and now Sam, had set off a new round. We heard that Chuck Yeager, who was famous for having broken the sound barrier and remained involved with military test piloting at George Air Force Base, near Edwards, went around laughing that the astronauts were going to have to sweep monkey mess out of the capsule before riding into space.

Our first flights were indeed going to be up-and-down parabolas such as Sam's. They would be launched by the Redstone missile, with its seventy thousand pounds of thrust. Around the end of the year Redstone inspection took us to Huntsville, Alabama, and the Army's Redstone Arsenal, where we met the chief of its Guided Missile Development Division, Wernher von Braun, and went to his home.

The German-born engineer, then in his late forties, was a handsome, broad-shouldered man with thick dark hair. Von Braun had been a devoted Nazi during World War II, but his rocket expertise was valuable to the United States, and when the war ended he and members of his team of German scientists were brought to this country. Other scientists went to the Soviets. Whatever anyone thought of von Braun's previous allegiances, he had a well-deserved reputation for heading an effective rocket team. He had led the development of the German V-2 guided ballistic missiles that had rained destruction on England and Allied-held Europe near the end of the war. In the new postwar equation, the V-2 became the basis for the Redstone, which gave the United States its first effective intermediate-range ballistic missile in the competition with the Soviets. Modifications would allow it to carry a Mercury capsule. Von Braun spoke of men riding the Redstone and other rockets into space and someday to the moon, of humans pitting themselves against enormous odds for the sake of discovery.

His library showed him to be a man whose interests were not confined to rocket science. I wandered into the book-lined room expecting to find nothing but tomes on engineering, astronomy, physics, and other technical matters. There were many of those. But I was impressed to find even more extensive collections in fields such as religion, comparative religion, philosophy, history, and government.

By the time of our trip to Huntsville, we were also doing parabolic flights at Edwards that gave up to a minute of weightlessness at the top of the parabola. We used two-place F-100 trainers. We'd take the rear seat while the pilot would go up to forty thousand feet, make a dive reaching Mach 1.4, and then head up again while pushing over to an angle that would make anything on the cockpit floor float up in front of you, holding that balance all the way over the top of the arc and back toward Earth again. It was like an extension, but much faster and farther, of that brief moment when your car goes over a rise in the road and you're lifted out of your seat. While we were strapped down and couldn't float in the cabin, it gave us the chance to try eating and drinking and manipulating equipment during weightlessness.

We had done something similar at Wright-Patterson

in a C-131, a cargo plane. This gave us only about fifteen seconds, but was more fun because we were unstrapped and could float and turn flips in the cabin.

I wrote Jim Stockdale at the end of the year, describing the travels, tumble runs, and parabolic flights of the training program. I said that I enjoyed the sensation of weightlessness, and that I thought I'd finally found the element in which I belonged.

We were traveling constantly. As I told Jim in my letter, Langley had gotten to be a place where we went back to get clean skivvies and shirts, and that was about all. We were supposed to have three hours a week to keep up our flying skills on the F-102s that NASA got from the Air Force, but we were always having to fly from here to there on commercial airliners. Since we usually flew from Washington, at least I was able to spend time at home at the beginning and end of each trip.

The number of new areas we were studying was almost unbelievable. At the rocket manufacturing plants—the Redstone Arsenal and the Convair division of General Dynamics in San Diego, where the Atlas booster was made—we met with the design engineers to learn more about the hardware. Our visits had another purpose, too. It was to encourage the workers on the line to tolerate no defects. Z-D—zero defects—became the watchword in our talks on the assembly line as NASA and the contractors moved toward man-rating boosters originally designed for weapons use. Soon the term Z-D appeared on posters, on bulletin boards, and in company newspapers.

B.F. Goodrich, in Akron, Ohio, had been selected to

design, test, and fit us with pressure suits, so we made trip after trip to Akron for precise fittings and to learn more about the suit.

Bob Gilruth had made good on his word to involve us in all aspects of Project Mercury. We each had specific areas of responsibility. Since I had probably flown more different types of aircraft than the others, I was handed cockpit layout and instrumentation, spacecraft controls, and simulation.

Scott's domain was communications equipment and procedures, periscope operation, and navigational aids and procedures. Gordo's area was the Redstone booster, trajectory, aerodynamics, countdown, and flight procedures, emergency egress, and rescue. Gus was responsible for the reaction control system, hand controller, autopilot, and horizon scanners. Wally had environmental control systems, pilot support and restraint, pressure suit, and aeromedical monitoring. Recovery systems, parachutes, recovery aids, recovery procedures, and range network were Al's job. And Deke had the Atlas booster and escape system, including configuration, trajectory, aerodynamics, countdown, and flight procedures.

We traveled and trained together as much as we could. When we traveled separately, it was to keep up with our specialty areas. This usually consumed one week a month, when we scattered to the winds, and then we reassembled for group briefings.

We made visits to multiple contractors and subcontractors, but our most important training centered on the Mercury spacecraft and its systems. That's what we had to know perfectly. No one else would be with us in space, and our lives might depend on how well we knew the spacecraft. It was here that we had the biggest effect, and initiated a major design change. On our first trip to St. Louis and McDonnell late in 1959 to discuss what they were doing with the capsule, we took on a glaring problem that we had noted in the blueprints. There was no window.

The capsule wasn't completely visionless. It had a small porthole, but it was located just off the astronaut's left shoulder, where it would be almost impossible to use during flight. There was also a periscope, which had a fisheye lens that covered 180 degrees but distorted everything around the edges. We saw the absence of a window as a serious problem in terms of both safety and imagination. With no possibility of visual orientation, we might lose the ability to correct the capsule's attitude. And one of us was likely to be the first human being to escape Earth's atmosphere and look back down upon our planet. To have no way to describe a sight that had been an ageless dream of humankind was unthinkable. We wanted a window.

In defense of the design team, they were well aware of the desirability of being able to look directly outside. But they faced some tough limitations. A window and its support structure added weight. Weight was critical, and to save weight they had designed the capsule for a cabin pressure of only 5 pounds per square inch for the vacuum of space. Air pressure at sea level is 14.7 pounds per square inch, which allows the body to absorb the necessary oxygen out of the sea-level atmospheric mix of 20 percent oxygen and 80 percent nitrogen. At 5 pounds per square inch, the engineers had to increase the oxygen level to 100 percent to give the body what it needed. Designing for a lower cabin pressure saved weight. It was dangerous; pure oxygen could turn a spark and flammable material into an uncontrollable inferno. But to design the capsule for triple the pressure meant it would weigh too much for the boosters we had.

Oddly, we were limited in the weight we could rocket into orbit because of our advanced technology. The Soviets had lifted far heavier satellites than ours. Both nations' boosters had been designed as ICBMs, but because we had done a better job of reducing the size of our nuclear warheads, we could use smaller missiles. The Soviets had failed at making smaller warheads, and so needed larger boosters. The tables were turned when it came to putting satellites, and eventually manned spacecraft, into space. Our Atlas rocket, with its 367,000 pounds of thrust, could deliver nuclear warheads from the United States to Moscow, but could barely lift the four-thousand-pound Mercury capsule into orbit; the Soviets could have orbited a house if they had wanted to.

Bob Gilruth and NASA's design team had agreed with us on the window if the weight problem could be solved. Max Faget, the capsule's original designer, attacked the problem. If the daddy of the spacecraft thought it could be done, it could be done. Eventually Max gave the new design his go-ahead, and soon afterward we got word that McDonnell was incorporating a window in the Mercury capsule.

We were nearing the end of our first year of training in February 1960 when NASA's scientists came up with an-

other what-if. They wanted to know how we would react
if we had an orbiting spacecraft that was tumbling out of
control.

This could be caused by one or more stuck thrusters
that caused the capsule to turn wildly on any of its three
axes. Roll was the capsule's sideways tilt. Pitch was its
nose position, up and down. Yaw meant the capsule's motion around its vertical axis. In their characteristic effort
to see how we would be affected by, and respond to, every
potential emergency, the scientists wanted to find out our
limits of control and debilitation under such circumstances.

The flight propulsion laboratory at NASA's Lewis
Research Center in Cleveland had designed the diaboli-
cally perfect machine for the job. It was the MASTIF—
Multi-Axis Space Training Inertial Facility. The MASTIF
was about nineteen feet across. It combined three gimbals, one inside the next, that looked like the skeletons of
globes. At the center of the three layers of rings was a
cockpit. Sitting inside, you could at the whim of the
MASTIF's programmers be placed into roll, pitch, or yaw,
two of them at a time, or all three simultaneously.

We sat tightly strapped in, facing a panel that contained eight instruments, and gripped a three-axis hand
control. The programmers set the machine in motion; once
moving, it continued to rotate on its own momentum. It
was up to the astronaut to get it under control and bring it
to a stop. Moving the stick triggered compressed-air bottles on each of the gimbals that gave the same thrust against
weight that we would have in the Mercury spacecraft.

They started us out slowly in a single-axis exercise. It
was easy enough to concentrate on the instruments and

use the hand controller to stop roll, for example. Then, once you got the hang of single-axis response, the technicians would throw in a second axis. Finally you'd be doing roll, pitch, and yaw all at once. Even slow, it wasn't easy. But they would move the speeds up until you were working that stick as if electricity were running through it and you *couldn't* let it go.

In what amounted to our graduation exercise, we were doing thirty revolutions per minute in all three axes simultaneously. What a ride that was! You could get it under control, but if you didn't, you were going to have a bad day. You came out of it reeling, and it took about half an hour on the cot the technicians thoughtfully provided before everything stopped spinning.

All of our travels and study of the various system components were sandwiched in between intense academic training. We studied astrophysics, trajectories and orbital mechanics, the basics of propulsion, and astronomy. This last was aimed at teaching us star recognition, so that if the automatic attitude control system of the capsule failed, we could set the attitude manually by aligning with the stars.

The Morehead Planetarium at the University of North Carolina had been chosen as the site for our stargazing lessons. It had reclining seats below a dome on which star projections could be set for any season or specific date. We could look at the star pattern for a given launch date, and the star track over any orbital route. Once we had begun the training, the planetarium added an enclosure that gave us the view of the stars through a simulated Mercury window.

The ability to recognize constellations and orient the capsule in relation to them could be critical in an orbital flight. The capsule had to be lined up just right at the moment we fired the retro-rockets that would slow us for reentry. Too steep a reentry would send the capsule into the atmosphere too fast, and it would burn up; too shallow, and it could skip off the atmosphere like a stone off the surface of a pond and not be able to return. The capsule's automatic attitude control would probably work perfectly, but knowing how to position the capsule by the stars was at the very least a good backup.

Here was an argument against both the few Air Force test pilots who claimed the astronauts would be just passengers and the heavy thinkers who thought that machines could learn as much about space as humans and for a lot less money. This was to be a different kind of flying, for sure. But machines failed, and only humans in the cockpit could take over when they did. NASA knew that from the start.

CHAPTER 15

As Project Mercury entered its second year, we began to spend a lot more time at Cape Canaveral. The Cape had been a military missile launching site since 1949, and had been chosen as the Project Mercury launch site. Its lineup of launch pads and gantries jutted up from the flat sand scrub of the Florida coastline between the Atlantic surf and the mosquito-infested palmetto and mangrove thickets along the Banana River and the Intracoastal Waterway. The only places to stay were to the south, in a little tourist town called Cocoa Beach and farther down at another barrier island town, Indialantic.

Cocoa Beach was in the early stages of a boom created by the various government contractors building the missiles that were launched from the Cape, and expanded by NASA and the manned space program. Its one motel, the Starlight, was home to the contractors, the military, some of the astronauts, and the reporters who were covering the space program. They relaxed at the motel pool after hard days at work and drank at the Starlight Lounge, and everybody who wanted to be where the action was converged at the Starlight. Henri Landwirth, a Belgian immigrant who had survived the Holocaust, managed the Starlight. When he moved to a new Holiday Inn, the astronauts moved with him.

People wanted to be where we were. It wasn't just at the Cape. It was everywhere. On our visits to the factories where the components of the Mercury system were being made, everyone wanted to show us a good time. Our military lodging and food allowances covered us adequately, but the program's contractors had big budgets and they used them. The contractors, local businessmen, chambers of commerce, and politicians couldn't do enough for us. The food was lavish and the liquor flowed, and all they wanted in return was their picture taken with the astronauts.

Any one of us who was looking for companionship while he was on the road would not have to look very far. Of course, what each person did in his private time was his business. But *Life* magazine was painting us all in glowing terms, and I thought that one story tarnishing that image was all it would take to put the whole program in jeopardy. The effectiveness of Project Mercury was already being questioned in the media, and was increasingly becoming a topic in the preliminaries of the 1960 presidential race. The proponents of unmanned flights, or of an increased number of primate flights, were still vocal, and now seemed to be gaining ground. If the program's most visible and highly regarded components—the astronauts—proved to have feet of clay, I was afraid it would erode popular support and provide further grounds to attack the program.

Not only that, I thought we owed it to people to be-have. It was now clear that, rightly or not, we had been placed upon a pedestal. We hadn't asked for that, and certainly we'd done nothing so far to deserve it. But that was our image, and I thought we had an obligation to live up to it. We owed it to the kids and to the country, because they believed in us.

I had mentioned this in passing at some of our "séances" at Langley. I'd also discussed it with John A. "Shorty" Powers, the honey-voiced retired Air Force colonel and veteran pilot—he had flown in World War II, the Berlin airlift, and the Korean War—who had become NASA's spokesman. Shorty agreed with me, but this was not a welcome subject of discussion within the group.

The matter came to a head late one night in San Diego, where we had gone for another meeting with the Atlas engineers. We were staying at the Kona Kai, a hotel overlooking San Diego bay. I was asleep at two in the morning when the phone rang. I answered it and heard a nervous Shorty Powers.

"What you've been talking about has happened," he said.

"What do you mean?"

Shorty told me that a leading West Coast paper had called him for reaction to a front-page story it was planning to publish the next morning. A reporter and photographer had trailed one of the astronauts and gotten compromising photos.

Shorty didn't think that anything could be done about it. I told him I was sure going to try. I called the reporter and photographer. I called the editor and got him out of bed. I talked about why it was important to the country and the program that the story not run. I talked about the godless Communists and how they were ahead of us, and how the press had to let us get back in the space race. I pulled out all the stops.

When my alarm rang in the morning, I got up and went straight to the hotel newsstand. The story wasn't in the paper. To this day, and knowing the press much better now, I'm still amazed that it didn't run.

That day I asked for one of our séances there at the Kona Kai. It's been described in practically everything that's been written about Project Mercury, most notably by Tom Wolfe in *The Right Stuff*. I was mad, and I read the riot act, saying that we had worked too hard to get into this program and that it meant too much to the country to see it jeopardized by anyone who couldn't keep his pants zipped. I said we had dodged a bullet because I had made some predawn phone calls, but sooner or later the press was going to decide the behavior was just too blatant to ignore. And that had the potential to kill it for all of us.

The response was that I was out of line, foisting my view of morality on everybody else, and that I should mind my own business.

My concerns weren't of a moralistic nature. I had firm ideas about how I wanted to live my own life, but I didn't want to meddle in anybody else's. I was completely and wholeheartedly devoted to Annie, but I understood that not every relationship could be like that. What I did care about was the potential for negative publicity and the impact it would have on the program. We were public figures whether we liked it or not, and we had to act like it.

My views were in the minority, but I didn't care. I had

made my point, and I didn't think being an astronaut was a popularity contest.

I would turn out to be wrong about that.

I looked forward to visiting the various plants and sites and meeting with the contractors. I enjoyed talking with the engineers and the workers on the assembly lines. I wanted to have a personal relationship with the people who were putting the boosters and the project's other components together. I wanted them to see my face and feel my handshake and know that I cared about what they were doing. If they cared about me in turn, they might take those extra precautions that would save my life. I knew that every personal contact was a step toward a man-rated, zero-defect vehicle.

Sometimes, when we were hurrying from one meeting to another at one of the plants, I'd stop among the workers and get so involved shaking hands and talking that it delayed our meetings. Project Mercury seemed to me to be a vast collaborative enterprise, its lofty goals bringing out the best in everybody from the assembly line on up. We astronauts may have had our differences, but they didn't affect the teamwork that was required. The idea that we were part of a team that required such a level of trust was exciting. The demands on all of us were great, from the scientists and engineers at NASA and the manufacturers to the men and women who were putting together the vehicles in and on top of—which we would ride into space. The high stakes gave the enterprise a focus and a spirit of all-for-one, one-for-all camaraderie.

We all took notes at our site visits. To keep Annie,

Dave, and Lyn up to date, I bought a big two-reel Ampex tape recorder and converted my notes to tapes that would brief them on what we were doing. Nobody else was going to explain it to them, and I thought they needed to understand as much as they could about the training and the amount of time it involved.

The increasing amount of time we were spending at the Cape allowed me to bring Annie and the kids down when school ended for the summer of 1960. Henri Landwirth gave the astronauts good deals on rooms, and I installed the family with me at the Holiday Inn at Cocoa Beach. It sat by itself in a field of sea grasses between Highway A1A and the ocean.

We started out the mornings together. I went running on the beach, and sometimes the whole family would join me. More often, Annie and Lyn would walk and pick up shells while Dave and I ran together. During the day, when I was at the Cape, they camped by the motel pool. The pool was a luxury, and they took full advantage. The kids swam and Annie sunbathed. Then they'd go somewhere for lunch. There was usually a thunderstorm in the early afternoon, which meant an after-lunch nap, and then they'd go back to the pool or to the beach. On weekends, we water-skied with Henri and his family from the inland beaches along the Banana River. We had a little sailboat, a Sunfish, that Lyn and Henri loved to sail among the schools of porpoises that cavorted in the river.

Dave and Lyn spent more time with Annie and me, and with each other, than they did with other kids, since they were older than the other astronauts' children. I included them in every part of the training that I could.

That summer, to gain some feel for weightless conditions in space, our training regimen sent us to Little Creek, Virginia, to work with the Navy's underwater demolition teams, and I took Dave along. He lived with me in the bachelor officers quarters, attended meetings and took notes, and joined in the scuba training we got in the pool. The divers put the astronauts through a range of underwater exercises, even simulating emergencies in which the divers would yank off our masks and breathing equipment to check our recovery procedures.

The divers took us a mile out into the Chesapeake Bay for our graduation exercise. Dave was along for that, too, but he stayed in the boat. They dropped us off two at a time—the safety-oriented buddy system—where the water was thirty or thirty-five feet deep, and we had to swim a compass course along the bottom to a marker near the shore, alternately leading and following. It was like dead reckoning in flying, with currents replacing wind speed and direction in your figuring. Most of the time you couldn't see five feet in front of you.

When we astronauts started surfacing, we were scattered along the beachfront. We didn't hit the marker in good frogman style at all. But even if we weren't good underwater navigators, we did get used to a different kind of breathing and operating in a hostile environment.

Another kind of hostile environment might occur after we landed. NASA wanted to prepare for any number of emergencies that might divert us from our designated landing sites. An aborted orbital flight could mean a landing in the Sahara Desert of North Africa, the outback of

Australia, or the dense jungle of Papua New Guinea. We trained for all the possible landing contingencies. NASA thought it could locate us fairly quickly no matter where we went down, but fairly quickly could mean as much as seventy-two hours. We had to be prepared to survive for that long.

Stead Air Force Base was outside of Reno, Nevada. NASA sent us out there for a few days of desert survival training in July 1960. Temperatures in the heat of the day rose above 120 degrees Fahrenheit. I hadn't been anyplace so hot since I was training to fly Corsairs at El Centro, California.

We spent two days in classrooms reviewing survival procedures and gear. Our equipment was dictated by the space limitations of the capsule, which was so closely designed for size and weight that there was no room for elaborate survival gear. A small survival kit, equipped mainly for a water landing, included a one-man life raft, minimal food and water, shark repellent and dye marker, sunglasses and zinc oxide, a whistle, a flashlight, two knives, waterproof matches, a location beacon and a signal mirror, medical injectors and a first-aid kit, a nylon lanyard, and soap. The routine NASA had designed would let us convert our parachutes to clothing, and this had particular value for a desert landing.

After our classroom work, Air Force helicopters flew us sixty miles east to a remote desert of hot sand, cactus, and sagebrush, and left us. We scattered to individual sites, where we were to survive for three days. We had the parachutes and shrouds and the survival kits. The first thing you did was take a knife and cut a section of parachute, make a hole to stick your head through, and drape it over you for sun protection. Repeated several times, you had a version of the layered, loose-fitting costumes worn in North Africa and the Middle East. Some more of the parachute fabric went into making a cover for your head. Still more of it became your shoes, sections that were rolled up and bound to your feet with the parachute shrouds. It was an ingenious system.

Bill Douglas oversaw the desert training. We were all on our own in the desert, but Bill knew where we were and checked on each of us regularly to see if we were okay.

The instructors had told us before we went out about the debilitating effects of dehydration. I was curious about what it was like, and told Bill I wanted to try it.

He told me he'd keep an eye on me, so for the first twenty-four hours I didn't drink any water.

By then I was getting pretty low. I had dug out a little area under some sagebrush, as we had been taught to do—the subsurface temperature was a little lower, and the sagebrush with parachute fabric over it provided shade—and I was lying there as debilitated as I have ever been. When Bill showed up again, I didn't even want to raise my hand and shade my eyes to look at him as he came walking up.

Bill had brought plenty of water, and told me to drink as much as I wanted. I drank fifteen pints over the next nine hours without passing a drop. I could not believe how quickly my body had dried out, and to this day I still carry extra water if I have to drive across the desert.

One instructor in the desert training was an Air Force master sergeant named Urbin. Among his specialties was snakebite. He put on a real show, with snakes as props, and described slashing the wound and sucking out the venom.

"Some people won't do it," he said. "They'll act like they're sucking it out but they'll fake it because they're afraid it will kill them. But it won't hurt you at all. It's nothing but protein."

He reached down into a snake cage on the table and brought out a big fat water moccasin. He caught it behind the head and milked its venom into a little shot glass, then swallowed the venom in one gulp.

"Of course, if you have an ulcer it's a different story," he added with a grin.

We looked at one another. None of us had an ulcer, but I don't think I was alone in hoping that any emergency would land us somewhere free of snakes.

In any case, it was more likely that we'd come down over water. There's a lot more water than land on Earth. Training in sea survival took us to Pensacola, Florida, where crews took us a couple of miles out in the Gulf of Mexico and put us overboard with only the capsule's survival kit. We practiced inflating and getting into the life raft. Once situated, we activated our ultra-high-frequency SARAH location beacons, put on our sunglasses and zinc oxide, and bobbed around for three or four hours until the crew came back to pick us up.

At the end of July NASA planned a launch of the final version of the Atlas missile that was going to propel the first orbital flight. The Atlas had continued to be plagued with problems. Its first twenty-five or thirty launches produced a failure rate of 45 percent. Even if we had been able to forget the spectacular explosion we had seen the previous

May, figures like those weren't reassuring to those of us who might be riding the Atlas.

It was most susceptible to failure at the point of max Q, the moment in a launch where air resistance and the vehicle's velocity combine to produce maximum aerodynamic force. For the Atlas booster, max Q came at around thirty-two thousand feet. It was moving at fourteen hundred feet per second and the air, while thinning, still produced resistance of 982 pounds per square foot of capsule surface. This buffeted and shook the rocket at the same time as its diminishing fuel supply permitted the thin shell to bend and flex. That was apparently what had happened to the one we saw explode. Past max Q, thinner air would smooth its flight.

The morning of July 29 was dark and rainy. The countdown proceeded anyway. NASA had assembled an audience that included the astronauts, its own officials, and executives and engineers from General Dynamics. The Atlas sat on the launch pad, topped with a simulated Mercury capsule, the package looking exactly as it would look on the day an astronaut was aboard ready to be lifted into orbit. We listened on the squawk box as the count went down. The stage was set for the debut of *Mercury-Atlas 1*.

The launch went perfectly. The rocket rose on a column of flame and disappeared into an orange halo in the clouds. A minute later the squawk box erupted with hurried, cryptic messages indicating that flight telemetry— the signals from the rocket to the ground—had been lost and the rocket had disappeared from radar. Another half a minute, and some people on the ground thought they heard an explosion.

The investigation that followed determined the rocket had failed structurally and blown up going through high Q at thirty-two thousand feet. The debris fell into the Atlantic. The only good news was that capsule telemetry had continued until it hit the water, and all of its shattered pieces were recovered.

I didn't know what to think. We were much closer now to a manned flight than when we had witnessed the earlier explosion. The first flights were going to be on top of the more proven, smaller Redstone rockets, but sooner or later an astronaut was going to be riding an Atlas because only the Atlas had the power to put a spacecraft into orbit. The failure of *MA-1* set NASA's launch schedule back by months.

The rocket's failure was well publicized. It triggered more criticism of NASA in the press, and concern among the families of all the astronauts. All I could say to Annie, Dave, and Lyn was, "They'll go back to the drawing board and get it fixed."

I know that didn't satisfy them, or me either, for that matter.

We astronauts were looking for a sign of good news after the Atlas explosion, and found it in the August launch of the *Echo* satellite. It went into an orbit a thousand miles high, but even at that distance—if you knew its schedule—you could see at night the glint of sunlight off the hundred-foot silvery Mylar balloon that had inflated and expanded once it was in orbit. *Echo* was the first communications satellite. It was also the first satellite of any kind that you could actually see. We all had copies of its schedule, and wherever we were, we would go outside to catch a glimpse of it at night.

That fall the astronauts began to experience the feeling and the sound as well as the task requirements of suborbital space flight. The programmers of Computer Typhoon at the Johnsville centrifuge input data that made the centrifuge act in imitation of the Redstone launch. We would climb into the cab wearing our pressure suits, and strap ourselves into the contour chairs that were fitted to our backsides. Then the voice of ground control took us through the countdown to launch. Vibration began, and our helmet headsets played recordings of the sound— muted because it would be far below us—of Redstone liftoff. The Gs would build up slowly to a maximum of eight times the force of gravity, then drop off suddenly as if the capsule were floating at the top of its arc, then return with the force of reentry. We would have the same buttons to push and switches to adjust as in actual flight, and the same responses to ground control to make.

As the experience of riding the centrifuge shifted from the brutal tossing around of "eyeballs-in, eyeballs-out" to a facsimile of actual flight, other newly developed simulators improved the feel of what we could expect in flight.

The ALFA trainer at Langley provided a more sophisticated taste of capsule control than the wildly gyrating MASTIF. The capsule's roll, pitch, and yaw would be controlled by one- and five-pound hydrogen peroxide thrusters. The ALFA trainer, unlike the MASTIF, rode on an air bearing. (Its initials stood for Air Lubricated Free Attitude. NASA-speak was full of acronyms.) Manipulating the hand controller not only adjusted the trainer's attitude at the same thrust-to-weight ratio as the actual, it produced from its compressed-air jets the same sibilant whoosh the hydrogen peroxide thrusters would make as they moved the capsule. A screen imitated the porthole and periscope, providing views from aerial cameras of the landmarks we would overfly. They moved with our attitude, as if we were actually changing pitch or yaw as we flew by. They gave us a visual basis for adjustments.

Procedures trainers, which amounted to sophisticated flight simulators, took the experience even closer to reality. By the fall of 1960 we had procedures trainers at Langley and at the Cape. These were exact replicas of the capsule cockpit, including all instruments, buttons, indicator and warning lights, switches, safety overrides, and levers for controlling the one- and five-pound thrusters. We had information about pressurization, oxygen supply, temperature, electrical and fuel supply, cabin environment, the gyroscopes that governed the capsule's automatic attitude control, a position fix via the Earth Path Indicator, a view through the periscope—everything as if we were in the capsule in space except the G loads. These controls and instruments were tied into a computer run by instructors who trained us on all the different systems.

We all spent long hours on our backs in the trainers as we ran through launch simulations again and again. We trained system by system, until we were capable of running them all together on simulated missions. After that, the training staff gave us emergencies of all kinds, in all systems, at any time during the flight. They seemed to take fiendish pleasure in thinking up new emergencies for us to react to—oxygen failures, launch aborts, electrical system failures, control system failures, early aborts, and emergency reentry.

By late 1960 we were testing systems in the spacecraft itself in Hangar S at the Cape. This meant one or another of the astronauts would have to spend hours at a time in his pressure suit in the cockpit, operating the controls while technicians checked and rechecked their operations. Occasionally there would be unplanned delays. Two-hour tests could stretch to four or five or more. This was how we all became members of the Wetback Club.

No urine collection device had been designed for use inside the space suit. The early missions were going to be only fifteen minutes long, and nobody had thought that the testing and other delays might stretch on as long as they did. To break for a pit stop was often impossible in the middle of a test. It meant getting out of the pressure suit, which was no quick and easy task, and then putting it on all over again. I forget who was the first to do it, but once we realized how much time and trouble was involved, we just shrugged and let it go. You got used to it after a while, lying on your back while your thick suit underwear with its air tubes sopped up the urine as it worked its way up your back. It was preferable to bringing the whole operation to a halt just to be fastidious.

The intensity of our training schedule made it hard to pay attention to much else. But I couldn't miss media reports that the effectiveness of the space program was increasingly an issue in the 1960 presidential race. Massachusetts Senator John F. Kennedy, the Democratic nominee, had placed his Republican opponent, Vice President Richard

M. Nixon, on the defensive by talking about a "missile gap" and lax oversight of the space program during the Eisenhower administration.

The fact was that on any number of levels, the United States was not behind but ahead. We had launched more satellites than the USSR, and were receiving scientifically valuable information from space. Nixon cited this American advantage in satellites and space probes. But the Soviet Union's successes were spectacular and its failures unpublicized, so the majority of voters believed that we were playing from behind.

In May the Soviets had launched the first of several space capsules into orbit containing human dummies they named Ivan Ivanovitch. In the fall, with the campaign in full swing, the Soviets sent up another well-publicized flight that carried two dogs, Belka and Strelka, in space suits, and brought them back to Earth in a parachute landing.

Meanwhile, Project Mercury, after the first flush of publicity and the ongoing bland and upbeat stories from *Life,* struggled for public support. The project had plenty of critics. Even before the *MA-1* failure, some scientists and members of Congress had argued that sending humans into space was too expensive, that robots and other machines were more efficient and effective. And some refused to see the value of going into space at all.

Khrushchev, whether he knew it or not, helped the Democrats every time he crowed about a Soviet advance. His antics had the effect over time of pushing the critics into the background while the burden on Project Mercury to get a man into space before the Soviets grew greater.

But after Kennedy was narrowly elected president on

November 8, another highly visible launch failure triggered a new round of criticism, even as it blasted morale at NASA and gave the astronauts a new feeling of us-against-them solidarity.

This time it was the test flight of *Mercury-Redstone 1*, the supposedly well tested Redstone rocket, topped with a Mercury capsule just as it would be for the first manned suborbital flight. Once again NASA assembled an audience of several hundred dignitaries and politicians at the Cape for the November 21 launch. The Redstone fired, rose about four inches, then cut off and settled back on the pad. The three small rockets of the capsule's escape tower worked perfectly, however. They lifted the tower—*without* the capsule—four thousand feet. The capsule stayed atop the rocket, but the parachutes that were supposed to bring it down activated. The drogue chute popped out and floated down, carrying the capsule's antenna canister. Then the main and reserve chutes billowed out, settled down over the capsule and booster, and floated gently in the breeze.

The astronauts, NASA officials, and Wernher von Braun and members of his Redstone team watched in consternation from the blockhouse. Then we couldn't leave. Von Braun was afraid that if a gust caught one of the parachutes, it would pull the rocket over, and it would blow up with its entire fuel load. It was several hours before we could scramble out.

The press again derided NASA. Reports said the sight of the escape tower popping from the top of the rocket looked for all the world like a champagne cork popping from a bottle.

In the wake of the *MR-1* failure, we trained harder

than ever. We were afraid that any lapse in our efforts could provide the president-elect an excuse to drop or delay the manned space program in favor of machines. The arguments that they would be cheaper and produce more knowledge rose in volume. So did the opinion that—with the continued rocket failures—an extended series of primate flights should precede a manned flight.

We knew it wasn't our decision. But we also knew that if we had anything to do with it, the astronauts weren't giving way to monkeys or machines.

CHAPTER 16

I thought I had a good shot at being named for the first flight. The competition to be first into space had been a component of Project Mercury from day one. I had worked and studied hard, dedicating myself to the program, and though I hadn't sought the publicity, I had emerged as a favorite in the press, if the number of times I was quoted was an indication. Not that good press was a factor. I was sure Bob Gilruth and NASA would make the decision based on objective criteria. Still, while we had all trained hard, I felt my performance was at least as good as anybody else's.

So it was an odd moment around the end of the year when Bob called us all together. "I know you all want to go first," he said. "What I want you to do is vote on which one should be first if you couldn't go yourself. List your choices in order."

It took a moment to sink in that he was asking for a peer vote. I was enormously disappointed and very upset.

The military often used peer votes, but not like this. Months of training were being reduced to a popularity contest. After my comments at the séance about everybody's needing to keep their pants zipped, I could imagine where I'd stand in a peer rating. My chances of being first in space were slim if a peer vote was even a small factor in the decision.

Nevertheless, we all voted. I wrote Scott's name at the top.

One day after the first of the year, when we were working at Langley, Bob called us together again. He reminded us that Keith Glennan had said that we wouldn't know who would be first until the morning the flight was scheduled. But he had rethought that and decided it wouldn't work. The pilot who was first, and his backups, should have priority on the procedures trainers to prepare. Therefore, he had made a decision. He had chosen the astronauts who would make the first three flights. NASA was going to release their names, and that was all that would be known. The first three were going to be Al, Gus, and me. The press could speculate about who would be first, but that decision also had been made. Al, he told us in confidence, was going to make the first flight.

Six of us swallowed our disappointment. And Al tried to play the part of a good winner as we all congratulated him.

But I didn't think that was necessarily the end of it. The next day, January 20, 1961, was inauguration day. Although it was a cold day and there was snow on the ground, Annie, Dave, and Lyn had attended the new president's parade from the Capitol to the White House. I drove from Langley to Washington in the Prinz with Loudon Wainwright, one of the *Life* writers, and dropped him off at National Airport before heading home to Arlington. We were trying to get the inaugural address on the car radio. Even when it came through the static, however, my mind was elsewhere. I kept thinking that Bob Gilruth wouldn't have asked for the peer vote unless he meant to use it, and it must have been a big part of his decision. When I got home, I sat down to write a letter.

The letter was to Bob. It pointed out the unfairness of the peer vote as a factor, and gave my reasons for why I might have rated low in the vote, as opposed to my training performance. I said I thought I might have been penalized for speaking out for what I thought was the good of the program. It was a strong letter, because I felt strongly.

I gave the letter to Bob when I returned to Langley, but I heard no more about it.

We were all following Al now. The decision was made, and we were a team, so we went back to work. My training routine was welded to his. I was to be Al's backup, and would make the flight if anything happened to him.

Outside the tight NASA circle, the country thought the candidates had been narrowed down to three but that the decision was still to be made. We all went along with the charade. It was hard enough for Gus and me, but at least we were part of the group the press now was calling "the first three." For Deke, Gordo, Scott, and Wally, it was worse. They were the forgotten four.

On January 31 a Redstone booster carried a thirty-seven-pound chimpanzee named Ham, strapped into a Mercury capsule, on a flight that reached an altitude of 157 miles and landed 418 miles downrange. Ham's flight was far from perfect. It had gone forty-two miles higher and 125 miles farther than intended, due to a malfunctioning control valve that let too much fuel into the engine and gave the rocket a faster burn and a steeper climb angle than programmed. Ham was subjected to more than the normal seven Gs on takeoff. On reentry, the retro-rockets jettisoned too soon and Ham came down too fast. He felt almost fifteen Gs pressing on him—one fewer than the level at which the astronauts had started to black out on the Johnsville centrifuge—and he had a rough splashdown that drove a bolt through a bulkhead and made the capsule leak. Since he was far from the expected landing site, it also took a while to find him. But the capsule's life support equipment worked, and he was in apparently good health when he was recovered.

But Ham was not as happy as the photographs taken of him seemed to indicate. He was supposed to be pulling levers to avoid shocks to his feet, as he had been taught to do in his operant training, but the capsule's electrical system had gone haywire, and instead of rewarding him with banana pellets for the correct response, it gave him shocks instead.

The public nevertheless viewed the flight as a success because the pictures seemed to show Ham smiling. But Wernher von Braun decided that another Redstone flight was needed before it was safe to put a man aboard.

This drove Al crazy. His flight had been scheduled for March, and he was ready to go.

The Mercury-Atlas team reassembled at the Cape on February 21 for the first test launch of the capsule-booster combination since July's disaster. To solve the problem that apparently had caused *MA-1* to self-destruct, the engineers had devised an eight-inch-wide "belly band" of stainless steel to stiffen the rocket at its flex point below the mating ring with the capsule. The test—held this time without an audience of dignitaries—was successful.

Then a rumor reached us that made Al, and the rest of us, even crazier.

We knew that conservative doctors and medical experts had approached the President's Science Advisory Committee with the argument that the few primate flights made so far had provided too little knowledge about space flight. They believed the primate program should be extended and expanded before an astronaut went up. The rumor we heard was that the PSAC, which helped shape national scientific policy, was preparing to recommend to President Kennedy that manned flight be delayed until two dozen or more additional primate flights were made.

Al wanted to deep-six this ridiculous idea. He said he was ready to have a chimp barbecue. The rest of us were behind him all the way. Bob and Christopher C. Kraft Jr., who was NASA's flight director, also were concerned. I had met Chris Kraft back when I was at BuAer and he was at NACA. He had figured out a solution to a serious fuselage flexibility problem in the Crusader, and I knew he would apply himself to the PSAC problem with equal tenacity. He, Bob, Al, and I decided to hit it head-on before the program was set back for years to come.

We issued an invitation through NASA headquarters, where a new director, James E. Webb, had been sworn in, for the committee to come down to the Cape to hear our side.

The committee was composed of leading scientists in a wide range of fields. It was a large group, and most of the members came. We had the new Mercury control center set up for them, and we pulled out all the stops with a full-day presentation against the center's impressive backdrop of world maps, tracking charts, and television monitors. We went through all the records of our testing, all the medical results from the voluminous studies that had already been conducted. We produced the primate medical data, as well as our own, to show them their concerns were unfounded. We told them we were trained and ready to go. The bigger picture was that we were in a race, and everybody knew it.

The PSAC returned to Washington, and we heard no more about an expanded cohort of primate flights. But we still had to wait for the additional Redstone test von Braun demanded.

It went up on March 24, the date originally scheduled for Al's flight, and functioned perfectly.

As the first flight neared, the psychologist who had headed our pre-selection testing reentered our lives. This time, I supposed, George Ruff was looking for signs of stress, fear, or overeagerness, anything that departed from the psychological profiles he and his colleagues had made of each of us during selection.

Some psychiatrists were concerned about what they called the "breakaway phenomenon." They thought that if you got up in space, you might feel such euphoria that you might not want to come back. Al wouldn't have the option, since he wouldn't achieve orbit, but the doctors wanted to know about any odd feelings that zero G might bring on.

I never doubted that I would want to come back. But the additional testing didn't bother me, because I continued to believe that whatever Ruff and his colleagues learned would be important. If the shrinks wanted to follow us around to learn something, that was fine with me. I held fast to my view that we were into something so new to the human experience that almost any testing could produce new knowledge, so my response, as usual, was, "Let's try it."

As Al was chafing at the delays and we were fending off the primate-flight advocates and satisfying the psychiatrists that we weren't likely to fly off into deep space, NASA announced that all the stations of its worldwide Mercury tracking network were operational.

I was now Al's alter ego, his virtual twin. He was extremely busy with his final training, honing his skills on the simulators, keeping physically fit and rested, and keeping in touch with his family. There were still engineering meetings, weather briefings, and updates on what the contractors and NASA were doing about any problems. I attended in Al's place when he couldn't be there. We needed to absorb the same information so as to produce the same results. I worked hand in glove with him. I don't think our work showed any sign of the disappointment I, and the others, still felt at not being selected to go first. We all threw ourselves into supporting Al's flight as seriously as if we had been the ones scheduled to fly.

Then came the stunning news that changed the equation.

It reached us first as a rumor that the Soviets had once again done something big. The something big proved to be Major Yuri A. Gagarin's 108-minute single orbit around Earth and his safe return. Not only had the Soviets

put a man into space first, they had put him in orbit—
something we didn't plan to do for several flights to
come.

Kennedy captured the national mood of fatigue and
disappointment in his regular news conference. He said,
"No one is more tired than I am" of seeing the United
States second to Russia in space flight. "We are, I hope,
going to be able to carry out our efforts . . . this year.
But we are behind. . . . The news will be worse before it is better, and it will be some time before we
catch up."

Reporters asked me what I thought. They still assumed I was the one who had been shot out of the saddle
by the Gagarin flight. I said, "They just beat the pants off
us, that's all, and there's no use kidding ourselves about
that. But now that the space age has begun, there's going
to be plenty of work for everybody."

The technological feat of making our nuclear weapons—and our boosters—smaller had allowed the Soviets
to put satellites into orbit first. Now, the prestige of orbiting a manned capsule was theirs as well, and for the same
reason.

Al was bitterly disappointed. He had prepared to be
the first man into space. Now Kennedy had signaled that
Al's suborbital flight would not compare to Gagarin's. He
had as much as said we would still be behind even after Al
went up.

The media's attention swung to him when his name
was announced on May 2, the first date scheduled for his
flight. By then, the disastrous Bay of Pigs invasion had further embarrassed the United States. Another Atlas launch,
this one trying to put a "mechanical astronaut" into orbit,

had failed because of guidance problems and had to be destroyed, although the escape system worked well. A Little Joe rocket carrying a Mercury capsule had also failed. The nation was in need of a tonic, and looked to Al to provide it.

A weather postponement moved his flight to May 5. I woke up ahead of him in the crew quarters at Hangar S, where we both were sleeping, and went to the launch pad to check out the capsule. All the systems were go. Before I went back down, I left an enticing piece of pinup art on the control panel that I was sure he would appreciate, and a sign saying No Handball Playing in This Area.

He provided some anxious moments before he got off. I went back up the gantry to help insert him into the capsule, then returned to the control center to monitor the countdown. The holds stretched on, and Al had the problem we had all faced a time or two—he had to urinate. After more holds, he did. "Well, I'm a wetback now," he said.

The Redstone finally fired around nine-thirty in the morning. The flight went perfectly. The capsule that Al had named *Freedom 7*—we had decided that each astronaut would name his own capsule, with a seven added to signify that we were a team no matter who was in the cockpit— reached an altitude of 115.696 miles and a speed of fifty-one hundred miles per hour. He pulled six Gs on launch, manually controlled the capsule's attitude while weightless at the top of its arc, and hit eleven Gs briefly on reentry. The Navy vessels in the target area three hundred miles downrange had been joined a few hours before liftoff by a Russian trawler bristling with antennas. *Freedom 7* splashed into the Atlantic after a flight that lasted almost fifteen minutes. Soon afterward, Al was aboard the USS *Lake Champlain* and talking to the president.

The United States had joined the space age.

I spent the night in the same room with Al that night on Grand Bahama Island. Being backup meant you virtually lived with the person. While his flight had gone perfectly, uncertainties remained. The doctors preferred not to leave the astronaut completely on his own even after the flight. Nobody knew what the delayed reaction might be.

Al's reaction was exuberance and satisfaction. He talked about his five minutes of weightlessness as painless and pleasant. He'd had no unusual sensations, was elated at being able to control the capsule's attitude, and was only sorry the flight hadn't lasted longer.

Al's flight was greeted as a triumph around the world because it had been visible. The world had learned of Gagarin's flight from Nikita Khrushchev. It had learned of Al's by watching it on live television and listening to it on the radio. That openness was as significant a triumph in the Cold War battle of ideologies as Gagarin's flight had been scientifically.

Kennedy used the momentum of Al's flight boldly. Now that men on both sides of the Iron Curtain had entered space one way or another, the president leapfrogged to the next great step. He went to Congress on May 25 and in a memorable speech urged it to plunge into the space race with both feet. He said, "I believe this nation should commit itself to achieving the goal, before this decade is out, of landing a man on the moon and returning him safely to Earth. No single space project in this period will be more impressive to mankind or more important for the long-range exploration of space; and none will be so difficult or expensive to accomplish."

I assumed after serving as Al's backup that I would be named to make the second flight. Again, I was surprised and disappointed when it went to Gus and I was named backup for the second time. When something has not gone right for me, when I've had a disappointment, I tend to clam up and avoid inflicting my feelings on others. I'm sure there was a lot of that in my attitude then, but I tried to lighten up and not take it out on Annie and the kids. They certainly didn't deserve the worst of me, since they saw me so little anyway.

Gus's flight was set for July 19, the day after my fortieth birthday. He would have a view Al didn't have. Al had ridden the Mercury capsule as originally designed, with a porthole and no window. We had discussed other changes with Max Faget and the engineers at McDonnell. Deke wanted foot pedals to make the capsule's controls more like a plane's. I had wanted to replace the gauges with tape-line instrumentation that would provide information at a glance. Both systems would have added too much weight. But Gus's *Liberty Bell 7*, as he had named his capsule, had a window.

One problem nobody had figured out the answer to, however, was the one that had plagued Al.

The night before Gus's flight, I was staying with him in crew quarters as his backup. There was a little medical lab next door. We went in and set to work trying to design a urine collection device. We got some condoms in the lab, and we clipped the receptacle ends off and cemented some rubber tubing that ran to a plastic bag to be taped to his leg. It seemed to work well enough, and Gus put it on in the morning before he suited up.

Gus's flight went as well as Al's had. He splashed down on target. Then things started to go wrong. The hatch blew off prematurely, the capsule started taking on water, and Gus had to get out. Water poured into his pressure suit through the area around the neck. The helicopter that was supposed to pick him and the capsule up together tried to retrieve the capsule, but it was too heavy with water, and the pilot had to let it sink. Meanwhile, Gus almost drowned as his water-filled suit lost buoyancy. He finally caught a horse collar lowered from a helicopter and was lifted to safety.

That night in the room we shared at the debriefing site on Grand Bahama, Gus talked about how close he had been to drowning. He hated the loss of *Liberty Bell 7*. But like Al, he was exhilarated by his flight and eager to go back up again.

More suborbital flights were scheduled. After playing backup twice, I was in line to make the next one. Then, barely two weeks after Gus's flight, on August 6, the Soviets launched another cosmonaut, Major Gherman Titov, into a seventeen-orbit flight around Earth that lasted twenty-five hours. NASA quickly decided the two suborbital flights had gone so well from launch to splashdown that we too could move to the next stage.

It was time to orbit.

CHAPTER 17

The early fall of 1961 was, for me, a time of self-doubt and anxiety. I was disappointed not to have been chosen for either of the first two flights. After it seemed certain I would make the third, the shift to orbital preparations threw the choice once again into confusion. NASA tantalizingly refused to announce its selection. I also worried whether, having turned forty, I had crossed onto the wrong side of some invisible barrier. Some of the pundits following the space program were saying I was too old. I knew I was fit to fly, but I could hear the clock ticking.

On my occasional weekends at home I water-skied with Annie, Dave, and Lyn, and Tom Miller and his family, on the Potomac near Quantico. With Annie and the kids and my best friends, all of us enjoying each other, I found an antidote to the disappointment that nagged at me. I had been a forward-looking optimist all my life, and now more than ever, I needed to keep that view. I struggled to remember that the space program was bigger than any one person.

NASA, meanwhile, intensified its preparations. Its goal, though still unstated, was to put an astronaut in orbit before the end of the year. No one knew at the time that Titov had suffered a prolonged bout of nausea in the middle of his flight, and that the Soviet space czars had grounded their orbital flights for a year to try to figure out the cause.

Al and Gus had come through their five-minute stints of weightlessness with flying colors. Gagarin seemed to have survived his hour and a fraction in good health, although no one really knew. With medical data on Titov's flight not forthcoming, NASA's doctors remained uncertain about some of zero gravity's effects. What it did to the vestibular system of the inner ear was one of the things that worried them. Would it throw the sense of balance so out of whack that the astronaut would suffer from debilitating nausea and be unable to function? Would his eyeballs, freed of the need for support from the lower eye structure, change shape and reduce vision? Would his eyes move uncontrollably, so that he couldn't focus on the instruments and perform the tasks necessary to get down— if he even wanted to get down? The shrinks still suspected that space held a mysterious allure that could cause the breakaway phenomenon.

Atlas testing moved into its final phase. A September 13 Atlas launch, *MA-4*, carried a dummy astronaut into orbit and back after circling Earth once, and the capsule landed on target in the Atlantic. At that point the system seemed ready. The Atlas had been strengthened not only by the belly band but with the use of thicker metal near the top. But Bob Gilruth and Hugh Dryden, NASA's deputy administrator, wanted to send a chimp into orbit before risking a man.

This time the chimp was named Enos, and he went up on November 29. Like Ham, he had been conditioned to pull certain levers in the spacecraft according to signals flashing in front of him. Like Ham's, his flight was not altogether perfect. The capsule's attitude control let it roll 45 degrees before the hydrogen peroxide thrusters corrected it. Controllers brought it down in the Pacific after two orbits and about three hours.

When Enos was picked up he had freed an arm from its restraint, gotten inside his chest harness, and pulled off the biosensors that the doctors had attached to record his respiration, heartbeat, pulse, and blood pressure. He also had ripped out the inflated urinary catheter they had implanted, which sent his heart rate soaring during the flight. It made you cringe to think of it.

Nevertheless, Enos's flight was a success, and he appeared unfazed at the postflight news conference with Bob and Walt Williams, Project Mercury's director of operations. All the attention was on the chimpanzee when one of the reporters asked who would follow him into orbit.

Bob gave the world the news I'd learned just a few weeks earlier, when he had called us all into his office at Langley to tell us who would make the next flight. I had been elated when, at last, I heard that I would be the primary pilot. This time I was on the receiving end of congratulations from a group of disappointed fellow astronauts. Now, as the reporters waited with their pencils poised and cameras running, Bob said, "John Glenn will make the next flight. Scott Carpenter will be his backup."

The first American orbital flight had taken on new importance now that the Soviets had orbited in their two flights. I tried to take the news in stride, but I couldn't help being excited. I was going to be the fifth human into space, and the first American to orbit. Of course, as the test pilots at Edwards liked to point out, I was going to be following a chimp.

NASA had hinted that the flight might occur before Christmas. Bob knew it wouldn't hurt the agency if the president could give a Christmas present to the country in the form of an orbital flight.

I set about naming the capsule. Al's *Freedom 7* had struck the right note. Gus, in *Liberty Bell 7,* had been inspired by both patriotism and the capsule's shape. I had several ideas, but I was trying very hard to keep Dave and Lyn involved and make them feel a part of my mission. I asked them if they would be willing to think about some names.

I said, "There's only one ground rule. The world is going to be watching, so the name should represent our country and the way we feel about the rest of the world."

They pored over a thesaurus and wrote dozens of names in a notebook. Then they worked them down to several possibilities, names and words including *Columbia, endeavor, America, Magellan, we, hope, harmony,* and *kindness.* At the top of the list was their first choice: *friendship.*

I was so proud of them. They had chosen perfectly. I told the artist at Hangar S that my capsule would be *Friendship 7*, and asked her to put it on in script to add a little individuality to the usual block letters. The name would be a closely held secret until the day of the flight.

I spent most of my time at the Cape, working on the procedures trainer or in the capsule itself. For many of the runs, I wore EKG and biosensor leads. Finally, I asked flight surgeon Bill Douglas if putting them in the same spot each time wouldn't provide more accurate readings. He agreed, and that's how I got tattooed. Bill roughed up my skin on each of the spots with a scalpel, found a bottle of India ink on somebody's desk, and applied a drop on each spot—one around my collarbone, one just below my sternum, one under and behind my right arm, and another under and behind the left. What people won't do in the interests of science—I still have those tattoos today.

Scott, as my backup, was taking care of the details. By now all of us had had so much training that we could have gone on a moment's notice. We kept working out anyway, and I kept jogging up and down the hard sand of Cape Canaveral to stay in shape. I had worked my run up from two to five miles, and it became a form of therapy. After a day of involved meetings, simulator training sessions, or equipment checkouts, I couldn't wait to get out on the deserted beach and run. My only companions were pelicans and seagulls, sandpipers skittering up and down the surf line, and crabs that scuttled away from my approach, and as I ran I heard the constant slosh of waves collapsing on the shore.

At first Scott and I could move around Cocoa Beach without too much trouble. Even in the flamboyant Shelby Cobra he was driving then, with exhausts that gurgled out generous portions of noise from the V-8 engine, we went unrecognized by most people outside the space community. We liked to go to the Kontiki Village, a Polynesian restaurant, and listen to the lounge singer do "Beyond the Reef."

Soon, however, the normal trickle of reporters at the Cape turned into a flood. I usually attended Sunday services at the Riverside Presbyterian Church on A1A. Reporters and photographers started crowding into the back of the church. The cameras made an awful racket, and after one Sunday of that I asked them to wait outside, which they did. I kidded them that one by-product of the space program was to get journalists into church, some of them for the first time in years.

Fan mail started arriving by the bagful. All of the astronauts had received a steady stream of fan mail and autograph requests from the beginning, and we had always been able to keep up with it. I believed that if people took the time to sit down and write a letter, they deserved a response. But now the volume was overwhelming. It was more than I could handle, and our secretary, Nancy Lowe, applied her considerable organizational skills to sorting through the letters and bringing the bills and other necessary mail to my attention.

I had moved to the crew quarters at the Cape by then. They were on the south side of Hangar S, the same area on the second floor where the suit room was and the doctors had their offices. Crew quarters was a room with blue walls, stacked metal bunk beds, and desks. It was like an aerospace summer camp. I slept in an upper bunk, and was grateful for the isolation. The Cape had its amenities. Bea Finkelstein, a dietitian from the aeromedical laboratories at Wright-Patterson, was assigned to Project Mercury. It was her job to feed us a healthy diet, and to put us on low-residue foods three days before a scheduled flight, since space toilets had yet to be developed. She had taken over a small building near the beach and turned it into a dining hall. It nestled in the palm trees, and it was fun to stop there for dinner after work or a run.

One of the things that led me to isolate myself was my concern about getting sick. I could see, in my worst dreams, coming up to flight day and catching some kind of a bug that would get me bounced off the flight at the last minute. Now that I was next in line, I was determined that no freak accident or sickness would interfere.

Henri Landwirth called from Cocoa Beach almost every day. Henri really liked working with the astronauts. He understood the pressure we were under and he always tried to find ways of relieving it. He'd go to elaborate lengths to play a practical joke. Once I checked into the Holiday Inn and found no towels in my room. I went to the desk and complained in mock anger. Henri came running out to see what was wrong, and I got my towels. The next time I checked into the Holiday Inn, I went to my room, and opened the door to a sea of white—towels stacked four feet high on the bed, towels on the chairs, towels on top of the TV, towels on the bedside tables around the lamps, towels stacked in front of the mirror in the lavatory, towels filling the bathtub. Then I heard Henri's Belgian accent behind me, inquiring innocently, "Do you have enough towels, sir?" He played jokes on all the astronauts, and we'd repay him in kind; Henri came by the swimming pool one morning to find Gordo sitting out there with a fishing pole angling for one of the coolerful of fish he'd let loose in the pool. Another time Al and a couple of the others put a live alligator in the pool.

Henri didn't always relieve the pressure with a joke. He knew I wasn't getting off the Cape, and it was in his nature to be concerned. One day he called and said, "I know you haven't been off the Cape for a few weeks. Why don't you come and have dinner with us at home?" It sounded like a good idea, typical of Henri's generosity, so I went and had a good time with him, his wife, Jo, and their three young children, Gary, Greg, and Lisa.

The next morning Henri called again. He said, "Hey, I hope you've had the mumps."

I laughed and said, "Who told you I never had the mumps?"

"No, I'm serious for a change," he said. "Greg's got the mumps."

I had assumed I was immune. Annie had had the mumps when we were in college, and so I had been more than adequately exposed. I didn't worry about it. But a couple of weeks later I woke up with soreness on both sides of my neck, and started to panic. It turned out that my neck was sore from work I had been doing the previous day, straining my head against my inflated pressure suit to establish my maximum vision pattern. It was a false alarm, but reinforced my instinct to stay at the Cape as much as possible.

I tried to picture the flight in advance. You wanted it to be "nominal"—the engineers' dull word for things going as expected. Shorty Powers had deftly substituted A-OK during Al's flight. It caught on with the public, but it was a term the first seven astronauts never used and objected to for reasons probably none of us could explain. I believe we just thought it was silly.

The training and simulations had produced a

movielike image of the flight that I could switch on at any time. It was more difficult to anticipate the unexpected. I knew that my orbital track would take me over desert areas for which we had prepared in survival training. I would also fly over remote jungle areas. I could see a distinct possibility, in an emergency, of coming down among people in central Africa, the Australian outback, or Papua New Guinea who weren't familiar with the space race. I could imagine they would be nonplussed at best, and seriously alarmed at worst, to see a heat-charred space capsule descending from the sky under a parachute and a stranger in a silver suit emerging. I thought I should be prepared to explain—if I even survived such a landing—that I was no threat to them, and I came up with a message that, unfortunately, sounded like a Martian saying "Take me to your leader." NASA asked linguists at the Library of Congress to translate it phonetically into some of the more remote languages and dialects. I noticed that in many of these dialects, the words for "stranger" and "enemy" were the same. I never forgot that, and realized that the better acquainted we become, and the more we know about people, the less likely we are to look at them as enemies and the more likely we are to become friends. It reinforced my belief that *Friendship* 7 was an apt choice of name.

I wanted some mementos of the flight for family members and the other astronauts, so I sketched a little capsule track circling Earth, and had Galt Jewelers in Washington make it into gold pins.

Several possible launch dates came and went. Finally, the Project Mercury brain trust decided that with Christmas approaching, rather than hold the whole operation—literally thousands of people, including engineers, technicians, the medical staff, ground control, teams at the tracking stations around the world, and more thousands who were watching and waiting, the press included—hostage to the whims of the weather and the mission's demanding technology, it was better and safer to stand down. I agreed with the decision. As much as I wanted to go, it was far better if everybody had their minds on the mission when the time came. NASA told everybody to go home for the holidays, and announced a date of January 16.

I was grateful for the chance to be at home in Arlington with Annie and the kids. The pressure had built up on them, too. The respite of Christmas let us stay quiet and homebound. We did a lot of visiting back and forth with the Millers next door, and with Les Brown and his family. Les was another Marine pilot, like Tom assigned to headquarters, who had built his house on the other side of Tom's a little after we built ours. We had a regular Marine compound there on North Harrison Street. The *Life* contingent was there, as usual, but for the most part it was just family and friends.

That Christmas was snowy and cold. I got one bad scare when I slipped on some ice and fell. I reacted instantly, curling my hands up against my chest and landing on my backside. I got up with everything but my dignity intact, and without the broken hand or arm that would have scrubbed me from the flight. Then one day between Christmas and New Year's, the bright sun emerged from the clouds, tempting us outdoors. Annie, Dave, Lyn, and I drove up the Potomac to Great Falls Park for a winter picnic.

We had had snowy winter cookouts before, all of us bundled up around a fire, no one else around, a few birds fluttering nearby for handouts, and the water sounds of the Great Falls muffled in the background. It was a wonderful family time.

I think Lyn brought it up first, something I knew had been on their minds but left unstated until now. Haltingly she asked, "What kinds of things could happen to you, Dad?"

This was something I probably should have brought up before, but in trying to keep the best outlook had not. Now was the time.

"You're not little kids anymore," I said, "and you understand more about this flight than anyone except those of us in it. You know I've always leveled with you two, and I will now. There will be some risks when I go up, just as we take some risks every day. Like coming out here, on the road, things can happen, but we're careful and take precautions. You work hard to make sure that it won't, but there's still some risk no matter what you do.

"I wanted this," I continued. "And I know you all wanted it for me. We've worked hard for this opportunity, not just me but all of us in this family. I'm sure nothing's going to happen, but it's possible, and I'm glad you brought it up, Lyn. If I believed something was likely to happen, I wouldn't want to go, and NASA wouldn't send me in the first place. But if anything did happen to go wrong, I don't want you to blame anybody, okay? Not NASA, not anybody. I'm doing something I really want to do because it's important for our whole country. And it's something we should keep doing, so don't let anybody tell you we should stop trying to get up into space.

"And don't blame God. He gave us the opportunity to be where we are in the first place. I'm thankful for that, and I hope you are, too. Nobody forced me into this. This is our choice, and it's going to be a great flight."

I paused, and the silence stretched into a minute or so as we all just stared into the fire. Then there were solid hugs for each, and a foursome family hug. It was an interlude we would all long remember.

This kind of thing had been understood between Annie and me when I went away to fly in combat, and when I was working as a test pilot. The kids were older now, of course, and with Lyn's question I knew they had needed to hear the things that would make it possible for them to go on if anything happened.

We welcomed 1962 by watching the Times Square crowd on television and listening to Guy Lombardo play "Auld Lang Syne."

Back at the Cape, when the January 16 launch date was postponed to January 23, the pressure ratcheted upward. Everybody seemed to feel it but me. I had plenty to keep me occupied. I was working like a demon on the procedures trainer, and eventually worked through seventy simulated missions and reacted to almost two hundred simulated system failures. I was also trying to convince NASA to let me take a camera into space.

At that stage of Project Mercury, cameras were considered too great a distraction to an astronaut. They would keep him from performing his array of scientific tasks. I would have a camera aboard, but it was an ultraviolet spectrograph with only a single roll of film for taking pictures of the sun and stars. Just as the designers had left a window out of the first capsule, the slide-rule-and-computer contingent lacked the imagination to see the value of photographs that would help translate an astronaut's experience for anyone who saw them. They had their checklist, and that was all that was important. They argued that people already knew what Earth looked like from space. After all, we had satellites in orbit taking pictures.

I finally went to Bob Gilruth and said, "This is ridiculous. I need to take some pictures, because people are going to want to see what it looks like to be an astronaut. I'm not going to let my own safety and running the spacecraft take second place, and I'm going to get through the checklist."

I knew from previous conversations that Bob was sympathetic to the idea. He agreed, and we started looking for a camera. It had to be small enough to operate with one hand, and adaptable so that I could advance the film with my thumb and snap the shutter with my forefinger while wearing a pressure glove, since I didn't want to take the glove off to take pictures.

NASA didn't have a camera section then. One of the *Life* photographers, Ralph Morse, was assigned to NASA, and he suggested trying to adapt a Leica and a couple of other kinds of camera. The machine shop tried all kinds of contraptions, but nothing worked very well.

One day I was down in Cocoa Beach getting a haircut, one of my rare trips off the Cape. I went into the drugstore next door to pick up some things, and saw a little Minolta camera in a display case. It was called a Hi-Matic. I picked it up and thought, "Jeez, it's got automatic exposure." That was brand-new at the time. Nobody had ever heard of that. You didn't have to fiddle with light meters and f-stops. I bought it on the spot for $45. When I took it back to the Cape, it turned out to be more readily adaptable than any of the others.

Once they had it rigged, I laid all the cameras out on my bunk, put on my pressure glove, and tried them. The Minolta was the easiest to use.

The date approached, along with a further buildup of tension. The anticipation became a story in itself. The long buildup allowed more aspects of the flight to be explored, such as the television coverage. ABC, NBC, and CBS all planned uninterrupted coverage, about which one network executive told *TV Guide,* "That man's life will be hanging in the balance all the time he's up there. How can we go back to quiz shows or soap operas while he's risking his neck for us?"

That set the tone for much of what was written and said, and contributed to a kind of soap opera in its own right: Will he be launched, or won't he?

January 23 came, and weather kept pushing the date back day after day. The flight had taken on far more importance than I had attached to it in the beginning. It seemed that every news media person in the world was descending on Cape Canaveral and the surrounding area. Those who weren't, Annie reported when we talked on the phone, were camped on the street in front of our house in Arlington. The delays distracted all of them more than it did me. I was disappointed each time, and continued to worry about being exposed to some kind of bug, but I rationalized that there would always be another day, and I just kept working on the trainers and running on the beach at the Cape.

On January 27 it looked as if we were ready to go. The day before, I had written a statement to the American people that I gave to Shorty Powers for release in case something happened to me on the flight. Even if something went wrong, I thought it was important for Americans to continue to support NASA and space exploration.

I got up at two o'clock, ate my low-residue breakfast of filet mignon, scrambled eggs, toast with jelly, orange juice, and Postum, had my EKG and biosensor leads stuck over my tattoos, and struggled into my pressure suit with the help of suit technician Joe Schmitt. I walked out of Hangar S carrying the portable blower that kept air flowing through my suit, waved at the assembled workers and ground crew, and rode in the transfer van to Launch Complex 14. I rode twelve stories up in the gantry elevator, crossed the catwalk to the capsule on top of the sixty-five-foot rocket, and, with Scott's help, climbed into the tiny cabin of *Friendship 7* to wait for liftoff. Then I lay there on my back in the contour couch for nearly six hours, wishing the gantry would pull back, signaling a break in the clouds and imminent liftoff. But they didn't break, and the launch window closed at twelve-thirty in the afternoon.

Back in Hangar S, I had just gotten out of my suit when a small assembly of NASA officials came into the conference room next to the crew quarters. I assumed they wanted to express regrets about the canceled mission and talk about the next time. But what they wanted was for me to call Annie. It seemed Vice President Johnson wanted to visit her at home, and she was refusing.

Annie wouldn't have refused to see the vice president without a really good reason. I called her, and she said Johnson wanted to bring in network television cameras and some of the reporters who were camped outside, and wanted Loudon Wainwright, who was there for *Life,* to leave. She said she was tired, she had a headache, and she just wasn't going to allow all those people in her house. Part of her resistance came from knowing that the reporters would ask her questions with lights and cameras aimed at her. Part of it was the vice president's insistence that Loudon not be there, when we had a deal with *Life* that had White House approval. I told her whatever she wanted to do, I would back her 100 percent.

One of the NASA group, after I hung up the phone, said pointedly that it would be possible to replace me on this flight. I saw red. I said that if they wanted to do that, they'd have a press conference to announce their decision and I'd have one to announce mine, and if they wanted to talk about it anymore, they'd have to wait until I took a shower. When I came back, they were gone and I never heard any more about it.

NASA set a new launch date two weeks later. While the agency regrouped after the big buildup, I took a few days at home with Annie and the kids. President Kennedy invited me to the White House, and we spent a few minutes talking about the flight. His questions were so detailed that I asked him if he wanted me to come back with models and blueprints that could explain things better. He said he did. When I returned a few days later, I was prepared with models, maps, and charts. We moved from the Oval Office into the Cabinet Room, where I laid everything out on the big table. We must have talked for an hour. I was impressed with his curiosity about everything that was going to happen on the flight.

After the January 27 postponement, a few psychiatrists around the country started to write letters urging that I be taken off the flight because they were certain the delays must be starting to get to me. Some journalists were saying the same thing. Even Yuri Gagarin weighed in, no doubt for propaganda purposes, saying the wait in the capsule had exerted "serious psychological and moral pressure."

With no flight to describe, some of the reporters were hard pressed. After one of the delays, Shorty Powers told me he thought we ought to have something for the ever-increasing mob of journalists so they wouldn't have to report more speculation about the effects of the flight postponements on my mental state. There wasn't much to say, but I recounted running along the beach at the Cape and seeing some sea turtle tracks coming up from the water. I said I had followed the tracks up to a turtle nest, looked it over, and kept running. A columnist for the *Miami Herald* reported that, and added that I had a good recipe for turtle soup. That brought a bombardment of letters from people taking me apart for messing with an endangered species. I hadn't seen a turtle there, and I didn't touch a single egg in that nest. It really made me mad, because it seemed so unnecessary in the midst of all the preparations.

The delays continued. Most often the problem was the weather—if not clouds at the Cape, then storms buffeting the Atlantic in the splashdown area. Sometimes it was some small but important problem with the millions of interrelated components of the mission, not just the hardware of the Mercury-Atlas system but each part of the incredibly complex web of personnel and communications that would track the flight around the world, monitor it for anomalies, and relay news of what was happening to a global audience.

Of all the people who were driven to anxiety by the continued delays, near the top of the list was Henri Landwirth. I didn't know this at the time, because he worked very hard to keep it a secret, but Henri had decided to celebrate my flight by creating the world's largest cake. He planned to present it to me and the world on my return from orbit.

Henri's cake was a monster. It was a replica of the Mercury capsule, decorated in great detail even down to the *Friendship* 7 in script across the side. No one outside of NASA and my family knew what the name was, but Henri had somehow managed to break security and get a photo of the capsule that he used in his instructions to the bakers. It had to be baked in sections and assembled outside the Holiday Inn kitchen, since its door was only three feet wide. The finished cake weighed nine hundred pounds. He put it together in a truck that he planned to use to take the cake out to the Cape for my return.

The problem, of course, was that once Henri had the cake baked and assembled, he still didn't have a flight to celebrate. He had to rig the rented truck up with an air conditioner and a generator that ran day and night in the motel parking lot to keep the cake from spoiling.

I'm sure Leo DeOrsey was anxious, too. At the time I didn't have a lot of life insurance, and Leo had been trying to find a company that would insure me for another $100,000. Nobody would do it. There were no actuarial tables for astronauts. Finally he called and said Lloyd's of London was willing to cover me for the six-hour period from launch to landing. The premium was $16,000.

"Leo," I said, "if you think we can afford it, do it. Maybe you can use my *Life* money."

He called me again a few days later and said, "I'm not going to buy that insurance. I'm not going to bet against you. Here's what I'm going to do. If something happens on your flight, I'll make sure Annie gets one hundred thousand dollars. I'll guarantee it." And he wrote a check for $100,000 made out to Annie, and gave it to Bill Douglas to hold until it could be returned to him, or had to be delivered.

CHAPTER 18

I woke up at about one-thirty on the morning of February 20, 1962. It was the eleventh date that had been scheduled for the flight. I lay there and went through flight procedures, and tried not to think about an eleventh postponement. Bill Douglas came in a little after two and leaned on my bunk and talked. He said the weather was fifty-fifty, and that Scott had already been up to check out the capsule and called to say it was ready to go. I showered, shaved, and wore a bathrobe while I ate the now-familiar low-residue breakfast with Bill, Walt Williams, NASA's preflight operations director, Merritt Preston, and Deke, who was scheduled to make the next flight.

Bill gave me a once-over with a stethoscope and shone a light into my eyes, ears, and throat. "You're fit to go," he said, and started to attach the biosensors. Joe Schmitt laid out the pressure suit's various components—the suit itself, the helmet, and the gloves, which contained fingertip flashlights for reading the instruments when I orbited from day into night.

I put on the urine collection device that was a version of the one Gus and I had devised the night before his flight, then the heavy mesh underliner with its two layers separated by wire coils that allowed air to circulate. I put the silver suit on— one leg at a time, like everybody else, I reminded myself. In a pocket, in addition to the small pins I had designed, I carried five small silk American flags. I planned to present them to the president, the commandant of the Marine Corps, the Smithsonian Institution, and Dave and Lyn.

Joe gave the suit a pressure check, and Bill ran a hose into his fish tank to check the purity of the air supply; dead fish would mean bad air. I was putting on the silver boots and the dust covers that would come off once I was through the dust-free "white room" outside the capsule at the top of the gantry when I said casually, "Bill, did you know a couple of those fish are floating belly-up?"

"What?" He rushed over to the tank and looked inside before he realized I was kidding.

I put my white helmet on, left Hangar S carrying the portable air blower, waved at the technicians, and got into the transfer van. In the van, I looked over weather data and flight plans. About two hundred technicians were gathered around Launch Complex 14 when I got out of the van. Searchlights lit the silvery Atlas much as they had the night we had watched it blow to pieces. I thought instead of the successful tests since then.

Clouds roiled overhead in the predawn light. It was six o'clock when I rode the elevator up the gantry. Scott was waiting in the white room to help me into the capsule. In addition to white coveralls and dust covers on your shoes, you had to wear a paper cap in the white room. It made the highly trained capsule insertion crew look like drugstore soda jerks. I bantered with Guenther Wendt, the "pad führer" who ran the white room with precision, before wriggling feet first into the capsule and settling in to await the countdown.

The original launch time passed as a weather hold delayed the countdown. Then a microphone bracket in my helmet broke. Joe Schmitt fixed that, I said goodbye to everybody, and the crew bolted the hatch into place. They sheared a bolt in the process, so they unbolted the hatch, replaced the bolt, and secured it back into place. That took another forty minutes. I still didn't believe I would actually go.

I heard a steady stream of conversation on my helmet headset, weather and technical details passed between the blockhouse and the control center in NASA's technical patois. A voice said the clouds were thinning. Up and down the beaches and on roadsides around the Cape, I knew, thousands of people were assembled for the launch. Some of them had been there for a month. The countdown would resume, then stop. I just waited. Then the gantry pulled away, and I could see patches of blue sky through the window. The steady patter of blockhouse communications continued. Scott, in the blockhouse, made a phone call and let me know he had patched me through to Arlington so I could talk privately with Annie.

"How are you doing?" I said.

"We're fine. How are you doing?"

"Well, I'm all strapped in. The gantry's back. If we can just get a break on the weather, it looks like I might finally go. How are the kids?"

"They're right here." I spoke first to Lyn, then to Dave. They told me they were watching all three networks, and the preparations looked exciting. Then Annie came back on the line.

"Hey, honey, don't be scared," I said. "Remember, I'm just going down to the corner store to get a pack of gum."

Her breath caught. "Don't be long," she said.

"I'll talk to you after I land this afternoon." It was all I could do to add the words, "I love you." I heard her say, "I love you, too." I was glad nobody could see my eyes.

In a mirror near the capsule window, I could see the blockhouse and back across the Cape. The periscope gave me a view out over the Atlantic. It was turning into a fine day. I felt a little bit like the way I had felt going into combat. There you are, ready to go; you know all the procedures, and there's nothing left to do but just do it. People have always asked if I was afraid. I wasn't. Constructive apprehension is more like it. I was keyed up and alert to everything that was going on, and I had full knowledge of the situation—the best antidote to fear. Besides, this was the fourth time I had suited up, and I still had trouble believing I would actually take off.

Pipes whined and creaked below me; the booster shook and thumped when the crew gimballed the engines. I clearly was sitting on a huge, complex machine. We had joked that we were riding into space on a collection of parts supplied by the lowest bidder on a government contract, and I could hear them all.

At T minus thirty-five minutes, I heard the order to top off the lox tanks. Instantly the voices in my headset vibrated with a new excitement. We'd never gotten this far before. Topping off the lox tanks was a landmark in the countdown. The crew had begun to catch "go fever."

There was a hold at twenty-two minutes when a lox

valve stuck, and another at six minutes to solve an electrical power failure at the tracking station in Bermuda. Then the minutes dwindled into seconds.

At eighteen seconds the countdown switched to automatic, and I thought for the first time that it was going to happen. At four seconds I felt rather than heard the rocket engines stir to life sixty-five feet below me. The hold-down clamps released with a thud. The count reached zero at 9:47 A.M.

My earphones didn't carry Scott's parting message: "Godspeed, John Glenn." Tom O'Malley, General Dynamics' test director, added, "May the good Lord ride with you all the way."

Liftoff was slow. The Atlas's 367,000 pounds of thrust were barely enough to overcome its 125-ton weight. I wasn't really off until the forty-two-inch umbilical cord that took electrical connections to the base of the rocket pulled loose. That was my last connection with Earth. It took the two boosters and the sustainer engine three seconds of fire and thunder to lift the thing that far. From where I sat the rise seemed ponderous and stately, as if the rocket were an elephant trying to become a ballerina. Then the mission elapsed-time clock on the cockpit panel ticked into life and I could report, "The clock is operating. We're under way."

I could hardly believe it. Finally!

The rocket rolled and headed slightly north of east. At thirteen seconds I felt a little shudder. "A little bumpy along about here," I reported. The G forces started to build up. The engines burned fuel at an enormous rate, one ton a second, more in the first minute than a jet airliner flying coast to coast, and as the fuel was consumed the rocket grew lighter and rose faster. At forty-eight seconds I began to feel the vibration associated with high Q, the worst seconds of aerodynamic stress, when the capsule was pushing through air resistance amounting to almost a thousand pounds per square foot. The shaking got worse, then smoothed out at 1:12, and I felt the relief of knowing I was through max Q, the part of the launch where the rocket was most likely to blow.

At 2:09 the booster engines cut off and fell away. I was forty miles high and forty-five miles from the Cape. The rocket pitched forward for the few seconds it took for the escape tower's jettison rocket to fire, taking the half-ton tower away from the capsule. The G forces fell to just over one. Then the Atlas pitched up again and, driven by the sustainer engine and the two smaller vernier engines, which made course corrections, resumed its acceleration toward a top speed of 17,545 miles per hour in the ever-thinning air. Another hurdle passed. Another instant of relief.

Pilots gear their moments of greatest attention to the times when flight conditions change. When you get through them, you're glad for a fraction of a second, and then you think about the next thing you have to do.

The Gs built again, pushing me back into the couch. The sky looked dark outside the window. Following the flight plan, I repeated the fuel, oxygen, cabin pressure, and battery readings from the dials in front of me in the tiny cabin. The arc of the flight was taking me out over Bermuda. "Cape is go and I am go. Capsule is in good shape," I reported.

"Roger. Twenty seconds to SECO." That was Al Shepard on the capsule communicator's microphone at

mission control, warning me that the next crucial moment—sustainer engine cutoff—was seconds away.

Five minutes into the flight, if all went well, I would
achieve orbital speed, hit zero G, and, if the angle of as-
cent was right, be inserted into orbit at a height of about a
hundred miles. The sustainer and vernier engines would
cut off, the capsule-to-rocket clamp would release, and
the posigrade rockets would fire to separate *Friendship 7*
from the Atlas.

It happened as programmed. The weight and fuel tolerances were so tight that the engines had less than three seconds' worth of fuel remaining when I hit that keyhole in the
sky. Suddenly I was no longer pushed back against the seat
but had a momentary sensation of tumbling forward.

"Zero G and I feel fine," I said exultantly. "Capsule is
turning around." Through the window, I could see the
curve of Earth and its thin film of atmosphere. "Oh," I ex-
claimed, "that view is tremendous!"

The capsule continued to turn until it reached its normal orbital attitude, blunt end forward. It was flying east,
and I looked back to the west. There was the spent tube of
the Atlas making slow pirouettes behind me, sunlight
glinting from its metal skin. It was beautiful, too.

Al's voice came in my earphones. "Roger, Seven. You
have a go, at least seven orbits."

That was the best possible news. I was higher than
space flight when I heard that. The mission was planned
for three orbits, but it meant that I could go for at least
seven if I had to. The first set of hurdles was behind me. I
loosened the shoulder straps and seat belt that held me to
the couch, and prepared to go to work.

The capsule was pitched thirty-four degrees from

horizontal in its normal orbital attitude, so I could see back across the ocean to the western horizon. The periscope had automatically deployed and gave me a view to the east in the direction of the capsule's flight. The worldwide tracking network switched into gear. I talked to Gus, who was the capsule communicator, or capcom, at the Bermuda station. "This is very comfortable at zero G. I have nothing but a very fine feeling. It just feels very normal and very good."

Over the Canary Islands, almost to the west coast of Africa, I could still see the Atlas turning behind me. It was a mile away now, and slightly below me, losing ground because I was in a slightly higher orbit. I did a quick check of the capsule's attitude controls in case I had to make an emergency reentry. Pitch, roll, and yaw were primarily governed by an automatic system in which gyroscopes and sensors sent electrical signals to eighteen one- and five-pound hydrogen peroxide thrusters arrayed around the capsule. In "fly-by-wire" mode, I could use the three-axis control stick to override the automatic system using its same electrical connections. A fully manual system provided redundancy in a variety of attitude control modes. All three systems worked perfectly.

Friendship 7 crossed the African coast twelve minutes after liftoff, a fast transatlantic flight. I reached for the equipment pouch fixed just under the hatch. It used a new invention, a system of nylon hooks and loops called Velcro. I opened the pouch and a toy mouse floated into my vision. It was gray felt, with pink ears and a long tail that was tied to keep it from floating out of reach. I laughed; the mouse was Al's joke, a reference to one of comedian

Bill Dana's characters, who always felt sorry for the experimental mice that had gone into space in rocket nose cones.

I reached around the mouse and took out the Minolta camera. Floating under my loosened straps, I found that I had adapted to weightlessness immediately. When I needed both hands, I just let go of the camera and it floated there in front of me. I didn't have to think about it. It felt natural.

Telemetry was sending signals to the ground about my condition and the condition of the capsule. The capcom at the Canary Islands station asked for a blood pressure check, and I pumped the cuff on my left arm. The EKG and biosensors were sending signals about my heartbeat, pulse, and respiration, and the ever-present rectal thermometer was reporting my body temperature. At 18:41 I reported, "Have a beautiful view of the African coast, both in the scope and out the window. Out the window is the best view by far."

"Your medical status is green," Canary capcom reported. I asked for my blood pressure. The capcom reported back that it was 120 over 80, normal.

I took pictures of clouds over the Canaries. At twenty-one minutes, over the Sahara Desert, I aimed the camera at massive dust storms swirling the desert sand.

The tracking station at Kano, Nigeria, came on. I was forty seconds behind on my checklist of tasks, and went to fly-by-wire to check the capsule's yaw control again. The thrusters moved the capsule easily, and I reported at 26:34, "Attitudes all well within limits. I have no problem holding attitude with fly-by-wire at all. Very easy."

Over Zanzibar off the East African coast, the site of

the fourth tracking station, I pulled thirty times on the bungee cord attached below the control panel. I had done this on the ground, and my reaction was the same: It made me tired and increased my heart rate temporarily. I pumped the blood pressure cuff again for the flight surgeon on the ground. I read the vision chart over the instrument panel with no problem, countering the doctors' fear that the eyeballs would change shape in weightlessness and impair vision. Head movements caused no sensation, indicating that zero G didn't attack the balance mechanism of the inner ear. I could reach and easily touch any spot I wanted to, another test of the response to weightlessness. The ease of the adjustment continued to surprise me.

The Zanzibar flight surgeon reported that my blood pressure and pulse had returned to normal after my exercise with the bungee cord. "Everything on the dials indicates excellent aeromedical status," he said. This was what we had expected from doing similar tests on the procedures trainer.

Flying backward over the Indian Ocean, I began to fly out of daylight. I was now about forty minutes into the flight, nearing the 150-mile apogee, the highest point, of my orbital track. Moving away from the sun at 17,500 miles an hour—almost eighteen times Earth's rotational speed—sped the sunset.

This was something I had been looking forward to, a sunset in space. All my life I have remembered particularly beautiful sunrises or sunsets; in the Pacific islands in World War II; the glow in the haze layer in northern China; the two thunderheads out over the Atlantic with the sun silhouetting them the morning of Gus's launch. I've mentally collected them, as an art collector remembers visits to a gallery full of Picassos, Michelangelos, or Rembrandts. Wonderful as man-made art may be, it cannot compare in my mind to sunsets and sunrises, God's masterpieces. Here on Earth we see the beautiful reds, oranges, and yellows with a luminous quality that no film can fully capture. What would it be like in space?

It was even more spectacular than I imagined, and different in that the sunlight coming through the prism of Earth's atmosphere seemed to break out the whole spectrum, not just the colors at the red end but the greens, blues, indigos, and violets at the other. It made *spectacular* an understatement for the few seconds' view. From my orbiting front porch, the setting sun that would have lingered during a long earthly twilight sank eighteen times as fast. The sun was fully round and as white as a brilliant arc light, and then it swiftly disappeared and seemed to melt into a long thin line of rainbow-brilliant radiance along the curve of the horizon.

I added my first sunset from space to my collection.

I reported to the capcom aboard the ship, the *Ocean Sentry,* in the Indian Ocean that was my fifth tracking link, "The sunset was beautiful. I still have a brilliant blue band clear across the horizon, almost covering the whole window.

"The sky above is absolutely black, completely black. I can see stars up above."

Flying on, I could see the night horizon, the roundness of the darkened Earth, and the light of the moon on the clouds below. I needed the periscope to see the moon coming up behind me. I began to search the sky for constellations.

Gordo Cooper's familiar voice came over the headset as Friendship 7 neared Australia. He was the capcom at the station at Muchea, on the west coast just north of Perth. "That sure was a short day," I told him.

"Say again, *Friendship Seven*."

"That was about the shortest day I've ever run into."

"Kinda passes rapidly, huh?"

"Yes, sir."

I spotted the Pleiades, a cluster of seven stars. Gordo asked me for a blood pressure reading, and I pumped the cuff again. He told me to look for lights, and I reported, "I can see the outline of a town, and a very bright light just to the south of it." The elapsed-time clock read 54:39. It was midnight on the west coast of Australia.

"Perth and Rockingham, you're seeing there," Gordo said.

"Roger. The lights show up very well, and thank everybody for turning them on, will you?"

"We sure will, John."

The capcom at Woomera, in south-central Australia, radioed that my blood pressure was 126 over 90. I radioed back that I still felt fine, with no vision problems and no nausea or vertigo from the head movements I made periodically.

The experiments continued. Over the next tracking station, on a tiny coral atoll called Canton Island, midway between Australia and Hawaii, I lifted the visor of my helmet and ate for the first time, squeezing some applesauce from a toothpastelike tube into my mouth to see if weightlessness interfered with swallowing. It didn't.

It was all so new. An hour and fourteen minutes into the flight, I was approaching day again. I didn't have time to reflect on the magnitude of my experience, only to record its components as I reeled off the readings and performed the tests. The capcom on Canton Island helped me put it in perspective after I reported seeing through the periscope "the brilliant blue horizon coming up behind me; approaching sunrise."

"Roger, *Friendship Seven*. You are very lucky."

"You're right. Man, this is beautiful."

The sun rose as quickly as it had set. Suddenly there it was, a brilliant red in my view through the periscope. It was blinding, and I added a dark filter to the clear lens so I could watch it. Suddenly I saw around the capsule a huge field of particles that looked like tiny yellow stars. They seemed to travel with the capsule, but more slowly. There were thousands of them, like swirling fireflies. I talked into the cockpit recorder about this mysterious phenomenon as I flew out of range of Canton Island and into a dead zone before the station at Guaymas, Mexico, on the Gulf of California, picked me up. We thought we had foreseen everything, but this was entirely new. I tried to describe them again, but Guaymas seemed interested only in giving me the retro sequence time, the precise moment the capsule's retro-rockets would have to be fired in case I had to come down after one orbit.

Changing film in the camera, I discovered a pitfall of weightlessness when I inadvertently batted a canister of film out of sight behind the instrument panel. I waited a second for it to drop into view, and then realized that it wouldn't.

I was an hour and a half into the flight, and in range of the station at Point Arguello, California, where Wally Schirra was acting as capcom. I had just picked him up and was looking for a sight of land beneath the clouds when the capsule drifted out of yaw limits about twenty degrees to the right. One of the large thrusters kicked it back. It swung to the left until it triggered the opposite large thruster, which brought it back to the right again. I went to fly-by-wire and oriented the capsule manually.

The "fireflies" diminished in number as I flew east into brighter sunlight. I switched back to automatic attitude control. The capsule swung to the right again, and I switched back to manual. I picked up Al at the Cape and gave him my diagnosis: The one-pound thruster to correct rightward drift was out, so the drift continued until the five-pound thruster activated, and it pushed the capsule too far into left yaw, activating the larger thruster there. The thrusters were setting up a back-and-forth cycle that, if it persisted, would diminish their fuel supply and maybe jeopardize the mission.

"Roger, Seven, we concur. Recommending you remain fly-by-wire."

"Roger. Remaining fly-by-wire."

Al said that President Kennedy would be talking to me by way of a radio hookup, but it didn't come through and Al asked for my detailed thirty-minute report instead. I reported at 1:36:54 that controlling the capsule manually was smooth and easy, and the fuses and switches were all normal. I paused to ask about the presidential hookup. "Are we in communication yet? Over."

"Say again, Seven."

"Roger. I'll be out of communication fairly soon. I thought if the other call was in, I would stop the check. Over."

"Not as yet. We'll get you next time."

"Roger. Continuing report." I ran through conditions in the cabin and added, "Only really unusual thing so far besides ASCS [the automatic attitude control] trouble were the little particles, luminous particles around the capsule, just thousands of them right at sunrise over the Pacific. Over."

"Roger, Seven, we have all that. Looks like you're in good shape. Remain on fly-by-wire for the moment."

As the second orbit began, I thought I could see a long wake from a recovery ship in the Atlantic. One of the tracking stations was aboard the ship *Rose Knot,* off the West African coast at the equator. I moved into its range and reported a reversal of the thruster problem. Now I seemed to have no low right thrust in yaw, to correct left-ward drift. I performed a set maneuver, turning the capsule 180 degrees in yaw so that I was flying facing forward. "I like this attitude very much, so you can see where you're going," I radioed. I also reported seeing a loose bolt floating inside the periscope.

I passed the two-hour mark of the flight over Africa, with the capsule back in its original attitude. The second sunset was as brilliant as the first, the light again departing in a band of rainbow colors that extended on each side of the sunfall. Over Zanzibar, my eyeballs still held their shape. I reported, "I have no problem reading the charts, no problem with astigmatism at all. I am having no trouble at all holding attitudes, either. I'm still on fly-by-wire."

The *Coastal Sentry,* in the Indian Ocean, relayed a strange message from mission control. "Keep your landing bag switch in off position. Landing bag switch in off position. Over."

I glanced at the switch. It was off.

I returned to ASCS to see if the system was working. But now the capsule began to have pitch and roll as well as yaw problems in its automatic setting. The gyroscope-governed instruments showed the capsule was flying in its proper attitude, but what my eyes told me disagreed. The Indian Ocean capcom asked if I had noticed any constellations yet.

"This is *Friendship Seven*. Negative. I have some problems here with ASCS. My attitudes are not matching what I see out the window. I've been paying pretty close attention to that. I've not been identifying stars."

The ASCS fuel supply was down to 60 percent, so I cut it off and started flying manually. Gordo, in Muchea, asked me to confirm that the landing bag switch was off.

"That is affirmative. Landing bag switch is in the center off position."

"You haven't had any banging noises or anything of this type at higher rates?" He meant the rate of movement in roll, pitch, or yaw.

"Negative."

"They wanted this answer."

I flew on, feeling no vertigo or nausea or other ill effects from weightlessness, being able to read the same lines on the eye chart I could at the beginning. I pumped up the blood pressure cuff for another check and gave the readings in the regular half-hour reports. Flying the capsule with the one-stick hand controller was taking most of my attention. The second dawn produced another flurry of the luminescent particles. "They're all over the sky," I reported. "Way out I can see them, as far as I can see in each direction, almost."

The Canton Island capcom ignored the particles and

asked me to report any sensations I was feeling from weightlessness. Then came an unprompted transmission. "We also have no indication that your landing bag might be deployed. Over."

I had a prickle of suspicion. "Roger. Did someone report landing bag could be down? Over."

"Negative. We have a request to monitor this and to ask if you heard any flapping when you had high capsule rates."

It suddenly made sense. They were trying to figure out where the particles had come from. I was convinced they weren't coming from the capsule. They were all over the sky.

Daylight again. I had caged and reset the gyros during the night, and did it again in the light, but they were still off. I reported to the capcom in Hawaii that the instruments indicated a twenty-degree right roll when I was lined up with the horizon.

"Do you consider yourself go for the next orbit?"

"That is affirmative. I am go for the next orbit." There was no question in my mind about that. I could control the capsule easily, and I was confident that even with faulty gyroscopes I could align the capsule for its proper retrofire angle by using the stars and the horizon.

I flew over the Cape into the third orbit. The gyros seemed to have corrected themselves. Al radioed a recommendation that I allow the capsule to drift on manual control to conserve fuel.

The sky was clear over the Atlantic. Gus came on from Bermuda and I radioed, "I have the Cape in sight down there. It looks real fine from up here."

"Rog. Rog."

"As you know."

"Yea, verily, Sonny."

I could see not only the Cape, but the entire state of Florida. The eastern seaboard was bathed in sunshine, and I could see as far back as the Mississippi Delta. It was also clear over the recovery area to the south. "Looks like we'll have no problem on recovery," I said.

"Very good. We'll see you in Grand Turk."

Gus faded as I let the capsule drift around again 180 degrees, so that I was facing forward for the second time. It was more satisfying, and felt more like real flying. I still felt good physically, with none of the suspected ill effects. When I turned the capsule back to orbit attitude, the problem with the gyros reappeared, indicating more pitch, roll, and yaw than my view of the horizon indicated. The Zanzibar capcom asked why.

"That's a good question. I wish I knew, too."

I saw my third sunset of the day, and flew over clouds with lightning pulsing and rippling inside them. The lightning flashes looked like lightbulbs pulsing inside a veil of cotton gauze. Over the Indian Ocean, I went back to full manual control because the automatic with manual backup was using too much of the thrusters' supply of fuel. There had to be enough left when the time came to achieve the proper reentry attitude. I pitched the capsule up for a look at the night stars. The constellation Orion was right in the middle of the window, and I could hold my attitude by watching it.

Over Muchea, approaching four hours since liftoff, I told Gordo, "I want you to send a message to the commandant, U.S. Marine Corps, Washington. Tell him I have my four hours required flight time in for the month, and request flight chit be established for me. Over."

"Roger. Will do. Think they'll pay it?"

"I don't know. Gonna find out."

"Roger. Is this flying time or rocket time?"

"Lighter than air, buddy." Gordo would appreciate that. He and Deke had led the charge for getting us some flying time while we were training.

I turned the capsule around again so I could face the sunrise. The light revealed a new cloud of the bright particles, and I was still convinced they weren't coming from the capsule. The flight surgeon at Woomera suggested that I eat again. But I had been paying too much attention to the attitude control, and now I was concerned about lining up the spacecraft for reentry. This was the next set of hurdles, another crucial change in flight conditions that would require every ounce of my attention. I was over Hawaii when the capcom there said, "*Friendship Seven,* we have been reading an indication on the ground of segment fifty-one, which is landing bag deploy. We suggest this is an erroneous signal. However, Cape would like you to check this by putting the landing bag switch in auto position and see if you get a light. Do you concur with this? Over."

Now, for the first time, I knew why they had been asking about the landing bag. They did think it might have been activated, meaning that the heat shield that would protect the capsule from the searing heat of reentry was unlatched. Nothing was flapping around. The package of retro-rockets that would slow the capsule for reentry was strapped over the heat shield. But it would jettison, and what then? If the heat shield dropped out of place, I could be incinerated on reentry, and this was the first confirmation of that possibility. I thought it over for a few seconds. If the green light came on, we'd know that the bag had accidentally deployed. But if it hadn't, and there was something wrong with the circuits, flipping the switch to automatic might create the disaster we had feared. "Okay," I reluctantly concurred, "if that's what they recommend, we'll go ahead and try it."

I reached up and flipped the switch to auto. No light. I quickly switched it back to off. They hadn't been trying to relate the particles to the landing bag at all.

"Roger, that's fine," the Hawaii capcom said. "In this case, we'll go ahead, and the reentry sequence will be normal."

The seconds ticked down toward the retro firing sequence. I passed out of contact with Hawaii and into Wally Schirra's range at Point Arguello. I was flying backward again, the blunt end of the capsule facing forward, manually backing up the erratic automatic system. The retro warning light came on. A few seconds before the rockets fired, Wally said, "John, leave your retro pack on through your pass over Texas. Do you read?"

"Roger."

I moved the hand controller and brought the capsule to the proper attitude. The first retro-rocket fired on time at 4:33:07. Every second off would make a five-mile difference in the landing spot. The braking effect on the capsule was dramatic. "It feels like I'm going back toward Hawaii," I radioed.

"Don't do that," Wally joked. "You want to go to the East Coast."

The second rocket fired five seconds later, the third

five seconds after that. They each fired for about twelve seconds, combining to slow the capsule about five hundred feet per second, a little over 330 miles per hour, not much but enough to drop it below orbital speed. Normally, the exhausted rocket package would be jettisoned to burn up as it fell into the atmosphere, but Wally repeated, "Keep your retro pack on until you pass Texas."

"That's affirmative."

"Pretty good-looking flight from what we've seen," Wally said.

"Roger. Everything went pretty good except for all this ASCS problem."

"It looked like your attitude held pretty well. Did you have to back it up at all?"

"Oh, yes, quite a bit. Yeah, I had a lot of trouble with it."

"Good enough for government work from down here."

"Yes, sir, it looks good, Wally. We'll see you back East."

"Rog."

I gave a fast readout of the gauges and asked Wally, "Do you have a time for going to jettison retro? Over."

"Texas will give you that message. Over."

Wally and I kept chatting like a couple of tourists exchanging travel notes. "This is *Friendship Seven*. Can see El Centro and the Imperial Valley. Salton Sea very clear."

"It should be pretty green. We've had a lot of rain down here."

The automatic yaw control kept banging the capsule back and forth, so I switched back to manual in all three axes. The capcom at Corpus Christi, Texas, came on and said, "We are recommending that you leave the retro package on through the entire reentry. This means you will have to override the point-zero-five-G switch [this sensed atmospheric resistance and started the capsule's reentry program], which is expected to occur at oh four forty-three fifty-eight. This also means that you will have to manually retract the scope. Do you read?"

The mission clock read 4:38:47. My suspicions flamed back into life. There was only one reason to leave the retro pack on, and that was because they still thought the heat shield could be loose. But still, nobody would say so.

"This is *Friendship Seven*. What is the reason for this? Do you have any reason? Over."

"Not at this time. This is the judgment of Cape flight."

"Roger. Say again your instructions, please. Over."

The capcom ran it through again.

"Roger, understand. I will have to make a manual point-zero-five-G entry when it occurs, and bring the scope in manually."

Metal straps hugged the retro pack against the heat shield. Without jettisoning the pack, the straps and then the pack would burn up as the capsule plunged into the friction of the atmosphere. I guessed they thought that by the time that happened, the force of the thickening air would hold the heat shield against the capsule. If it didn't, *Friendship* 7 and I would burn to nothing. I knew this without anybody's telling me, but I was irritated by the cat-and-mouse game they were playing with the information. There was nothing to do but line up the capsule for reentry.

I picked up Al's voice from the Cape. He said, "Recommend you go to reentry attitude and retract the scope manually at this time."

"Roger." I started winding in the periscope.

"While you're doing that, we are not sure whether or not your landing bag has deployed. We feel it is possible to reenter with the retro package on. We see no difficulty at this time in that type of reentry. Over."

"Roger, understand."

I was now at the upper limits of the atmosphere. I went to full manual control, in addition to the autopilot, so I could use both sets of jets for attitude control. "This is *Friendship Seven*. Going to fly-by-wire. I'm down to about fifteen percent [fuel] on manual."

"Roger. You're going to use fly-by-wire for reentry, and we recommend that you do the best you can to keep a zero angle during reentry. Over."

I entered the last set of hurdles at an altitude of about fifty-five miles. All the attitude indicators were good. I moved the controller to roll the capsule into a slow spin of ten degrees a second that, like a rifle bullet, would hold it on its flight path. Heat began to build up at the leading edge of a shock wave created by the capsule's rush into the thickening air. The heat shield would ablate, or melt, as it carried heat away, at temperatures of three thousand degrees Fahrenheit, but at the point of the shock wave, four feet from my back, the heat would reach ninety-five hundred degrees, a little less than the surface of the sun. And the ionized envelope of heat would black out communications as I passed into the atmosphere.

Al said, "We recommend that you . . ." That was the last I heard.

There was a thump as the retro pack straps gave way. I thought the pack had jettisoned. A piece of steel strap fell against the window, clung for a moment, and burned away.

"This is *Friendship Seven*. I think the pack just let go."

An orange glow built up and grew brighter. I anticipated the heat at my back. I felt it the same way you feel it when somebody comes up behind you and starts to tap you on the shoulder, but then doesn't. Flaming pieces of something started streaming past the window. I feared it was the heat shield.

Every nerve fiber was attuned to heat along my spine; I kept wondering, "Is that it? Do I feel it?" But just sitting there wouldn't do any good. I kept moving the hand controller to damp out the capsule's oscillations. The rapid slowing brought a buildup of G forces. I strained against almost eight Gs to keep moving the controller. Through the window, I saw the glow intensify to a bright orange. Overhead, the sky was black. The fiery glow wrapped around the capsule, with a circle the color of a lemon drop in the center of its wake

"This is *Friendship Seven*. A real fireball outside."

I knew I was in the communications blackout zone. Nobody could hear me and I couldn't hear anything the Cape was saying. I actually welcomed the silence for a change. Nobody was chipping at me. There was nothing they could do from the ground anyway. Every half minute or so, I checked to see if I was through it.

"Hello, Cape. *Friendship Seven*. Over.

"Hello, Cape. *Friendship Seven*. Do you receive? Over."

I was working hard to damp out the control motions, with one eye outside all the time. The orange glow started to fade, and I knew I was through the worst of the heat. Al's voice came back into my headset. ". . . how do you read? Over."

"Loud and clear. How me?"

"Roger, reading you loud and clear. How are you doing?"

"Oh, pretty good."

The Gs fell off as my rate of descent slowed. I heard Al again. "Seven, this is Cape. What's your general condition? Are you feeling pretty well?"

"My condition is good, but that was a real fireball, boy. I had great chunks of that retro pack breaking off all the way through."

At twelve miles of altitude I had slowed to nearly subsonic speed. Now, as I passed from fifty-five thousand to forty-five thousand feet, the capsule was rocking and oscillating wildly and the hand controller had no effect. I was out of fuel. Above me through the window I saw the twisting corkscrew contrail of my path. I was ready to trigger the drogue parachute to stabilize the capsule, but it came out on its own at twenty-eight thousand feet. I opened snorkels to bring air into the cabin. The huge main chute blossomed above me at ten thousand feet. It was a beautiful sight. I descended at forty feet a second toward the Atlantic.

I flipped the landing bag deploy switch. The panel light glowed green, just the way it was supposed to.

CHAPTER 19

The capsule hit the water with a good solid thump, plunged down, submerging the window and periscope, and bobbed back up. I heard gurgling but found no trace of leaks. I shed my harness, unstowed the survival kit, and got ready to make an emergency exit just in case.

But I had landed within six miles of the USS *Noa,* and the destroyer was alongside in a matter of minutes. Even so, I got hotter waiting in the capsule for the recovery ship than I had coming through reentry. I felt the bump of the ship's hull, then the capsule being lifted, swung, and lowered onto the deck. I radioed the ship's bridge to clear the area around the side hatch, and when I got the all-clear, I hit the firing pin and blasted the hatch open. Hands reached to help me out. "It was hot in there," I said as I stepped out onto the deck. Somebody handed me a glass of iced tea.

It was the afternoon of the same day. My flight had lasted just four hours and fifty-six minutes. But I had seen three sunsets and three dawns, flying from one day into the next and back again. Nothing felt the same.

I looked back at *Friendship 7*. The heat of reentry had discolored the capsule and scorched the stenciled flag and the lettering of *United States* and *Friendship 7* on its sides. A dim film of some kind covered the window. *Friendship 7* had passed a test as severe as any combat, and I felt an affection for the cramped and tiny spacecraft, as any pilot would for a warplane that had brought him safely through enemy fire.

The *Noa,* like all the ships in the recovery zone, had a kit that included a change of clothes and toiletries. I was taken to the captain's quarters, where two flight surgeons helped me struggle out of the pressure suit and its underlining, and remove the biosensors and urine collection device. I didn't know until I got the suit off that the hatch's firing ring had kicked back and barked my knuckles, my only injury from the trip. My urine bag was full. A NASA photographer was taking pictures. After a shower, I stepped on a scale; I had lost five pounds since liftoff. President Kennedy called via radio-telephone with congratulations. He had already made a statement to the nation, in which he created a new analogy for the exploration of space: "This is the new ocean, and I believe the United States must sail on it and be in a position second to none."

I called Annie at home in Arlington. She knew I was safe. She had had three televisions set up in the living room, and was watching with Dave and Lyn and the neighbors with the curtains drawn against the clamor of the news crews outside on the lawn. Even so, she sobbed with relief. I didn't know then that Scott had called to prepare her in case the heat shield was loose. He had told her I might not make it back. "I waited for you to come back on the radio," she said. "I know it was only five minutes. But it seemed like five years." Hearing my voice speaking directly to her brought first tears, then audible happiness.

After putting on a jumpsuit and high-top sneakers, I found a quiet spot on deck and started answering into a tape recorder the questions on the two-page shipboard debriefing form. The first question was, *What would you like to say first?*

The sun was getting low, and I said, "What can you say about a day in which you get to see four sunsets?"

Before much more time had passed, I got on the ship's loudspeaker and thanked the *Noa*'s crew. They had named me sailor of the month, and I endorsed the fifteen-dollar check to the ship's welfare fund. A helicopter hoisted me from the deck of the Noa in a sling and shuttled me to an aircraft carrier, the USS *Randolph,* where I met a larger reception committee. Doctors there took an EKG and a chest X ray, and I had a steak dinner. Then from the *Randolph* I flew copilot on a carrier transport that took me to Grand Turk Island for a more extensive medical exam and two days of debriefings.

At the debriefing sessions, I had the highest praise for the whole operation, the training, the way the team had come back from all the cancellations, and the mission itself—with one exception. They hadn't told me directly their fears about the heat shield, and I was really unhappy about that. A lot of people, doctors in particular, had the idea that you'd panic in such a situation. The truth was, they had no idea what would happen. None of us were panic-prone on the ground or in an airplane or in any of the things they put us through in training, including underwater egress from the capsule. But they thought we might panic once we were up in space and assumed it was better if we didn't know the worst possibilities.

I thought the astronaut ought to have all the information the people on the ground had, as soon as they had it, so he could deal with a problem if communication was lost. I was adamant about it. I said, "Don't ever leave a guy up there again without giving him all the information you have available. Otherwise, what's the point of having a manned program?"

It emerged that a battle had gone on in the control center over whether they should tell me. Deke and Chris Kraft had argued that they ought to tell me, and others had argued against it, but I didn't learn that until many years later. I don't know who had made the decision, but they changed the policy after that to establish that all information about the condition of the spacecraft was to be shared with the pilot.

I was describing the luminous particles I saw at each sunrise when George Ruff, the psychiatrist, broke up everyone in the debriefing by asking, "What did they say, John?" (The particles proved to be a short-lived phenomenon. The Soviets called them the Glenn effect; NASA learned from later flights that they were droplets of frozen water vapor from the capsule's heat exchanger system, but their fireflylike glow remains a mystery.)

George also had a catch-all question tagged on at the end of the standard form we filled out at the end of each day's training. It was, *Was there any unusual activity during this period?*

"No," I wrote, "just the normal day in space."

Grand Turk was an interlude, in which morning medical checks and debriefing sessions were followed by afternoons of play. Scott, as my backup, stuck with me as I had with Al and Gus after their flights. We went scuba diving and spearfishing, and Scott rescued a diver who had blacked out eighty feet down, giving him some of his own air as he brought him to the surface. There were no crowds, since the debriefing site was closed.

Annie had tried to give me some idea of the overwhelming public reaction to the flight. Shorty Powers had said there was a mood afoot for public celebration. But I was only faintly aware of the groundswell that was building.

Vice President Johnson flew to Grand Turk to accompany me back to Patrick Air Force Base, south of Cape Canaveral, on February 23. Annie, Dave, and Lyn were waiting, along with a crowd that sent up a loud cheer. I hugged Lyn and Dave, and then hugged and kissed Annie. She was gorgeous, wearing a choker necklace, an orchid corsage, and a pillbox hat. I couldn't hold back tears. Annie, most of all, had been at the back of my mind the entire flight.

Johnson told the crowd, "It's a great pleasure to welcome home a great pioneer of history." I quickly learned that his accolade was only a prelude to a tidal wave of attention.

The drive twenty miles to the Cape along A1A through Cocoa Beach turned into a parade that I found out later Henri Landwirth and city officials had helped organize. Annie and I, and Dave and Lyn, shared a car with the vice president. Each of the other astronauts and his family—except for Gordo, who was still in Australia—was in a separate car. Bands were playing. Thousands of people lined the route, waving and clapping. A sign over the road said, Welcome Back, John. Another said, Our Prayers Have Been Answered. Johnson saw a boy wearing a toy space helmet riding on his father's shoulders. He asked the driver to stop, and beckoned the man over. People surged around the car, asking for autographs. They were saying things like, "Thank you," and "God bless you." Dave and Lyn looked at each other, their mouths open and eyes wide and shining with astonishment.

I saw Leo DeOrsey along the road near the Holiday Inn. I stopped the parade and called him over. "Boy, am I glad to see you," he said. The $100,000 check he'd written to Annie wasn't going to be cashed.

When the motorcade approached the guard gate at the Cape, I took my NASA identification card out of my wallet. Johnson thought I was serious. He leaned over and gave me an amused look from underneath his eyebrows. "I think they know who you are out here," he said.

The motorcade went to the Cape landing strip where President Kennedy was arriving aboard Air Force One. He landed, and toured the NASA facilities, including Launch Complex 14 and Hangar S, with me and Annie and the kids. *Friendship 7* was on display outside Hangar S. Somebody told me a decision had been made to send it on an around-the-world tour—the "fourth orbit" of *Friendship 7*—and then to the Smithsonian for permanent display alongside the Wright brothers' first plane and Lindbergh's *Spirit of St. Louis*. The little flag I had carried for the

Smithsonian was being displaced by my twenty-four-hun-
dred-pound spacecraft.

The president awarded me and Bob Gilruth NASA's
Distinguished Service Medal. He talked about going to the
moon and said, "Our boosters may not be as large as some
others, but our men and women are."

That set off another great cheer. Afterward, two thousand people gathered at a reception and ate pieces of the
huge cake made to look like *Friendship* 7. Henri Landwirth
took me to one side and asked me how the cake tasted.

"It tastes fine, Henri," I said. "Why?"

A look that combined mischief and relief spread over
his face. "It's almost a month old," he said conspiratorially.
"I made it when everybody thought you were going up before. You wouldn't believe what I had to do to keep it
fresh."

"I didn't say it tasted fresh," I said.

Amazing reports kept coming in. Shorty Powers, who
had monitored the coverage, told me that in New York's
Grand Central Station, thousands of commuters had simply stopped on their way to work to watch the launch on a
large-screen television set in the main concourse. Subway
conductors told their passengers when the rocket took off,
and asked them to say a prayer. Walter Cronkite lost his an-
chorman's objectivity and said, "Go, baby, go." The whole
country apparently had held its collective breath as it sat
glued to television sets and radios.

I didn't at first try to analyze this outpouring.
Gradually it came to me through all the hoopla what had
happened. The flight was dramatic, but it wasn't just that;
Al's and Gus's flights had been dramatic, too. But the many

372 J O H N   G L E N N :   A   M E M O I R

delays produced the first suspense. When it finally got off, the flight sustained the drama for nearly five hours, all of which were the subject of live television and radio coverage. It had kept the nation on the edge of its seat. There had been the control problems, and then the question of the heat shield, which apparently the entire country knew about before I did, so the radio blackout brought on a riveting level of suspense. There was a national sigh of relief when I could be heard again, and another when I was safely down and aboard ship. Most important, there was the Cold War, and the sense the flight gave to the nation that we were back in the race and competing. We had put five times as many satellites in orbit as the Soviets, and three men into space to their two, but it took the successful end of a movie-length (and then some) drama to restore the nation's pride.

After the welcome-home ceremonies at the Cape and a news conference the next day in which I described my feeling at being inside the reentry fireball as "cautious apprehension," Annie, Dave, Lyn, and I spent the remainder of the weekend in seclusion on the naval base at Key West.

Having their company and no one else's for a change was a pleasure. We walked on the beach and swam, and I caught up on everything that had been going on at home and in school in the weeks leading up to the flight. But it wasn't a trip given entirely to leisure. The president had told me I had been invited to address a joint session of Congress the following Monday, after a parade in Washington.

I worked on my speech, and on Monday morning we

flew to the airport at West Palm Beach to meet the president for the trip back to Washington on *Air Force One*. Jackie was going to be staying in Palm Beach with Caroline and their infant son, John Jr. Jackie wanted Caroline to meet us, so they came out to the airport with the president to see us off.

We were on the plane, and the president boarded, and behind him came Jackie with little Caroline, holding her by the hand, and Jackie said, "Caroline, this is the astronaut who went around the Earth in the spaceship. This is Colonel Glenn."

Caroline looked at me, and then all around the plane. Finally she turned back to me, her face disappointed, and said in a quavering voice, "But where's the monkey?"

Somebody had typed my speech the night before, and as we flew north I asked President Kennedy to review it. He looked it over, said it was fine, and gave it back to me, and I put it in an inside pocket of my suit.

I had learned to like the president. He was well briefed on the space program, and talked about it with passion. He believed, as I did, that it was not just a scientific journey, but a source of inspiration that could motivate Americans to pursue great achievements in all fields.

I, in turn, was motivated by his example as a political leader. His vision set an inspiring example, and I saw that the Kennedy charisma could move millions to contribute to something I thought was vital—a democracy of energized participation in which people shared their talents with the nation and kept it improving and evolving.

Washington was cold and rainy. NASA officials and the astronauts and their families—except for Gordo, still not back from Australia—attended a brief reception at the

White House before we all piled into open cars for the parade to the Capitol. Thousands of people lined the streets despite the dismal winter weather. Annie and I rode, again with the vice president, wearing overcoats and gloves against the chill.

At the Capitol, the various requirements of protocol brought the senators into the House chamber, then ambassadors, ministers, and chargés d'affaires of foreign governments, then the Supreme Court, then Al, Gus, Deke, Scott, and Wally, then the Cabinet, and then Vice President Johnson as the Senate's presiding officer. At last I was escorted into the chamber. It was a moment of intense gravity to me. Through the standing applause, I was conscious that the opportunity to address a joint session of Congress was rare, a privilege reserved for royalty and heads of state. All the same, I felt at ease. I felt I was there not just as a test-pilot-turned-astronaut who happened to have made America's first orbital space flight, but as a patriotic American with something to say about the space program and its embodiment of our nation's persistent quest to expand our knowledge and press forward our frontiers.

I felt as natural standing at the podium in the great hall of our democracy as I had felt floating weightless around the globe six days before. The reverence I felt echoing the phrases of taps on Memorial Day at the cemetery in New Concord and the energizing civics lessons of my high-school teacher Harford Steele had not been lost on me.

"I am certainly glad to see that pride in our country and its accomplishments is not a thing of the past," I said. "I still get a hard-to-define feeling inside when the flag goes by. Today as we rode up Pennsylvania Avenue . . . I got this same feeling all over again. Let us hope that none of us ever loses it."

I introduced my parents and Annie's in the front row of the balcony. The senior Glenns and Castors were going to be staying overnight at our house when the festivities were over. Annie or I would have to remember to tuck away the bottles of vodka and gin we kept on hand for company, since our parents were hard-core teetotalers who frowned on even social drinking. I introduced Dave and Lyn, and then "the real rock in my family, my wife, Annie."

There was a lot of applause, and I was feeling very much in command of the situation when I reached inside my jacket and took out the speech I had prepared. I opened it as the round of clapping died to an expectant hush. I took a breath, looked down—and my heart stopped. I was staring not at page one of my speech but at page fifteen. At that moment I was closer to panic than I had been at any time in orbit.

I whipped off the page, and there, to my immense relief, was page one. The president just hadn't placed the last page on the bottom when he had finished reading.

In the speech (which now had all the pages in order, as I discovered by riffling through them every time there was so much as a brief round of applause) I addressed what I thought—and still do—was a key principle. The benefits of scientific inquiry, or any form of exploration, cannot always be known when the first steps are taken. "What benefits are we gaining from the money spent?" I asked rhetorically. "They are probably not even known to man today. But exploration and the pursuit of knowledge have always paid dividends in the long run—usually far greater than anything expected at the outset." I had in my mind a vision of pioneers setting out from the eastern seaboard of what would become the United States, not knowing what lay ahead but having the courage to make the investment and take the first step.

I wanted the Congress to have the courage to continue to support NASA's agenda and President Kennedy's bold and expensive requests for the space program. I closed by saying, "As our knowledge of the universe increases, may God grant us the wisdom and guidance to use it wisely."

More parades followed. Four million people, according to police estimates, turned out on a frigid March 1 in New York City for a ticker tape parade for all the astronauts. The city sanitation department swept up thirty-five hundred tons of paper. I spoke to the United Nations a day later. And on March 3 I returned home for a parade that even in tiny New Concord turned out seventy-five thousand people on another cold day.

All the while, mail was piling up at home. The Postal Service didn't put letters in the mailbox anymore. The deliverymen came out in special trucks and left letters by the bagful in the carport. Annie and I thought we could answer them all, and for a while we tried. It was impossible, and NASA took over the task with a special mail room at headquarters.

Gradually it began to dawn on me and Annie that our lives were on a trajectory as steep as the launch of *Friendship* 7. Only now it didn't seem to be in anyone's control. My speech to Congress had been well received.

Suddenly columnists such as Arthur Krock of *The New York Times* were speculating about my political potential. James Reston, also in the Times, had written that I embodied "the noblest human qualities." A group of Republicans in Nevada called on me to run for president. Jim Webb, at a news conference at which he announced the expansion of the astronaut corps, had to answer questions about my political plans. He predicted I would stay out of politics, and that was my intention. Nevertheless, obviously I could help persuade politicians to support the space program.

I told Jim and Bob Gilruth that I was devoted to the program, and wanted to stay with it as long as I could be useful. Two- and three-man missions were coming up as part of the Gemini and Apollo programs. I held out hope that, despite my age, I might be assigned another flight.

My age made me something of an example for middle-aged workers. The International Union of Electrical Workers said in an editorial in the union's publication that I had struck a blow against job discrimination due to age. Although I had passed "this magic milestone of maturity," I had shown that "a man did not become obsolete" at forty. I thought with amusement back to Project Bullet, when I had just turned thirty-six and the *Times* had placed me near the age limit for "piloting complicated pieces of machinery."

I found that the hardest thing during this time was to hang on to some semblance of perspective. Annie and I both felt buffeted by the winds of attention. We worked harder than ever to try to maintain some kind of normal family life. Our dinners by candlelight, with Dave and Lyn, all of us describing our days and talking among ourselves about what we had accomplished that day, grew in importance. I also, apart from the family, felt the need to hang on to something of the person I had been before the flight. I felt I was the same person I had always been, and what was so difficult was dealing with the notion people had that I was some kind of a hero.

The astronauts had made several visits to New York, and we always had a security detail of police detectives. I was curious about what their work was like, and my questions led to an invitation to make some rounds with them. On a trip to New York in April, I had the time to accept. *Life* had just run a story titled "The Toughest Block in the World." Probably some of the territory I had visited in Shanghai would have qualified, but somehow the magazine had decided that no place held a candle to New York's West Eighty-fourth Street between Columbus and Amsterdam Avenues for family fights, gun violence, robberies, and auto thefts. This sounded interesting to me, so I asked the chief of detectives if he could arrange a ride-along. He told me to show up at the Twentieth Precinct station house on West Eighty-second Street.

I turned up around eight at night, and before long I was out riding the streets of the Twentieth with a couple of young patrolmen, Kevin M. Hallinan and Henry Buck Jr. The precinct covered New York's Upper West Side from Central Park West to the Hudson River, between Sixty-fifth and Eighty-sixth Streets, and its reputation as a tough area seemed well deserved. I tagged along in my suit, but with a baseball cap and sunglasses, as the two cops responded to calls from burglaries to robberies; to domestic fights. We jumped across rooftops and climbed down fire escapes and up the dank stairwells of transient hotels. Nobody recognized me. I suppose they thought I was a detective. But I got one big shock on a domestic-violence call.

Hallinan explained on the way up the stairs of a West Eighty-third Street building how he and Buck dealt with fights between husbands and wives. "We each pick a partner and 'dance' with that partner," he said. "After a while they wear themselves out and calm down. Most of the time."

The husband and wife were both big people. Hallinan and Buck got them apart and started talking to them separately. Then Hallinan tapped me and pointed out a photograph of me from one of the newspapers, tacked to the kitchen wall. It was a picture of me on the *Noa* wearing nothing but the urine collection apparatus. Shorty Powers wasn't supposed to release any shots like that. I was about to get mad when I noticed that the tack that held the picture to the wall was placed through my head. Here was one place I definitely was not a hero, and I realized that the space program, for all its Cold War import and future scientific benefits, and for all the excitement it generated in many quarters, remained remote from people whose hard-pressed lives didn't allow them to appreciate it.

CHAPTER 20

That May 3, the Committee on Space Research of the International Council of Scientific Unions came to Washington for its third International Space Symposium. Gherman Titov and I were on the program of the Life Sciences session to present papers about our respective flights. The State Department had asked me to be his host.

Titov was a ruggedly handsome, small-framed man of about five feet four inches. He was quite young, only twenty-six. We got together first thing in the morning at the National Academy of Sciences, and Annie and I gave him and his wife, Tamara, a tourist's-eye view of Washington. He commented in the elevator on the way to the top of the Washington Monument that this was the first joint Soviet-American space venture. It was an inauspicious beginning, since fog interfered with the view.

As the day went on, we spent fifteen minutes in the White House with President Kennedy, and visited the Capitol, the Lincoln Memorial, the Tomb of the Unknowns at Arlington National Cemetery, and the Smithsonian, where Al's *Freedom 7* was on display. With the help of an interpreter, we talked informally about our flights, our training regimens, our reactions to weightlessness, and the as-yet-unexplained luminous particles that I had described as looking like fireflies. We speculated that they were frozen droplets of fuel from the thrusters. I spoke in specifics, and hoped he would respond with some of his own since we knew little about the Soviet space program. But he answered only with generalities and Soviet dogma.

We made our presentations in the State Department auditorium before an audience of scientists from thirty nations. Most of the scientists on hand wanted to explore the period of sickness and nausea he had experienced during his twenty-five-hour flight. A consensus developed that space sickness was a version of motion sickness, like seasickness, that a person can adapt to.

Titov was cordial but forceful and thoroughly indoctrinated. Charts and photographs had supplemented my presentation, but not his; he followed the Soviet line that disarmament would have to precede full sharing of information, since both sides used military rockets as launch vehicles. He also professed the official Soviet policy of atheism, as I learned when we fielded questions at the end of our presentations. Someone asked, "In Communism you don't believe there is a God. Did your space flight alter that?"

"Not at all," the cosmonaut said. "Only now there is proof for the Communist position. I went into space and didn't see God, so that must mean God does not exist."

"Did you see God in space, Colonel Glenn?" the questioner asked.

"I didn't expect to," I said. "The God I believe in isn't so small that I thought I would run into Him just a little bit above the atmosphere."

We attended a news conference later on. By then the State Department had breathed a sigh of relief. I had asked Titov and his wife if they wanted to come to Arlington the following evening to have a typical American dinner at home with me and Annie and the kids. He and Tamara thought that was a good idea, and said they'd check with the delegation head. But I hadn't run it through State, and when I mentioned it to one of our department escorts, he was aghast. He cited the likelihood of pro-Hungarian pickets and anti-Communist groups descending on our front yard. But Titov's handlers said they couldn't accept.

The next afternoon, however, Annie and I were attending a reception for the visiting group at the Soviet embassy. As we came off the end of the receiving line, the State Department man rushed up to me with the news that the Soviets had reversed themselves and accepted our invitation.

I turned to Annie and said, "The Russians are coming."

At that point it was late in the afternoon, and we were going to have to scramble to get dinner ready not only for the Titovs but for the head of their delegation and one of their chief scientists.

Al and Louise Shepard were at the reception, and I huddled with Al and said, "Titov's coming to dinner after all. I don't have anything ready. Can you bring them out to the house? Stall things, make a mistake, take a wrong turn to Baltimore or something. Give us as much time as you can."

Al grinned and said he thought he could figure a way
to get lost on the way to Arlington.

Annie and I headed home with a National Park Police
motorcycle escort, sirens screaming. When we got there,
we sent one of the policemen to a 7-Eleven for some
frozen peas and other dinner items. Another one called the
Arlington police to arrange security in case there were
pickets. We started canvassing the neighborhood for
steaks.

Between the Millers, the Browns, and some other
neighbors, we corralled enough steaks. The policeman returned from the 7-Eleven, and we had the makings of dinner. I moved our grill and the Millers' out into the middle
of the carport, lit the charcoal, and put a fan on the flames
to speed the process. I draped a towel around my neck and
brought out the steak sauce. I wanted to have everything
ready when they got there.

The steaks were sizzling when, suddenly, my fire-
building efforts went awry. The fan-aided flames had gotten so hot they burned through one grill's center post and
dumped the steaks onto the charcoal. The fat blazed up, I
frantically dumped water on the fire, and smoke and steam
were billowing from under the carport roof when I looked
up and saw the limousines arriving. I rescued the steaks,
brushed them off, wiped my hands on the towel and put
the best face on the situation as the Titovs and Al and
Louise led the delegation up the driveway. Al was laughing
up a storm. Bringing up the rear were a bunch of people
who looked like the Soviet version of the Secret Service,
and some reporters and photographers.

"I don't know how you do it in Moscow," I said to

Titov, "but sometimes over here you have to work for your dinner. Come on and give me a hand."

He took his coat off and joined me. When he and Al and I finished cooking the steaks on the second grill, we carried them into the kitchen. Annie and Louise and Tamara were at work, Tamara grinding peanuts from a Planter's can over the top of the salad. All three of them had their shoes off.

The steaks survived the disaster and were delicious. Afterward, Titov, Al, and I headed off for a joint television appearance. Titov later told me that evening was the most fun he and Tamara had had in Washington. My first stab at international diplomacy had been harried but effective.

Project Mercury returned to space on May 24, with Scott's three-orbit flight in *Aurora 7*. Deke had been scheduled to make the flight, but the doctors had grounded him after detecting a slight heart murmur. The country—and Rene and their children—had some bad moments at the end of a productive flight, when Scott's capsule overshot the splashdown area by 250 miles and he was lost to radar and out of radio range. A plane spotted him and the capsule thirty-six minutes after he splashed down, but nobody at Mercury Control knew for even longer if he was alive or dead. As the minutes ticked by, I kept in touch with Rene by telephone to update her. Meanwhile, out in the Atlantic, Scott had climbed out of the capsule and was lolling on his life raft talking to two frogmen who had parachuted into the water to assist him. It was three hours before he finally was

picked up. We shared a heartfelt hug when he reached
Grand Turk for his debriefing.

The summer of 1962 was a crossroads in many ways.
NASA's Space Task Group had announced plans to move
from Langley to a new Manned Spacecraft Center south of
Houston, Texas. While we were preparing for our move, I
was experiencing a growing friendship with Bobby
Kennedy.

While I liked President Kennedy and felt comfortable
with him, I found a deeper level of compatibility with
Bobby and his wife, Ethel. They invited us to their home,
Hickory Hill, which was only a short drive from where we
lived in Arlington.

Hickory Hill, with its grounds, qualified as a country
estate, but there was nothing sedate or formal about it
with Bobby and Ethel as hosts. Bobby collected people,
and in many cases they seemed to be people who had performed feats of great daring or physical prowess.
Roosevelt Grier, the football player, was part of his circle.
So was Jim Whittaker, the first American to reach the
summit of Mount Everest, in 1957. Andy Williams, whose
singing I admired, the humorist Art Buchwald, the
Olympic gold medal decathlete Rafer Johnson—all would
find a place in Bobby's orbit. It was a lively group of people who had had interesting experiences and a lot to relate
and talk about. I guess he thought I qualified on similar
grounds.

Ethel enjoyed giving pool and garden parties, and this
was the occasion the first time Annie and I were at

Hickory Hill. Ethel had set up an arched bridge across the pool, with a table for two set in the middle over the water. This, it developed, was for me and her. Ethel was a consummate hostess, and took delight in pairing people. As the evening went on, people started bouncing on the bridge. A lot of people ended up in the pool, including Ethel, in her party clothes, and Arthur Schlesinger, who came up with his glasses askew on his nose. I managed to stay dry.

Ethel was full of life and conversation and ideas. She frequently paired the two of us, and Annie and Bobby consequently spent a lot of time talking together. While the family image was of people in constant motion, Bobby was quiet and gentle with Annie. He was absolutely patient when her stutter intruded in her conversation. The four of us became good friends.

Later that summer, after the pool party, they invited us, and Dave and Lyn, for a weekend at the fabled compound at Hyannis Port. The whole family was there: Bobby and Ethel; the president and Jackie; Teddy, who was then running for the Senate from Massachusetts; old Joe Kennedy, who was in a wheelchair after a paralyzing stroke. We spent a day water-skiing in Lewis Bay; I skied with Jackie and Bobby, but the real stars were Dave and Ethel, who slalomed together and managed several crisscross maneuvers without falling in the drink, as Jackie and I both had. We also talked aboard the *Marlin*, the fifty-two-foot family sailing yacht. I had to defend the *Life* contract to the president, who was under mild pressure to cancel it. I explained that the astronauts weren't funneling scientific information about the space flights exclusively to *Life*, only personal family information. Kennedy agreed.

Bobby's political views were to the left of mine, but they still were similar in many ways. I strongly supported the military and its role in national security. But I had never forgotten the lessons of my Depression childhood, and like Bobby, I saw government as a tool with vast resources for improving people's lives without impinging on their liberties.

In those first meetings, we began a political dialogue that would continue for some time.

The attention that had followed my flight continued unabated. Every move I made—and some I didn't—was chronicled. A coat-of-arms specialist emerged with the information that the Glenn family coat of arms contained the Latin words for "ever higher" and "to the stars." Even a tussle I had with some drunken teenagers outside a church party had made the papers back in March. And the *Times* carried the story in May when my mother received a World Mother award from the American Mothers Committee. She told the group in her acceptance remarks that I was "just a normal American boy" who "always seemed to finish what he started."

Of course, I didn't mind the coverage of Mother's award, nor of Muskingum's granting my bachelor of science degree at last. Muskingum had awarded me an honorary doctorate in 1961, but now, with my additional credits applied to the credits I'd earned at Muskingum more than twenty years earlier, I was finally a full-fledged, legitimate college graduate despite my lack of residency in recent years.

Dave and Lyn, when we were all together in public, had taken to walking behind me and Annie just to watch people's reactions. I thought for a time they were embarrassed to be seen with their parents. They told me later they delighted in the open mouths and double takes. But it did make normal family life difficult.

A chance remark during a television interview led to an opportunity to escape for a while. I had said how nice it would be to get away with the family, and the host, Mark Evans, said he could arrange it through a friend of his. Near the end of the summer, after we had sold our house in Arlington and shipped our household goods to Houston, we drove in a borrowed camper to Jackson Hole, Wyoming, where we met Eileen Hunter, whose ranch overlooked the main peaks of the Grand Teton Mountains.

After a day or two in one of her comfortable guest houses, we moved up to Burt Turner's Triangle X Ranch, where we picked up horses and supplies and packed into the Pacific Creek Wilderness Area just east of Yellowstone National Park. There we found a campsite beside a trout stream where we lived for a week in blissful solitude, just the four of us in our tents. We had loved camping together when the kids were younger. This was our first camping trip in over three years, since I had joined Project Mercury.

We spent our days riding in the mountains. We fished together, and cooked our catch over an open fire. Wild strawberries grew in a field across the stream from our campsite, and we picked some for our dessert.

At the end of our trip, we headed south to Houston, where our new house was being built.

The move to Houston produced trauma on a number of levels. We had enjoyed our house in Arlington, and hated to break up the mini-neighborhood of Marine pilots we inhabited with the Millers and the Browns. Dave was sixteen that summer, and Lyn fifteen, difficult ages at which to leave friends and a familiar school, and move someplace new. We were a military family, after all, and they'd experienced several moves, but I couldn't pretend it was going to be easy for them, especially since they would have to deal with the added attention of being John Glenn's kids.

NASA and the astronauts were undergoing scrutiny as well. Leo DeOrsey, who had evolved from the architect of the *Life* deal to our de facto business advisor, had gotten us a small interest in a motel called the Cape Colony in Cocoa Beach. But he didn't like the way the place was being run, and pulled us out of it. Nevertheless, the episode focused attention on our business dealings.

Sometimes it was hard to keep my head. Just after my flight, Leo had called with a tempting question. "Do you want to be a millionaire?" he said.

I gulped and said, "What?"

"General Mills wants to put you on the Wheaties box. Their first offer is a million dollars. I'm sure I can get it higher."

The money was attractive. My four hours of flight time in space had added $245 to my monthly $904.68 paycheck. But I had to turn it down. I just didn't think it was right. I was still part of the space program, still a Marine, and I just couldn't do it.

But that kind of fervor, the idea that an astronaut could with his picture or name alone sell a box of cereal, reached new heights when we moved to Texas. The Manned Spacecraft Center, being built on a thousand-acre tract near the little town of Seabrook, west of Galveston Bay, already promised millions of dollars and thousands of jobs to the local economy. But some business people couldn't resist the idea that their ventures needed the astronauts themselves.

A home developer named Frank Sharp wanted to give us sixty-thousand-dollar homes in the Houston area. Sharpstown, as his development was modestly called, was a little far west and north, closer to Houston proper than was strictly convenient for getting to and from the space center. But a free house, furnished to boot, wasn't something any of us could afford not to think about. Sharp had in mind to put us all along one street in his still-undeveloped subdivision. He was going to donate the lots, builders would build houses, and local furniture and appliance stores would furnish them.

The NASA hierarchy was noncommittal on the deal, but when Leo said we might accept it, the press furor made NASA call a halt. This was the right thing to do. It would have created too many unstated obligations, when in fact our business was still about returning to space, extending the length of our missions and moving to two- and three-man orbital flights and space walks to prepare for an eventual moon landing.

Annie and Rene Carpenter had picked out lots side by side in a quiet development of nice homes called Timber Cove. The lots backed up on a canal leading to Taylor Lake, which in turn led to Clear Lake and Galveston Bay. Annie and I, and the Carpenters, contracted for houses that focused on the water outside our back doors. We trailered the boat from Arlington, and I could be water-skiing five minutes after I got home.

Project Mercury's schedule of one-man orbital flights continued. I manned the Point Arguello, California, tracking station for Wally Schirra's six-orbit, nine-hour flight in *Sigma* 7 on October 3.

Before the end of the month, President Kennedy had revealed that the Soviets were installing offensive missiles in Cuba, and he forced Khrushchev to back down with a sea and air blockade. He was putting federal troops behind the aspirations of the civil rights movement in the South. The astronauts continued training. I still ran to keep my weight down, and one day I was running at Vandenburg Air Force Base in California when a pickup truck passed me, stopped, and waited for me to catch up. As I passed, the driver leaned out the window and called, "Hey, buddy, who you fighting?"

Meanwhile, a new crop of nine astronauts joined the training schedule. This time, the NASA-devised routines took Project Gemini's two-man missions into consideration. We went to Panama for jungle survival training. After two days in the classroom learning what snakes to avoid and which leaves, roots, and bugs were edible, we spent three days in Choco Indian country, high-canopy rain forest, with the spacecraft's survival kit and jungle hammocks. My training partner was Neil Armstrong, who I found to be a great companion.

I always got a kick out of Neil's theory on exercise: Everyone was allotted only so many heartbeats, and he didn't want to waste any of his doing something silly like running down the road. Actually, he stayed in better shape than that would indicate.

Neil had a sly sense of humor. After we had built our two-man lean-to of wood and jungle vines, he used a charred stick to write the name Choco Hilton on it. It rained every day. We used the jungle hammocks to stay off the ground. They were tented to keep off the rain, and had mosquito netting. We caught a few small fish and cooked them on a damp wood fire. At the end of the three days, the astronauts assembled from their scattered sites and followed a small stream to a larger river. There we put on life vests and floated downriver to one of the feeder lakes to the Panama Canal, where a launch picked us up to end the exercise.

I was aboard the *Coastal Sentry* near Kyushu, Japan, in May of 1963 when Gordo made his twenty-two orbit flight in *Faith 7*. He had to come down early after his spacecraft lost orbital velocity, and I helped talk him through the retro fire sequence. He fired the retros "right on the old gazoo," as I reported, and came down in the Pacific near Midway thirty-four hours and twenty minutes and 546,185 miles after liftoff, ending what proved to be the last and most scientifically productive flight of Project Mercury.

That summer, Bobby Kennedy called and said he wanted me to come to dinner at Hickory Hill the next time I was in Washington. I, and Annie and the kids, had toured Japan for NASA after Gordo's flight. The agency also dispatched me on a lot of trips east to congressional and other meetings. Annie joined me on a trip late in the summer, and we went out to McLean for dinner with Bobby and Ethel.

Bobby wanted to talk to me about my running for the Senate from Ohio. Stephen Young, the first-term incumbent, was an old-line Democrat who was in his seventies. He was up for reelection the next year. So was President Kennedy. He had failed to carry Ohio in 1960, and wanted to strengthen the Democratic ticket there to ward off any Republican challenge since Ohio was an important swing state. I also sensed that the president wanted a stronger voice for his agenda in the Senate. Bobby said he suspected Young would retire from the Senate in exchange for an appointment to an as-yet-undecided ambassadorship.

I told Bobby I was flattered. I was certainly interested in politics and contributing to public life. Rumors that I would make a political run already had surfaced the previous summer. I had held a news conference on my forty-second birthday to respond. I said I was not thinking of going into politics, but I refused to rule it out, along with any other career that might follow active participation in the astronaut corps.

Articles at the time speculated about my politics. The Democrats thought I was one of them, and Republicans said they were convinced I was a Republican. I found it a good sign that neither party knew my politics. I wasn't that much of a Kennedy-phile, despite our friendship. And although my parents were both Democrats and I had grown up convinced that Roosevelt's New Deal had saved the country, I had never registered as a Democrat. I had never registered with either party, and had split my votes according to candidates' individual qualities. I had followed the

practice of many military officers, describing myself as an
independent.

But while I was pleased to be approached and felt I had
a contribution to make, I thought it wasn't the right time.
I thought that I should stay at NASA. There still were very
few people who had space flight experience. Manned flights
were on hold while the two-man Gemini spacecraft was
prepared. NASA also was beginning to get into the design
of the Apollo spacecraft. Grumman had the contract to
build the Lunar Excursion Module. Engineers there already
had produced the first plywood mockup, and I was a part
of the advisory group focusing on the cockpit layout. We
were moving around paper cutouts of the instruments and
flight controls. I thought that my experience was important to NASA's ongoing program, and that it was my duty
to stay on.

I also hoped to get another flight assignment.

I told Bobby I hoped he and the president would still
back me if the time ever came when I was ready to enter
politics.

But a flight assignment didn't come. I began to get
frustrated sitting at a desk, so I asked Bob Gilruth when I
might expect to get another flight. Bob said headquarters
didn't want me to go up again, at least not yet. Later, author Richard Reeves wrote that President Kennedy had
decreed I was too much of a national asset to risk on a second flight, but I had no inkling of that at the time, and have
never known whether it was true. I kept asking Bob
Gilruth about a flight, and he gave me the same answer
while hinting that he wanted me to manage some of the
astronaut training. That wasn't something I particularly
wanted to do. Meanwhile, I continued as a kind of ambas-

sador for NASA, going where the agency sent me to appear with members of Congress and other politicos until, finally, I was irked and told Jim Webb I didn't want to do it anymore. He said it was important to the program, in order to keep the funding coming.

Bob Voas, the psychologist who had helped design the astronaut selection and training criteria, had moved to Houston with the rest of the Langley NASA contingent. He and I had talked politics since before my flight. He had told me after my address to the joint session of Congress that if I ran for anything, he wanted to help. Now our discussions intensified.

Then came that unforgettable November day when President Kennedy was assassinated in Dallas. I was coming home from the NASA flight line at Ellington Air Force Base when I heard the news on the car radio. Annie was at a department store in Houston. I called Bobby and Ethel that night, and later represented the astronauts at the president's funeral.

In the days following, as the initial shock and grief receded, Annie and I sat back as we had after Pearl Harbor and assessed our responsibilities to each other and to the country, and what we might do. It was a time for soul-searching. We discussed three basic questions: Was I going to leave NASA? Should I go straight into the business world, in which many opportunities had opened since my flight? Or was now the time to make an entry into politics?

I had received no indication that I would get another flight. I had been stonewalled every time I asked about the possibility. I had always believed that serving in high public office and having the opportunity to help determine the future of the country was one of the greatest positions that anyone could aspire to. Now, with JFK's assassination, it was more important than ever for good people to enter public life.

We decided that if I was ever going to get into politics, now was the time.

Near the end of the year, I went back to Bobby and told him of my decision. He thought it was too late to put together a campaign, and recommended against it. I thought I could still do it. I told Bob Gilruth I wanted to be relieved of my assignment as an astronaut, and notified the commandant of the Marine Corps that I intended to resign my commission.

On January 17, 1964, at the Neil House hotel in Columbus, I announced that I would challenge Young for the Democratic nomination for the Senate seat from Ohio.

PART FIVE

PUBLIC LIFE

CHAPTER 21

My campaign was complicated from the start. NASA had relieved me of my astronaut assignment, but I was still a Marine. I had a block of unused leave time, so I took it to go to Ohio to start planning and organizing. The retirement papers I had filed were working their way through a molasses-slow bureaucracy, and until I was officially retired and no longer a government employee, the federal Hatch Act prevented me from engaging in direct political activity.

When the delegates to the Ohio Democratic convention assembled three days after my announcement, they broke with the party leaders, who were backing Young. Officially, the convention made no endorsement, but not to endorse the incumbent was a vote in my favor.

"You're going to the U.S. Senate," Bob Voas exulted. Bob had resigned from NASA to manage my campaign. Ford Eastman, a NASA public affairs officer assigned to the astronauts, became my press secretary.

We were all neophytes. I had little money and less organization. Young came out swinging. I wasn't running as a hero, but he mockingly treated me as if I were, saying that Able, Baker, Ham, and Enos were the true space pioneers. He sounded like some of the group at Edwards. The Ohio primary was in May, and we settled in for a fight.

My campaign was barely a month old when it ended.

I was staying in an apartment in Columbus that a supporter had contributed to the campaign. Annie had gone back to Houston to pick up some things to make the apartment look more like home. One morning I was shaving and getting ready for a briefing on some campaign issues when a problem I have had all my life—a compulsive tendency to fix things—reared up and bit me. The large, heavy cabinet mirror in the bathroom was on a sliding track, and it wouldn't slide. I removed it to clean out the track. When I went to put the mirror back, it slipped from my hands and fell toward me. I instinctively threw my hands up to the right and ducked to the left to protect my head, and when my weight shifted, the bath rug I was standing on skittered out from under me. I fell and hit my head full force on the metal track on the rim of the bathtub that held the shower door.

I apparently was out for a few seconds, because the next thing I knew I was on one knee looking at a pool of blood on the floor. The mirror had broken over my head and cut my scalp. I didn't feel the cuts, but the pain over the left ear where I had hit my head was excruciating.

Fortunately, the people waiting to brief me heard my crash landing and came running. They took one look and called an ambulance. At the hospital the left side of my head swelled out beyond my ear. Later, X rays showed damage to the vestibular system from a concussion, and blood and fluid collecting in the structure around the inner ear. The concussion itself was mild, but the swelling and bleeding disrupted the functioning of my inner ear so that I suffered disabling dizziness and nausea. Any but the slowest of head movements would make the whole world spin. I was virtually immobile.

Annie rushed back from Houston, and if the news was not already bad enough, she brought word that Tom Miller's oldest child, Donny, had died in a motorcycle accident. He was a month older than Dave. They had been classmates and very close friends in Arlington, and Donny was like a member of our family. A few days later, as I was being transferred on a medevac plane to Wilford Hall Air Force Hospital at Lackland Air Force Base in San Antonio, Annie was flying to Washington to Tom's son's funeral.

While I was confined to bed at Wilford Hall, Annie returned to Ohio to try to keep the campaign alive. Her stutter had been so aggravated by the phone call telling her about my accident that she had had to summon a friend to call the children home from school. But now she and Rene Carpenter went out on the campaign trail and stumped around Ohio. Annie would make brief introductory remarks, then Rene would start pounding the pulpit. Their theme was that politics didn't have to be a dirty business, that good people could enter politics and make it better. Annie called me every night to tell me what had happened that day, and the next day, before Rene took over, she would tell the crowds in a few words what I had said.

As for friends and supporters outside the family, Bobby and Ethel Kennedy called every couple of days.

They were the exception. An initial flurry of calls dropped off to almost nothing. I was left to contemplate the irony of having been unscratched while flying combat missions in two wars and being launched into space atop a rocket, only to be brought down by a slippery bath mat on a tile floor.

The doctors called it "traumatic vertigo." They said I would improve as the blood and fluid in the inner ear were reabsorbed. There was no treatment other than time, they told me, and as many as a third of the people with similar injuries were impaired to some extent for the rest of their lives. Eight to twelve months was the best estimate they could give me for whatever recovery I would achieve. I could walk, but only slowly and only by keeping my feet wide apart and restricting my head movements.

My advisors told me I could beat Young anyway. But after a month I had had enough. I didn't want to be nominated as a hospital candidate, unable to present myself to the people and let them make a decision based on something besides the memory of my flight in *Friendship 7*. And even if I won the primary, it wasn't at all clear that I would be well enough to run a vigorous campaign before the general election in November. My first priority had to be to get better. I wanted to heal and start over. On March 30, 1964, I called a news conference in my hospital room in San Antonio. Annie was with me, again demonstrating her courage by facing the microphones and introducing me while I sat back in the bed in my pajamas. I said, "I do not want to run as just a well-known name. No man has a right to ask for a seat in either branch of Congress merely because of . . . orbiting the Earth in a spacecraft."

With that, I withdrew from the race.

• • •

Dropping out was a hard decision to make, not least because of people like Bob Voas and Ford Eastman, who had staked their futures on the possibilities of my campaign. And the months following were full of the trauma not only of my slow recovery, but of the aftermath of my decision.

When I had fallen I was still on leave, and while I recovered I withdrew my application for retirement from the Marines. My Marine salary then was around $15,000, not counting flight pay. I wasn't getting any of that, since I could barely walk, let alone fly. I had no immediate prospects for future income, and retirement pay was only 55 percent of my base pay. It was something we could count on, but it wasn't much. I was also worried about losing my medical benefits before I was fully recovered.

We were living amazingly close to the bone. The campaign owed $16,000 when I withdrew. I was determined to repay it from my own pocket, even though it would take almost everything we had. I couldn't attend a political fund-raiser, and the thought of asking people to bail me out of campaign debt from a hospital bed repelled me. It reminded me of my boyhood, and how much I liked delivering newspapers but hated collecting from my customers. And there was another expense on the horizon. Dave was graduating that spring from Clear Lake High School and had been accepted at Harvard. While he had won a $1,000 scholarship, there was a lot of difference to make up. Doing that, and paying the campaign debt, took most of what we had set aside from *Name That Tune* and

Life. Just making the house payments was going to be hard.

I was released from the hospital after nearly two months. At home I continued the process of recovery. Its slow pace was the most aggravating part. I'd always been able to address most problems I had in my life with a plan I could carry out step by step. But now I could only wait—and hope.

Each morning I would move from the bedroom to the living-room couch. It took me almost ten minutes to make my way along the hall with one hand on the wall to keep from falling. Some days I would sense notable progress, and then I'd go for a week or two without noticing any improvement. The pattern seemed to be a series of plateaus. My symptoms resembled those of Ménière's disease, a chronic disruption of balance and equilibrium. What I had was triggered by traumatic injury, however, while the causes of Ménière's can be mysterious. Lyn said later, "That may have been the first time that I realized my father was human."

I knew that Lyn and Dave had stepped back from the attention aimed my way after the flight. And of course, for all my efforts to be close and involved when I was home, I had been away a lot. The last thing I had wanted was to be some remote, mythic, superhuman father figure. To have me home, struggling, and very, very human was perhaps a good reminder for us all.

To make use of the time when I was confined to the house, Annie, Rene, the kids and I, along with some of the neighbors, went through some of the letters I had received after

A relaxing moment in an astronaut's life: at home with Annie, Arlington, Virginia, 1959.
[RALPH MORSE/ *LIFE* MAGAZINE © TIME INC.]

I had to run every day to keep in shape. I'm not sure how much Annie and the kids liked it.
[RALPH MORSE/ *LIFE* MAGAZINE © TIME INC.]

In the desert, astronaut survival training meant wearing your parachute on your head.
[RALPH MORSE/*LIFE* MAGAZINE © TIME INC.]

The Mercury 7 astronauts. *Left to right:* Scott Carpenter, Gordo Cooper, me, Gus Grissom, Wally Schirra, Al Shepard, Deke Slayton. [NASA]

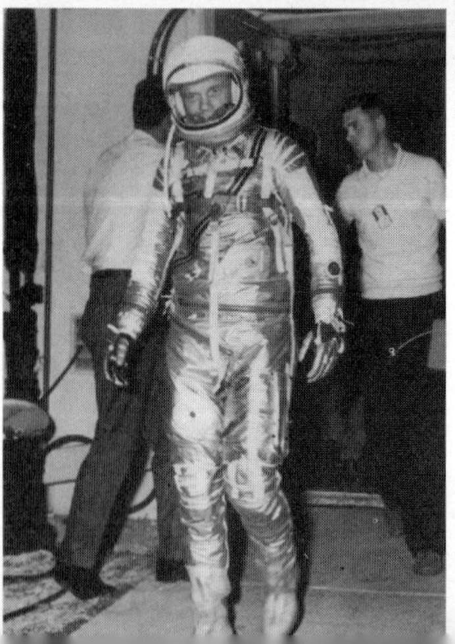

Suited up on my way to a flight simulation test ten months before my flight. [NASA]

After many delays I finally went up in *Friendship 7* on February 20, 1962. [NASA]

Close-up in space from the instrument panel camera. [NASA]

I called Annie after I landed. [MICHAEL ROUGIER/*LIFE* MAGAZINE © TIME INC.]

Debriefing aboard the destroyer *Noa* an hour after splashdown. [NASA]

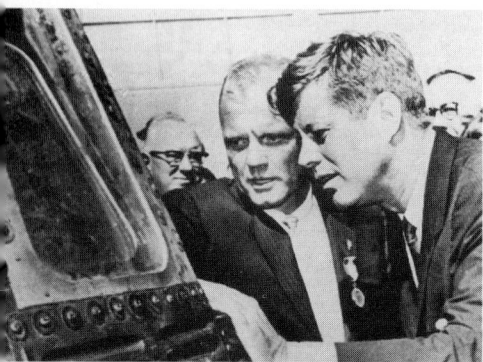

President Kennedy came to Cape Canaveral, where we inspected *Friendship 7*. [NASA]

In high spirits with Vice President Lyndon Johnson, on the way to the Capitol. [UPI]

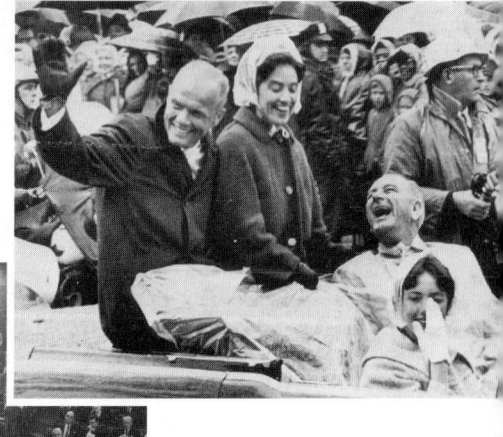

February 26, 1962. I spoke to a joint session of Congress about my flight and my love for America. [NASA]

Little New Concord threw a big parade. Dave and Lyn are in the front seat.

After all the attention, we treasured our moments together.

June 1963: jungle survival training in Panama with the new crop of astronauts. *Left to right:* Neil Armstrong, Edward White, me, guide Gertrudio Arauz, and Scott Carpenter. [NASA]

Bobby Kennedy was a good friend. Here we're with Willy Schaeffler and Jim Whittaker on a raft trip in Idaho.

Annie and me with Hillary and Bill Clinton in Arkansas in 1983, during my presidential bid.

Flying to Cape Canaveral with Vice President Bush and Senator Jake Garn after the *Challenger* disaster, January 1986. [WHITE HOUSE PHOTO]

April 1998: training to go back into orbit. This was a simulated parachute jump into water. [NASA]

We had to learn to rappel from the space shuttle in case of emergency. [NASA]

I'm with Chiaki Mukai in launch position on the mid-deck. The Terminal Countdown Demonstration Test was one of the last steps in our training. [NASA]

I'm wired in the head net I used in a sleep-monitoring experiment. [NASA]

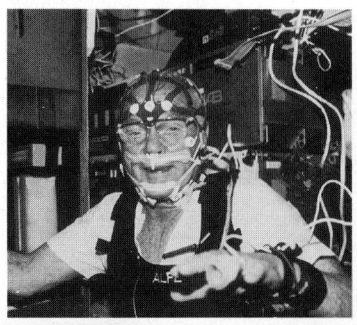

The crew of STS-95 in flight. *Clockwise* from the bald man: Curt Brown, Steve Lindsey, Steve Robinson, Pedro Duque, Chiaki Mukai, and Scott Parazynski. [NASA]

The Glenn family on Earth, where Dave, Lyn, and Annie hope I'll stay for the rest of my life. [NASA]

the *Friendship* 7 flight. There had been over 350,000 in a six-month period, so many that NASA had had to create the "Glenn mailroom." Steve Grillo was in charge. Most of the letters were requests for information or autographs, but I had asked Steve to set aside ones that were particularly poignant, funny, or unusual. They ran the gamut, and we found ourselves sometimes moved to tears or dissolved in laughter.

One woman wrote describing a torrid all-night tryst she and I had had on Morningstar Beach in the Caribbean. When she continued to write, with lurid details she foresaw for our next visit and demanding that we continue our relationship, I turned the letters over to investigators. The postal inspectors learned that her partner on the beach had been a Navy lieutenant out of Norfolk. He had red hair and freckles, and told her he was me, in the islands to get some relaxation after the flight. The night she and the lieutenant had together sounded far from relaxed.

Now, rereading the letters and struck by the range of feelings they expressed, I decided to select some to compile into a book that we would try to have published.

Most of the people who wrote just wanted to share how they had felt on the morning of the flight. They let me see how the country had bonded, and how many others around the world had found common cause with the United States and its mission into space. Sentiments like these helped me fight off the despair of my confinement and the snail-like pace of my recovery. Others, whether they were barely literate, grammatically garbled, or utterly clear, were just plain entertaining.

One, forwarded to me from the White House, was from the owner of a rooming house in Newport,

Kentucky. He complained that since "Mr. Glen ho orbited the Earth 3 times has caused so much impurity in the Air that every One in this House has bin coffing and snoting i believe this kind of stuf should bee stoped."

A young Hoosier wrote: "I am in the fourth grade and I watched your flight through space. I wish I was with you and some times my mother wishes I was too."

From Oklahoma a young aspiring scientist wrote: "Bert Buffington and I are building a rocket and we would like to have some blueprints of the Atlas. Could we have some blueprints of what is in the capsule and of what is in the boster? Could we have the ingredients that you put in the RP-1 and liquid oxygen?"

"Roses are red, vetols are blue," wrote a California boy. "When you went around the world, we did to. We are thankful for all the sinestists who made this flight secesfully."

And from a fifth grader: "The day you were shot I watched you go up on TV in school."

Letters had come from students studying science, from children and adults in every corner of the globe, from young people who wanted to follow in the astronauts' footsteps, from people named Glenn, from others who had been inspired to name their children—or dogs or cows or hamsters—after me, from people ranting about one thing or another, and from women who thought we were married and boys who thought they were in love with Lyn. I received offers of pets and other gifts, letters from people who were ill or down on their luck and needed consolation, or who wanted money, tokens, or other souvenirs. Some added drawings, most notably a rendition of Annie doing the twist. Another one I especially liked came from a fifth-grade class in California: "To celebrate your great achievement we were all served cake for lunch today. We'd like for you to go again!"

Another letter presaged those I would receive in a similar context half a lifetime later. This one was from a woman who was feeling old at thirty-six: "I feel like I'm part of the world again, and I want to thank you for that. Now, instead of sitting around waiting for the bomb to drop, we are going to start doing the things we've put off in our lethargy. . . . We can do anything we set our minds to. We're young!"

More recent letters expressed, in turn, opinions about my entry into politics, and sympathy for my fall and wishes for a quick recovery.

Sorting through the letters and picking out ones that had a special appeal was a tonic for the disappointment I had felt in having to drop out of the race. As I lay on my back in bed—lying on my side or stomach would sometimes send me into fits of nausea—or felt my way along the wall as I made my slow recovery, I didn't yet know what I was going to be able to do with the rest of my life. But these reminders that the flight of *Friendship 7* had struck a chord among people all around the world were an impetus to shake off the doldrums and renew my efforts.

The book was published later that year, with a title inspired by a letter from a twelve-year-old sixth grader who ended her letter with a reference to an aspect of my flight that was broadcast on television and radio: "P.S. I listened to your heart beat."

And in most of the letters the writers had let me hear theirs.

. . .

I healed in fits and starts. Gradually I began to be able to move my head. Then I could ride in a car without getting nauseated. In the May Ohio primary I got over two hundred thousand votes and could only imagine what might have happened had I been able to campaign.

During this time my father was diagnosed with prostate cancer. He had been retired for only two years after a life of labor as a plumber. He and my mother had some money saved for retirement. It wasn't a lot, but the house was paid for and they had looked forward to hitting the road together. Dad had never lost his love of driving, and they had a long list of places around the country they wanted to visit together, traveling by car.

But Dad's cancer was well along when it was diagnosed. The money that was to have gone for a modest retirement of carefree travel began to go instead toward the treatment of his cancer. His insurance didn't nearly cover all the costs, and their nest egg diminished.

By the summer my balance had recovered enough for me to walk outside onto the patio at the back of the house at Timber Cove. The slim canal that led to Clear Lake beckoned, but that would have to wait. I still couldn't even handle bouncing around in a boat, much less water-ski. Next I was able to go out with the family for a meal. Everything I did, however, was predicated on my not having to move my head too fast or too abruptly. And it still was not clear, to me or to the doctors treating me, whether I would recover fully. The doctors in San Antonio had said that I had suffered no brain damage and that my injury was not related to the orbital flight. That was the limit of their certainty.

In July I wrote a letter to the commandant of the Marine Corps reiterating my intention to retire. I was on the list to be considered for promotion to full colonel, but I asked to be removed. Marine Corps promotions are based on both past performance and future promise, but only so many candidates would make the grade, and I knew that if I was promoted, another candidate would be passed over. I didn't think it was fair to make somebody else wait another year when I was going to retire anyway.

With my career in politics on hold, my astronaut career behind me, and my military career drawing to a close, some of the endorsement opportunities that had come my way in the wake of my flight began to reappear. I could have endorsed a line of automobiles and lent my name and face to a brand of sporting goods. But I didn't want to be simply a pitchman for some product or some company. I wanted a job in which I could actually learn something about business, where my opinion counted for something and I could be involved in making decisions.

Royal Crown Cola, the third leading cola maker behind Coke and Pepsi, approached me in August with the offer of a public relations job. I told their representatives that was just what I didn't want—no billboards, no box tops, no six-packs. The offer then went in a different direction. By October Royal Crown elected me to its board of directors and amended its offer to vice president for corporate development. The company, headquartered in Columbus, Georgia, was willing to announce that I would not serve in any advertising, promotion, or public relations capacity. The position came with a $50,000 salary and good stock options. Other jobs had offered more, but

I was comfortable with this arrangement. I accepted, pending my retirement from the Marine Corps.

By the end of October I had finally recovered, or at least felt I had. I knew I could pass the medical exam that signaled my eligibility for retirement, but I wanted some finely tuned measurements. I returned to the Naval Air Station at Pensacola to walk the balance beam and repeat tests on the disorientation devices I had done in astronaut training. If anything would reveal residual problems, this was it. I was elated when the results showed no change from my Project Mercury measurements.

Flying was the remaining acid test I needed to convince myself I was really well again. There was a short jet refresher course at the Marine Air Corps Station at El Toro, California, where I had once been stationed. It was designed for onetime pilots who had been on other duty, to reacquaint them with day and night flying, instruments, tactics, and acrobatics before they returned to their operating squadrons. I made one last request of the Marine Corps and was assigned to the course.

It felt unbelievably great to be strapped into the cockpit of a Panther again. I put it through its paces, and myself as well. The snap of the turns, the press of the G forces, swiveling my head to keep aware of my surroundings—it all felt good. I deliberately sought head positions that would cause dizziness, and found none. I was back! I flew twenty-six hours during fifteen flights over eleven days and never felt better. It was a good end to a bad year.

President Johnson, meanwhile, had proposed a plan to promote me separately, without passing over another eligible lieutenant colonel, and Congress approved it. I received my promotion at a White House ceremony after

Johnson's overwhelming election victory. In the Ohio senate race Steve Young was elected to a second term on Johnson's coattails, squeaking in by sixteen thousand votes.

I retired from the Marine Corps on January 1, 1965. Hanging up the uniform after twenty-three years was something I did with great reluctance, but genuine pride. Annie and I went to the commandant's office, where I saluted, accepted a handshake, and stepped into my future. I knew it was time to move on. But the Marines had given me some of my deepest friendships, my warmest memories, and, most important, a fundamental code of conduct. The Marines weren't the only source of this, but they brought together the other influences in my life in a way that I had come to revere.

I had learned a set of virtues from the time I was a child, from my parents, my church, my teachers, and all the other components of the community that was New Concord, Ohio. The Marine Corps embodied those same virtues. They included hard work, religious faith, and personal discipline. They embraced love of country, love and respect for one's family, a willingness to accept responsibility for one's actions, and a combination of feelings in which brotherhood, courage, and loyalty were all bound up together. Those convictions are reflected in the Marine Corps' simple but far-reaching motto, *Semper Fidelis*— Always Faithful. That was what Pete Haines, the skipper of my Corsair squadron back in World War II, had meant when he said the Marines were different, that you could trust your life to another Marine. You'd risk getting hurt yourself before you'd let your buddy down. When you had that kind of faith and loyalty and discipline, the next steps were that much easier. You'd do anything before you let your loved ones, or your country, down.

I confess that I had trouble controlling my emotions at leaving the Marines. Whatever I did in the future, I knew the Marine way of life would always be a part of me.

A former governor of Georgia, another Marine whom I've met and respect, summed all this up in a book called *Corps Values*. Zell Miller's book is on the shelf of every Marine I know because it so eloquently reflects our feelings.

Dave's life as a Harvard freshman was far different from the kind of life I was leaving. The unquestioning loyalty to country and to authority, which I had accepted as a given, was not a given in the halls of the Ivy League, nor in many of the other colleges and universities around the country. Students who were for the most part as loyal as any Americans had ever been were seriously questioning the way authority over their lives was being exercised. Vietnam, a divided sliver of Southeast Asia on the east coast of the Indochina Peninsula, was the issue. Congress had passed the Gulf of Tonkin Resolution the year before, authorizing the president to wage war in Vietnam. American involvement was escalating, and the draft was sweeping increasing numbers of young men into service.

In the months leading up to his departure for college, Dave and I had talked about his plans. He told me he had been impressed with how much I enjoyed life as a fighter pilot. He thought he would get his degree, join the

Marines and go to flight school to fly fighters, and after serving, either stay in or take up a career with one of the airlines.

But the ferment that was beginning to roil the universities, not just over Vietnam but over many other issues, including the struggle for civil rights and the relevance of religion, challenged Dave's assumptions. This was normal for any college student, but I don't suppose it made it any easier for him that my own beliefs had been so visible. I knew he was struggling to anchor his beliefs when a letter arrived from Frank Irwin, our pastor back in Arlington. It was a copy of a letter he had sent to Dave, in response to Dave's request for a reading list of writers who argued on behalf of Christianity. Obviously Dave felt the need for some backstopping of the faith he had grown up with. Harvard was a far different place from Muskingum, and the 1960s were a far different time from the 1930s, when I had accepted without question the role of the church in my life.

Frank suggested an interesting mix, including C. S. Lewis's *The Case for Christianity* and *The Screwtape Letters* and Edward W. Bauman's *The Life and Teaching of Jesus*.

Lyn seemed relatively untouched by the turmoil that was going on at the next step up the educational ladder. She was enjoying her senior year at Clear Lake High School, where she was a cheerleader and one of the school's most vivacious and popular girls, and had been accepted at Mount Holyoke College, in western Massachusetts.

I could now afford all this schooling without worrying where the next tuition checks were going to come from. I could also—and it was this for which I was most grateful—help my father as he struggled with prostate cancer.

Mother and Dad's savings and insurance had long since run out, and they would have been in dire circumstances otherwise.

My duties at Royal Crown took me far afield. I continued to make a few public appearances, not only for NASA, which had begun the two-man flights of Project Gemini in the spring of 1965, but also for the Boy Scouts of America and other public service organizations. I was named to the editorial board of *World Book Year Book*, published by the *World Book Encyclopedia*. Another subsidiary, the World Book Science Service, had published *P.S. I Listened to Your Heart Beat*. I wondered, with Lyn and Dave both away at college in the fall and me traveling a lot, how Annie would cope with the occasional problems presented by her stutter. But Scott and Rene Carpenter's presence next door was a source of reassurance. And in September Lyn left to start her freshman year and Dave returned to Harvard.

When the family reassembled for the Christmas holidays at the end of 1965, Annie was quite concerned about her health, but she hid the full extent of her worry. She had always been anxious not to trouble me. Now she was downplaying chronic internal pain that was sometimes so severe, it kept her in bed. Her doctor finally diagnosed endometriosis, a condition in which uterine tissue grows outside the womb. At that time it was suspected to be a potential precursor of malignancy. Her doctor recommended surgery.

I was worried as I had not been in nineteen years—not since she was rushed back to the hospital with an infection after Lyn was born and I flew home from Okinawa. But the hysterectomy went well, and follow-up tests showed no signs of cancer. Annie's few days in the hospital after the New Year were lightened by Lyn's drawings and cartoons. Mike DeBakey, the noted Houston heart surgeon, was a friend, and he looked in every day. When Annie came home, I stayed off the road while she recuperated.

My dad's six-year battle with cancer ended in New Concord that February 1966. He had made me promise, soon after he was diagnosed, that when the final days came we would make no heroic efforts to prolong his life in a hospital. He wanted to be at home. Annie was able to travel by the time his days were dwindling, and she and I went home to be with him and Mother. Dad had been an elder in our Westminster Presbyterian Church ever since I could remember. I was with him when he died. He was at peace with himself and his God, and was ready to go.

CHAPTER 22

Dave came home from Harvard at the end of 1966 with his hair much longer than the crew cut he had worn in high school. I didn't exactly approve, but I kept quiet. Then one day, Annie and I were Christmas shopping with the kids—Lyn, who had transferred to Stanford, was also home—at one of the malls that were springing up around Seabrook and Clear Lake with the ongoing development of the Manned Space Flight Center. We had just sat down to lunch when Dave broached a subject that obviously had been on his mind.

"Mom, Dad," he said, "you wouldn't believe some of the things going on at school. Almost everybody I know is against the war in Vietnam. Some of the guys are even talking about going to Canada or Sweden. And I've got to tell you, I find it hard to take the other side. I know how you feel, but I'm having trouble with it, too.

"That's not all. There's even a movement saying God is dead. And to be honest with you, it's made me think. I have trouble explaining why I believe in God. How can anybody prove that God exists?"

The sixties were not an easy time for parents and their college-age children. The times were a-changin', as Bob Dylan memorably said. The critical questioning of religious beliefs, the increasingly unpopular war in Vietnam, the use of drugs, and the perception of inequities that had spurred the civil rights movement combined into an incendiary mix. Campuses around the country, full of students like Dave and Lyn who were forging their own beliefs and ways of thinking, were centers of dispute and debate and also forums for civil disobedience like sit-ins, boycotts, and occasional riots.

I went with Vietnam first. "Nobody was for the war in Korea, either, Dave," I said. "But we fought it, and it sent a message to the Communists that we weren't going to sit back and let them take over. Look what they've done since the end of World War Two. It's still a threat."

"Dad, Korea was an invasion. This is a civil war. We've got, what, four hundred thousand troops over there, we're trying to bomb North Vietnam back to the Stone Age, and it's not working. That's what kids are saying. They're saying it's only poor kids who get drafted, and we've got problems at home we need to deal with."

"I'll agree with you on one thing, Dave," I said. "If we're going to be there—and I think we should—we should have put enough forces in to win. We could have been out of there already, instead of upping the ante and then upping it again. It's not the war, it's the gradual escalation that's wrong. That's why all those body bags are coming back."

"I don't know, Dad. I don't know if we should be there in the first place," he said.

This was very different thinking from what I was used to. The allegiance to authority exercised to good purpose that I had grown up with, the unquestioning pursuit of national goals as articulated by leaders the people respected, had given way to doubts and questions as goals became less clear, especially among those whose lives would be on the line. I realized that I would have to take this kind of thinking into account both as a parent and as a would-be politician. I decided to move on.

"How can anybody prove there's a God?" I said. "I can't. There's no mathematical formula or chemical composition that adds up to God, just like there's no formula for love, or hope, or honesty. I don't believe that God is dead. I can't look around this world and believe that it came out of chance encounters of cosmic debris. But you know, God doesn't have to be believed in to exist."

Dave looked at me, not accepting what I said so much as accepting my belief in my version of things. And I knew I would accept his beliefs as they evolved. He and Lyn and Annie and I had had a number of discussions like this in recent years, as had many parents and children, and we would have more. The tumult of the sixties was making them more intense. But if the bedrock was quaking, it was also forming anew. I knew Dave and Lyn would reach beliefs that were not necessarily the same as Annie's and mine, but that would be equally strong in respect and caring for others.

My responsibilities at Royal Crown increased. The board named me president of Royal Crown International, which gave me the opportunity to put together deals that would get RC and its products distributed overseas.

Morgan Cramer, the former head of P. Lorillard, was brought aboard. He had long experience in international business, and for the next couple of years he and his wife, Marian, and Annie and I literally roamed the world negotiating business agreements and making new contacts. It was a real education, not only in the world business arena, but in the opportunity to see the varied ways of life around the globe.

The new job with Royal Crown took us from the Houston area to New York City. Annie and I lived on the twenty-seventh floor of an apartment building at 58 West 58th Street, just a few blocks from the company's Madison Avenue offices.

Annie had a habit of always waving goodbye when the kids left for school or I left for work. One morning after the move to New York I kidded her about never waving to me anymore. She took up the challenge, and when I got downstairs that morning I walked a short way down the street, turned, and looked up. There was Annie on the balcony, waving with big sweeps of her arm. I responded in kind and was chuckling to myself when I heard an amused voice behind me. "Glenn, what in the world are you doing?" Standing there was Robert Goulet, whom I had met after one of his Broadway performances in *Camelot*.

A little embarrassed, I tried to explain our tradition of waving, as Goulet responded, "Yeah, uh-huh, um, uh-huh, yeah, yup," as though I had to be humored. Then Goulet stood there with me in the middle of the 58th Street pedestrian traffic, both of us waving at a nearly invisible figure twenty-seven stories up while passersby looked at us with the special tolerance New Yorkers reserve for the harmlessly addled.

I forgot about that until some months later, when Royal Crown had a convention in Las Vegas and Goulet was appearing at a nearby hotel. RC chairman Bill Young suggested we attend his show and made the reservations, which turned out to be ringside. Goulet was in the middle of his big opening buildup when he noticed Annie and me in the audience. He stopped singing, silenced the band, and proceeded to relate the whole silly story, complete with body language and exaggerated arm waves, while the audience roared and applauded and Annie and I sat there red-faced that one of our corny habits had been brought to light.

During this same time I was able to take a major step toward financial independence—a source of great relief for the future. I joined with my old friend Henri Landwirth and a real estate broker named John Quinn in a Holiday Inn franchise near the site of the new Disney World south of Orlando, Florida, securing a construction loan from a Cleveland bank. Quinn had a superb property and Henri was a consummate hotel manager, so we were able to expand that Holiday Inn three times over the years as Disney World grew ever more popular. In the years since then, we have built three more Holiday Inns in the same area, all of them successful, and almost everything we have done has been on the basis of a handshake. Our partnership has been an unusual example of trust in a rough-and-tumble business world.

The year 1967 began with a tragedy in the U.S. space program. Gus Grissom, who had become the first person to fly into space three times, was killed in a fire on the launch pad during a test of the Apollo spacecraft, along with the newer astronauts Ed White and Roger Chaffee. The first deaths in the manned space program, coming as they did, not during a spectacular launch or splashdown, but in a relatively innocuous procedure, highlighted the sometimes dangerous technological trade-offs. Fire had swept through the spacecraft as quickly as it had because the need to save weight by keeping the interior cabin pressure low meant that pure—and fire-feeding—oxygen was still inside the capsule.

Gus was buried at Arlington National Cemetery. The service, the caisson rolling slowly past the sea of gravestones, the missing-man formation flyover, the rifle volleys in salute, the ceremonial folding of the flag that had draped his coffin, and the finality of taps underscored that nothing could be taken for granted. The space program had gotten off Earth and was reaching for the moon, but as always, courage was a cornerstone. Tragically, our friends were gone, and with Gus's loss the original seven were now six.

But a test of the huge Saturn rocket that was the engine of the Apollo program, the rocket that would take men to the moon and back, went perfectly in November and returned momentum to the program.

By the end of 1967, with Dave a senior at Harvard and wearing his long hair tied with a leather headband, and Lyn still at Stanford, the United States had almost half a million troops in Vietnam. Poverty, hopelessness, and nagging frustration had produced riots in the urban ghettos of

Chicago, Detroit, Newark, and Watts in Los Angeles. For a course in urban sociology, Lyn lived with a black family in Oakland for a week and discovered at least one source of the anger: Grocery prices in the ghetto were much higher than in upscale neighborhoods. Sometimes she heard gunshots echoing in the night. Antiwar sentiment had spilled out of the colleges and universities into the streets, and increasingly into the living rooms where Americans gathered around their television sets and made their political decisions. These matters loomed in the upcoming election season.

Annie and I had kept in touch with Bobby and Ethel Kennedy. We had skied with them and their kids at Sun Valley and rafted with them down the middle fork of the Salmon River in Idaho. I had grown to consider Bobby a good friend, a person with whom I could share my deepest thoughts. Our friendship was unaffected by the scandal sheets that featured lascivious lies, like the one that reported that Ethel and I were having an affair, accompanied by a picture of the two of us skiing that had cropped out Bobby and Annie on either side. It was a friendship that allowed us to talk forthrightly, over a campfire in Wyoming, about how many true friends we really had, friends to whom we would be willing to entrust our money and the most intimate of family matters. Not many, we all agreed. In the aftermath of his brother's death, Bobby had been hurt that people he thought were stalwart friends had stopped calling, now that he was no longer the president's brother. I recalled my similar experience after falling and withdrawing from the Senate race. When Ethel challenged us to name the people we felt close enough to to carry the casket at our funeral, we were all hard-pressed to come up with six names.

Bobby had been elected to the Senate from New York in 1964 and had spoken against the war in Vietnam. I didn't question the purpose of the war, as Bobby did. But I continued to question the way we were fighting it. Every military instinct I had said that if we had to be there, we had to go all out to win it and as fast as possible. That saves lives in the long run. In my judgment, the problem was that the politicians were refusing to let the generals win in Vietnam. Gradual escalation, expecting the enemy to capitulate each time, is not the recipe for a short war.

I also saw Bobby's deep compassion for people—sympathy for the underdog greater than that of anyone I had ever known. He also believed that no matter how large a problem was, a way could always be found to make it better. It was this quality, more than any other, that attracted me to him, as a person and as a politician.

Eugene McCarthy, the U.S. senator from Minnesota, was outspokenly against the war, and the only Democrat with the nerve to challenge President Johnson in the early going for the Democratic nomination in 1968. Bobby thought about it but vacillated, despite Lyn's urging him on during a ski trip at Waterville Valley with a quote from Shakespeare's *Measure for Measure*: "Our doubts are traitors, and make us lose the good we oft might win, by fearing to attempt." He mentioned that later as he weighed his decision. After McCarthy virtually tied the president in the New Hampshire primary vote and captured most of the state's delegates, Bobby decided to enter the race. He called and asked if I would help.

Joining Bobby's campaign brought me up against a faction of the board of Royal Crown, on which I still served. Some of its members were conservative southerners, and they disapproved of Bobby's stand against the war and his support for civil rights. This faction was drafting a resolution prohibiting political activities by members. I got wind of it and told them that if they did, I would be forced to resign and make a public statement. The resolution was never formally proposed.

During the next three months Annie and I traveled frequently with Bobby's campaign. I made speeches for him. He liked to have his dog, Freckles, a black and white spaniel, with him. Sometimes we would ride on the back of a convertible with Freckles in the middle, as if he were the candidate and Bobby and I were his bodyguards. When Johnson dropped out at the end of March with his memorable statement that he would not seek, and would not accept, his party's nomination (and at the same time halted the bombing of North Vietnam), Bobby's chances to take the nomination looked good.

The assassination of Dr. Martin Luther King Jr. on April 4 shocked people of goodwill around the world. It deepened the festering sense of neglect among black people across the country, especially in the cities, and made the election of a healer like Bobby even more urgent. He won in Indiana and Nebraska, suffered a setback in Oregon, and moved on to California. It was an exhilarating time, and as I saw Bobby feed off the crowds that supported him and listen to those who wanted to condemn him, it whetted my appetite for politics again.

I went with him to a late-night meeting of Black Panthers and other militants at a church in Oakland, where we arrived around midnight and heard the audience hurl angry accusations about how their community was treated. Rafer Johnson, who was also part of the campaign, tried to intervene, but Bobby let them talk. The next day some of the same angry people were helping to keep order at a rally in West Oakland. Bobby's willingness to listen, and his visible caring, had a way of converting people to his side.

Annie and I were among the people with Bobby and Ethel in their suite the night of the California primary on June 4. As the results rolled in and it became clear that he had won, Bobby went down to make his victory speech. I was supposed to go along and be on the podium with him, but a lot of California people wanted to share the spotlight, so we decided to stay in the room.

We heard his speech on television and continued to watch after midnight as the cameras followed him from the ballroom through the hotel kitchen. Then came the madness and chaos of his shooting. Annie and I stared at each other in horrified disbelief. I rushed downstairs in time to follow him and Ethel to the hospital. Early in the morning, when after three hours of surgery Bobby was still in a coma but alive, Ethel asked if Annie and I would take their children—five of the eleven were with them in California—back home to Virginia. They had been asleep when Bobby was shot. When they woke up, we told them what had happened.

We flew back across the country in an Air Force jet dispatched by the president. We took the kids out to Hickory Hill and stayed the night with them. I was in Bobby's study when I saw on his desk a collection of Ralph Waldo Emerson's poems and essays. Leafing through it, I saw he had marked in the margins passages he liked. Two of them have stayed with me. One defined Bobby's approach to life: "Always do what you are afraid to do." The other was another challenge: "If there is any period one would desire to be born in, is it not the age of revolution, when the old and the new stand side by side and admit of being compared? When the energies of all men are searched by fear and by hope, when the historic glories of the old can be compensated by the rich possibilities of the new era, this time like all times is a very good one if we but know what to do with it."

During the night a call from California told us that Bobby had died. The next morning it fell to me and Susie Markham, a neighbor who knew the children well, to sit on their beds as they were getting up and deliver, as gently as we could, even worse news than they had heard the day before. It was one of the hardest things I've ever had to do.

I helped carry Bobby's casket at his funeral at St. Patrick's Cathedral in New York. Andy Williams sang "The Battle Hymn of the Republic." An honor guard stood along the sides of the casket, its members shifting constantly as a new person stepped up to take another's place, so that over time it included students, the poor, high government officials, civil-rights workers, military officers, and personal friends. Dave, then twenty-two, joined in the guard with me. We had stood there for several minutes when someone tapped me on the shoulder, and I moved aside for U Thant, the secretary general of the United Nations. Annie and I rode with Ethel on the train bearing the casket from New York to Washington, and accompanied the cortege to Arlington. There I helped carry his casket to the grave, folded in military fashion the flag that had draped it, and presented it to Ethel.

Dave graduated from Harvard soon afterward, on June 13. Annie and I attended and could not have been more proud. We visited at Hickory Hill often during that summer of the party conventions, which were marked by violent clashes at the Democratic National Convention in Chicago and continued protests over Vietnam. Richard Nixon went on to narrowly defeat Hubert Humphrey in November. Expanded Vietnam peace talks convened in Paris after the first of the year.

Meanwhile, draft-eligible upon his graduation, Dave chose to enlist in the Navy's Officer Candidate School. He had learned that his eyesight would prevent him from being a pilot, and after OCS he served aboard a guided missile cruiser, the USS *Albany*. After his discharge and a year of studying at Stanford, he would enter medical school, a path I might have chosen myself had war not intervened.

Stephen Young announced that he was retiring from the Senate at the end of his second term in 1970. I followed his announcement with one of my own—that I would run for the Democratic nomination to succeed him.

Soon afterward Young's former campaign manager, Howard Metzenbaum, a labor lawyer and former state legislator from Cleveland, proclaimed his own candidacy. Metzenbaum was close to the state party. He had made a great deal of money in airport parking lots and rental car franchises, and he spent almost three-quarters of a million dollars on the primary. My supporters told me the primary would be a snap. I was taking nothing for granted, but I had trouble convincing them that I needed a war chest for the primary. They didn't think I would need much money until the fall and what promised to be a tough general election fight against the likely Republican senatorial nominee,

Robert Taft. When the polls tightened during the primary campaign, it was too late. Metzenbaum was better funded and better organized. He ran well-produced TV spots and had fifteen offices around the state. Toward the end of the campaign he was outspending me by four to one.

I wasn't running as the first American to orbit the Earth. The space program had gone on to greater accomplishments than mine. Neil Armstrong, my partner in jungle survival training in Panama, became the first man on the moon in July 1969 as I watched from mission control in Houston. But I was still well known for my flight, and one of my campaign planks was more money for the space program.

However, other events were claiming the nation's attention.

The day before the May 5 primary, Ohio National Guardsmen, called out to quell antiwar demonstrations in the wake of the U.S. and South Vietnamese incursion into Cambodia, killed four students on the campus of Kent State University. The anguish was symbolized by a single photograph: a young woman, her arms outstretched, kneeling beside one of the victims. The nation's reaction was intense. Metzenbaum was vocally against the war, while my questions had focused mainly on its conduct. I don't know if the Kent State shootings made the final difference in the outcome, but I lost the primary by fewer than thirteen thousand votes.

While I toted up the political lessons I still had to learn, Metzenbaum went on to lose to Taft.

My political loss, however, paled next to the losses in our family shortly after the first of the year. Two of our parents, my mother and Annie's father, died within ten weeks of each other, in January and March, 1971. Twice more we made sad journeys to New Concord and to the cemetery where the sound of taps always echoed in my ears.

Annie was watching from the stands on Memorial Day 1971 when the pace car in which I was riding at the Indianapolis 500 slid and crashed into a flatbed trailer crowded with photographers and journalists. The four of us in the car were only bruised, but nine people were seriously hurt.

For all of my exploits, riding in a car and shaving seemed to be my most life-threatening activities. To have died at that point would have been a shame. It would have deprived me of the opportunity, two years later, to witness the beginning of an amazing and inspirational transformation.

Annie and I were watching the *Today* show one morning early in 1973 when a guest came on from a small college in Roanoke, Virginia. Ron Webster, a psychology professor at Hollins College, was doing research into why stutterers stutter. His program was having remarkable success in its early stages.

Annie had tried various treatment programs before. They had functioned primarily as feel-good sessions; they eased the psychological burden of stuttering, but didn't solve the problem. Webster's program was different. He approached stuttering as a physical problem, not a psychological one. A couple of his graduates were shown in before-and-after film clips. Before the treatment they sounded much like Annie. Afterward, they—and 80 percent of the students in the program—showed improvement. I remembered a dinner party next door at Scott and Rene Carpenter's in Timber Cove. Scott had a new stereo, and after dinner, as a parlor game, he had asked everybody to try reading from a newspaper while wearing headphones, with the stereo set to delay the reader's voice by half a second. Nobody could do it except Annie. The delay that confused the rest of us somehow allowed her to read perfectly.

"That's a new approach," I said when the *Today* segment ended. "It looks good. Do you want to try it?"

Annie was excited at the prospect.

Soon she was enrolled in a three-week course at Hollins that had her and other stutterers working intensively eleven hours a day. The theory behind the exercises was that speech patterns are set to a large degree by audio feedback, sounds and vibrations that let us pick up continuity and rhythm. If the nerve, brain, and muscle combinations that do this had some congenital weakness or for some reason got out of sync during childhood, stuttering could develop. Sometimes the stutterer's hearing could block out information given the brain by the sound of the voice, causing the speech muscles to jump out of control. The goal was to retrain the system and reestablish control.

Students relearned alphabetical sounds from the ground up and spent hours practicing them, then spoke syllables, dragging them out to two seconds, and progressed from there. The final exercise, at the end of the three weeks, had the students doing something Annie had never been able to do before. I'll never forget her phone call.

"John," she said on the line from Virginia, forming her words slowly and carefully as the muscles worked, "today we went to a shopping center and went shopping. And I could ask for things. Imagine that."

I had never heard Annie speak that many words without a single pause. It was all I could do to reply, "That's wonderful!"

"I think so, too," she said slowly. "It's a start."

Annie grasped the gift of speech and held it tight. Our lives were transformed. "John," she said when she got home, hiding an impish smile, "I've wanted to tell you this for years: Pick up your socks." Our phone bill increased as she started calling friends around the country. She had never been able to read children's stories to Lyn and Dave when they were little. People had always mistaken her reluctance to speak for shyness, but she loved people and was no wallflower. Now, she said, she felt like a butterfly that had been let out of a cage.

Annie's wonderful progress stood in contrast to a turbulent year in American life, one that was a watershed for much that had gone before. The Watergate investigations drew closer to President Nixon's Oval Office. American involvement in the Vietnam War ended with agreement in the Paris peace talks, and with it came the end of the draft and the release of more than six hundred prisoners of war.

The repatriation of the POWs had an intensely personal resonance for us because one of the bravest of them had remained a good friend ever since our days together at Pax River. I had taken Jim Stockdale on his first night jet cross-country checkout down to Opa Locka, Florida, turned the controls over to him on the way back, and after we landed at two in the morning, celebrated by scrambling eggs for breakfast at his house. Later I had written him about the welter of training activities in Project Mercury. Jim had been a Navy pilot flying from carriers when he was shot down over Vietnam in 1965. In the almost eight years of his captivity, he organized other prisoners, established a tap code for communication, and refused to submit to interrogation and torture by his captors. Once he hit himself in the face and head with a camp stool to disfigure himself with cuts and bruises rather than appear in a North Vietnamese propaganda film. He later wounded himself, almost fatally, to convince the North Vietnamese at Hoa Lo Prison in Hanoi that he would die before capitulating to their demands. All this came out later, when Jim was awarded the Congressional Medal of Honor. Annie and I had kept in touch with his wife, Sibyl, during Jim's long captivity, which had included years at a time in leg irons and solitary confinement, and it was a relief to know that he had not only survived, but prevailed.

Then politics reared its head again, a consequence of the developing Watergate crisis. As 1974 approached, a sudden vacancy from Ohio occurred in the U.S. Senate. Republican William Saxbe had been elected to his six-year term in 1968. But before he completed it, President Nixon's firing of Watergate special prosecutor Archibald Cox caused his attorney general, Elliot Richardson, and Richardson's deputy, William Ruckelshaus, to resign in protest. After this "Saturday night massacre" on October 20, 1973, Nixon appointed Saxbe attorney general.

It was up to John Gilligan, Ohio's Democratic governor, to fill the Senate vacancy. My view was that he should appoint me. Metzenbaum also lobbied for the appointment.

But Gilligan had his own agenda. He had an eye on the 1976 presidential nomination, or being tapped as somebody's vice president, but the first thing he needed to do was to get reelected governor. He asked me to run with him for lieutenant governor. I would ascend to governor when he made the national ticket, and could look to a future Senate race.

I had learned by then that politics was sometimes a waiting game, but lieutenant governor wasn't what I had in mind. I turned the offer down. I told Gilligan the least he could do was appoint somebody neutral, so that Metzenbaum and I could fight it out on equal footing for the 1974 Senate nomination, with neither of us having the advantage of incumbency. But the Metzenbaum machine was poised to back Gilligan for reelection in exchange for the interim appointment, and Gilligan named Metzenbaum.

That set up a battle royal in the 1974 primary campaign.

In addition to most of the unions, Metzenbaum had the backing of the state Democratic party, to which he had decades-old ties. With his own money, and hundred-dollar-a-plate fund-raisers the county chairpeople helped him put on throughout Ohio, the Metzenbaum campaign was flush.

One hundred dollars a plate was a lot of money to most people. I campaigned at places like the Flat Iron Cafe in Cleveland, where I put on a 99-cent corned beef and cabbage dinner and Rosey Grier and I stood on tabletops and spoke to the crowd. *The Plain Dealer* ran a picture of us the next day, opposite a photo of tux-clad guests at a Metzenbaum dinner in the ballroom of an uptown hotel. Steve Kovacik, a Columbus attorney who had money of his own, was the architect of my campaign. After starting out with a statewide tour in a converted bus, we concentrated on the state's fourteen largest counties. I had carried seventy-six of Ohio's eighty-eight counties in 1970, but Metzenbaum had carried most of the big ones. We wanted to reverse that outcome.

Metzenbaum's money became an issue in the campaign. While I was by now financially independent, I had paid hefty federal income taxes on the money I earned. Metzenbaum, by contrast, had paid very little, though there was nothing illegal about this. The difference came to light when we released our tax returns in the course of the campaign. I had paid more in taxes for 1972 than my opponent had paid for 1968 through 1972 combined, and my income was less than a third of his. Moreover, he was releasing detailed financial information in dribs and drabs, and only under pressure.

The facts about Metzenbaum's financial dealings, I said, were "dribbling out just like the facts about Watergate."

Toward the end of the campaign, when Ethel Kennedy was not able to repeat the appearance she had made in 1970, Jackie Kennedy stepped in with a radio spot endorsement.

I was leading in the polls. Then my opponent made a fundamental mistake that sealed the election in my favor. He put me on a war footing. He said I had never held a job.

When I first heard about the comments he had made at a rally in a Toledo union hall, I thought they were so ridiculous that I wouldn't dignify them with a reply. But over the next couple of weeks he kept repeating them at other stops, and the press was beginning to report his remarks as a factor in the campaign. In two elections his campaign had accused me of everything from voting for Richard Nixon over John Kennedy in 1960 to having an affair with Andy Williams's wife, Claudine Longet. But this was more than just a political sally, this was an assault on everyone who had ever served the country. Metzenbaum had never been in the military.

I waited two weeks, until the final debate of the campaign at the Cleveland City Club, before responding. We went back and forth in the debate, and when it came time to summarize, I was the last to speak. I said, "Howard, I can't believe you said I have never held a job." Then I gave what became known as the "gold-star mother speech."

I began by recounting my twenty-three years in the Marine Corps and my missions in two wars. I talked briefly about the space program, where it "wasn't my checkbook, it was my life that was on the line."

Then, turning to him, I said, "I ask you to go with me, as I went the other day, to a veterans hospital, and look those men with their mangled bodies in the eye and tell them they didn't hold a job. You go with me to any gold-star mother and you look her in the eye and tell her that her son did not hold a job.

"You go with me to the space program, and you go as I have gone to the widows of Ed White and Gus Grissom and Roger Chaffee, and you look those kids in the eye and tell them their dad didn't hold a job.

"You go with me on Memorial Day coming up and you stand in Arlington National Cemetery—where I have more friends than I like to remember—and you watch those waving flags and you stand there and you think about this nation and you tell me that those people didn't have a job.

"I tell you, Howard Metzenbaum, you should be on your knees every day of your life thanking God that there were some men—some men—who held a job. And they required a dedication to purpose and a love of country and a dedication to duty that was more important than life itself. And their self-sacrifice is what has made this country possible.

"I've held a job, Howard."

The audience of six hundred rose in a standing ovation. The speech attracted national attention. In the primary four days later, I won by almost a hundred thousand votes. I won the general election in November by a two-to-one margin, carrying every county in Ohio.

At last I was going to the Senate.

CHAPTER 23

It was January 14, 1975. Gerald Ford had become president of the United States following Nixon's resignation. I stood in the well of the Senate, raised my hand, and took the oath of office from Vice President Nelson Rockefeller. Ceremonial pictures followed, with Annie holding the Bible on which I placed my hand. I had had an informal swearing-in earlier, on December 24, 1974, after Metzenbaum resigned, to give me the benefit of a few weeks' seniority, but this made it real.

Looking around the Senate chamber, thinking of the people who had served there and the Senate's role, so recently exercised in the Watergate hearings, I couldn't help but remember Harford Steele, under whose tutelage my dream of public service had first glimmered.

I vowed to take a hard look at the usual way of doing things. I knew that lawmaking required compromise, but I wanted to avoid the cynical wheeling and dealing that too often sacrificed the country's well-being to the demands of special interests. I thought I could make reasoned, principled decisions on the issues by learning the facts, keeping in close touch with my constituents, and finding a balance between their needs and those of the country as a whole.

Lyn was in the gallery with Annie for my swearing-in. Later she appeared in my temporary office in the Russell building, where a group of well-wishers had gathered with my staff members, who were trying to create order out of chaos. Lyn had come from the National Gallery of Art with a gift for me: a print of Peter Paul Rubens's painting of Daniel praying in the lions' den. She said, "I thought this might be the way you're feeling today." She wasn't far wrong. I remembered Harry Truman's comment about his entry into the Senate. On his first day he couldn't believe he had gotten there. After six months he couldn't believe that many of the others had. I hoped, like Truman, to be a quick study.

Dave's view of my entry into the Senate was less dire than Lyn's, perhaps because of what had happened to him during the campaign. He, too, had worked in it whenever he could spare the time from medical school at Case Western Reserve in Cleveland—and had met his future wife, Karen Fagerstrom, who was a campaign volunteer.

The Democrats were the majority party in the Senate, and Mike Mansfield of Montana was the majority leader. He was a remarkable man who had dropped out of the eighth grade to enter the Navy during World War I, after that served in the Army and the Marines, and then completed school and earned his bachelor's and master's degrees. He went on to become a professor of Latin

American and Far Eastern history, then a member of the House of Representatives, and finally a senator in 1952. Now in his seventies, Mansfield became both my mentor and a close friend as I settled in.

I asked for assignments to the Government Operations and Foreign Relations committees. Government Operations oversaw the organization and efficiencies of government. Its jurisdiction was the broadest of any on Capitol Hill, allowing it to look into any phase of federal activity. Its Permanent Subcommittee on Investigations had a history of rooting out waste, fraud, and corruption, in and out of government. It offered many approaches to improving the ways things worked. And I wanted Foreign Relations because I thought it was an avenue for pursuing peace; I had seen enough of war. I got Government Operations right away, but had to wait for a seat on Foreign Relations.

During my first year I turned down most of the speaking invitations I received and stayed put in Washington. I maintained a good attendance record, learned the ropes, worked quietly for my positions, and was successful at adding floor amendments to legislation on energy, civil rights, and foreign policy. Being an effective senator was my agenda.

Within a year into my term, however, the 1976 presidential campaign season approached, and I saw my name among those mentioned as a vice presidential candidate. Henry "Scoop" Jackson, the Democratic senator from Washington, was making presidential noises, and my name cropped up with his as a potential vice president.

But Jimmy Carter, the former Georgia governor, had

gotten started early and had done an amazing job of organizing and making good showings in the Iowa caucuses and the early primaries. Once he won the Ohio primary in May, he had the nomination sewn up. Before the Democratic National Convention opened in New York in mid-July, I and several other vice presidential possibilities were summoned to Plains to make our beauty-pageant walks. Senators Walter Mondale of Minnesota and Edmund Muskie of Maine were among the group.

I got along fine with Carter. He was a former Navy submariner, and we had technical and scientific experience in common. I had some appreciation of the part of his background he seemed to like to bring up most—that he was a peanut farmer. I had plowed enough fields in my time. He obviously was a man of great intelligence and personal rectitude. I thought I could work well with him if chosen.

Still, I knew I was a long shot going in. Both Mondale and Muskie had more experience in government. But other factors may have been involved as well.

Robert Strauss, the Democratic party chairman, had named two keynote speakers to open the convention. I was one. I had always been more comfortable with the straightforward talk of engineering than with the rhetoric so dear to politics. So I worked on my speech long and hard, writing and rewriting and practicing with a TelePrompTer.

On the first night of the convention the other keynoter went first. Barbara Jordan was a black congresswoman from Texas and an electrifying speaker. She was a political evangelist who had the delegates jumping out of their seats as if to say "Amen!" Her speech set off a standing ovation and applause that went on and on, while Strauss pounded the gavel and finally gave up. When he shrugged and waved me to the podium, I knew I was in trouble. Jordan was a very hard act to follow, and my speech did not come off nearly as well as I had hoped. It set out the principles to which the Democratic party was committed in that first post-Watergate election, but apparently I failed to suffuse those principles with passion. I could see the attention of the delegates wandering as I spoke.

The television commentators didn't like it, either. Walter Cronkite came right out and called it dull.

Another possible factor was out of my control.

Carter's wife, Rosalynn, was a big part of the Carter brain trust. She sat in on the interviews with all the hopefuls. She had met Annie and seemed to like her, but afterward it was reported that Mrs. Carter was put off by Annie's stutter. Annie still had a vestige of it. She always will; it worsens when she's tired and is unable to muster the concentration to apply the physical effort necessary for her to speak smoothly. Stutterers never break through into that territory where speech comes perfectly naturally and without thought; it always requires focus and physical exertion. If the report was true, Mrs. Carter thought that Annie's less-than-perfect speech would be a handicap to the campaign. It shocked us and it hurt, and it was never denied to us directly.

Carter called our suite to tell me his decision. He said he'd picked someone else, that he had enjoyed our visits, that he wanted to remain friends, and that he hoped he could count on me in the campaign. I assured him he could, and he stayed at our house in Columbus later when he campaigned in Ohio. Mondale was his running mate.

"Well," I told Annie when I hung up the phone, "we wondered who was going to cut the grass at home this weekend. It's going to be me."

Carter carried Ohio by eleven thousand votes. It was a closely fought election, but the nation was ready to consign Watergate to history, and he won a narrow victory, with the Democrats retaining control of Congress.

Before the election, Mike Mansfield had surprised me with an invitation. He was planning a fact-finding trip to China, which he had first visited as a young serviceman, and asked if Annie and I wanted to accompany him and his wife, Maureen, and a group of Senate staffers.

It is unusual for a majority leader to ask a freshman senator to go along on such a trip. But China had fascinated me ever since I served there between the wars. I was still waiting for a seat on Foreign Affairs, and Mike had assured me it was pending. This was an opportunity not only to renew my interest, but to expand on the steps Nixon had made in opening China to the West. Then, soon before we left, Chairman Mao Tse-tung died. We offered to defer the trip, but the Chinese urged us to come anyway, and it seemed even more important to be able to see the country and meet informally with its officials in the wake of its revolutionary leader's death.

We traveled for more than a month, between mid-September and mid-October of 1976. China was in mourning. We entered the country at Shanghai, and visited communes near Nanjing where garbage was converted into methane gas for heating and cooking. Pursuing my lifelong interest in medicine, I asked to tour training centers for the "barefoot doctors," paramedics who serve medical needs in rural areas, and to observe operations where acupuncture was used for anesthesia. In Beijing our informal talks covered such areas as trade, the absence of a legal system on which businesses ready to enter China could rely, and a host of China-Taiwan issues, such as cross-strait family visits and direct trade and mail between the two. After Beijing we flew a Soviet-built Chinese Airlines plane to central Asia and the polyglot crossroads of the old Silk Route at Urümqi, then to the Turpan Depression to see irrigation projects designed to bring agriculture to the eastern Taklimakan Desert. I knew the Chinese tested nuclear bombs at Lop Nor, and asked to visit it, but our guides professed not to know where Lop Nor was.

As this suggests, the Chinese were very circumspect. My questions about their expanding rural electrification, agricultural plans, and other scientific matters were usually answered with the stereotyped response: "We are guided by the thoughts of Mao and the will of the people." But I knew there was more than that to the beginnings of massive electrification projects, for example. And when, after repeated requests, I visited a plant where huge electrical generators were being built, I saw at a meeting that one of the engineers understood English. In the conversation that followed—reluctant on his part—I learned that he had a degree in electrical engineering from the University of Illinois and that he was the designer of the generators. No matter what the post-Mao rhetoric said, China was clearly relying on education obtained in the West to move ahead in some key areas.

In the throngs of people in drab Mao jackets, the

hordes of bicycles that provided the bulk of China's urban transportation, the electric lines that were beginning to stretch into the countryside, you could see a land of vast and complex energies and ambitions. It was awakening, and how it viewed the West was a matter on which history might hinge. Mike's report to the Senate and the Foreign Relations Committee foresaw a need to expand relations and seek areas of common ground.

Back in the Senate I got my assignment to Foreign Relations and moved from the Committee on Energy and the Environment. I was made chair of the Subcommittee on East Asian and Pacific Affairs. And I found that the work of a third committee also interested me.

I had never gotten over the shock of seeing my father's retirement nest egg consumed by the costs of his cancer treatment. My parents would have lost their house if Annie and I had not been able to help them. Dad would have died destitute, and Mother would not have been able to live out her remaining years in the place that had been her lifelong home. Annie's father, suffering from Parkinson's disease and unresponsive to treatment, had also experienced difficult final years. When Annie was named honorary chairperson of the Nursing Home Association during the two years before I was elected to the Senate, she had visited nursing homes and stayed overnight in several of them to learn more about the treatment patients in long-term care received, and told me what she had seen. So I asked for a seat on the Special Committee on Aging. It couldn't write legislation and primarily served an informational purpose. But the demographics were changing as increasing numbers of Americans were growing older, and I thought I could use what I had learned from our own family's experience to improve the future for others. I remained committed to this goal throughout my years in the Senate.

In July 1978 President Carter tapped me to head the U.S. delegation to the ceremonies marking the independence of the Solomon Islands, and I returned to the Pacific.

The Solomons had resonance for me. They were where the Japanese tide of World War II was turned with the bloody advance of the Marines across Guadalcanal and the heroics at Henderson Field by Joe Foss and the "Flying Circus," fighting that had inspired me to become a Marine. They were where John F. Kennedy's PT-109 was sunk by a Japanese destroyer. A former British protectorate east of Papua New Guinea, the Solomons are a wide stretch of ocean away from the Marshalls, where I had served. But I thought the entire area—remote, palm-fringed, and lightly populated—was too often ignored. U.S. policies for the newly independent South Pacific islands were among the oversight responsibilities of the Foreign Affairs subcommittee that I chaired, so I accepted the assignment eagerly.

Richard Holbrooke, who twenty-one years later would become the U.S. ambassador to the United Nations, headed the East Asian desk in the State Department. We had worked closely together. He was part of the delegation, along with a group of Carter's personal representatives. One of them was an elderly South Carolinian, the Reverend I. Dequincey Newman.

The route to the Solomons took us through Port Moresby in Papua New Guinea, where we had an extra day and the memorable chance to visit a remote village. We flew in a small plane to Popondetta, and from there to Sila in the interior. From an impossibly short, uphill landing area, we hiked along a narrow jungle trail to Numba, so isolated that it and its people had been discovered by a missionary only seventeen years earlier.

Marija, the tribal leader, was a wizened eighty-eight. As the missionary acted as interpreter, he spoke of the time before the village was discovered. "When I was a young man, we fought all the time," he said, pointing to scars on his legs. "Sometimes when we stopped fighting, we would eat parts of each other. And then the missionary came, and we don't do that anymore. It is much better now."

I thought back more than sixteen years to my orbital flight, and the "take me to your leader" message I had composed in case I landed in a jungle like the one surrounding Numba and encountered cannibals. I wondered if, in Marija's case, it would have worked.

My gift to Marija was a gold bicentennial medal on a red ribbon, and I hung it around his neck with care. He, in turn, took from around his neck the medallion of sinew, bone, and feathers that signified his leadership, and I bent down as he placed it around my neck. The missionary had mistakenly told Marija I had been to the moon. Numba's people believed that when humans and animals died, their spirits went to the moon. That was why it was white. No one had come back from the moon in their experience, and they wanted to know if they would be able to touch me. They seemed surprised that they could. I had to shake hands, or otherwise be touched, by almost everyone in the village.

Reverend Newman was a black man, and his presence with our group startled and then engaged the Aborigines. We were among the first Americans they had seen, and they had thought all Americans were white like Europeans. When Newman stopped to rest on our hike back to the airstrip, a group of young people stopped with him, would not proceed until he did, and then surrounded him protectively as he walked. They treated him as if he were a venerated grandfather.

The visit to Numba wasn't one of the most important moments in a twenty-four-year Senate career, but it was one of the most unforgettable.

The Camp David Accords in September 1978 were President Carter's finest hour. They brought about the landmark peace treaty between Israel and Egypt the following March. Carter, seeking new avenues of peace, courted other Mideast leaders, including King Hussein of Jordan.

I had met Hussein during an earlier trip to the Middle East, when three years after the most recent Arab-Israeli war Senator Abraham Ribicoff had led a delegation seeking to understand all sides in the conflict. My dominant memories were of the obscene luxury of the shah of Iran's palace in Tehran, and a night on the Nile at the summer residence of Egypt's president, Anwar Sadat, at which Sadat spoke repeatedly, and fervently, of the region's need for peace. In Amman, our second stop after Israel, the king met our delegation, and he and I had an instant rapport.

Hussein was a pilot whose idea of relaxing after a hard day was to go to one of his air bases and fly a fighter jet for an hour or two. We talked with great enjoyment about the joys of flying.

Since then, Hussein's wife, Queen Alia, had died in a helicopter crash. He had remarried by the time Carter held a state dinner in his honor in June 1980.

Hussein's American-born new wife, Queen Noor, was the former Lisa Halaby. Annie and I had known her as a young teenager, when we skied at Sun Valley with Bobby and Ethel Kennedy, and she was there with her father, Najeeb Halaby, and the rest of her family. Jeeb had been a Navy test pilot and then head of the Federal Aviation Administration, and had gone on to an executive career with Pan American Airways. Lisa's marriage to King Hussein meant that we were acquainted with both sides of the family.

I knew that Hussein had hoped to visit the Smithsonian's National Air and Space Museum on one of his trips to Washington. It had been arranged, and postponed, more than once because his schedule was too full. During a Senate Foreign Relations Committee luncheon on this latest visit, I asked the Jordanian ambassador if the king might like to see the museum after the state dinner. He thought it sounded like a good idea, and eventually we got word that the trip was on, with me as the tour guide.

We had a delightful time at the dinner. Toward the end, the ambassador told me that Hussein would like to have us ride over in his limousine. We went to the king's waiting car while the postdinner entertainment was winding down. Soon the front door of the White House opened and the president and the First Lady ushered out their royal guests, the last and brightest stars in the evening's glittering social and diplomatic crown. Carter and Rosalynn walked the king and queen and two of his sons to the door of the car. When the president saw Annie and me in the backseat, his jaw dropped. "John, what are you doing here?" he asked.

"The king wanted to go through the Air and Space Museum, Mr. President," I said. "Would you like to join us?"

"No, I don't think so," said Carter, laughing. He and Rosalynn retreated into the White House while Annie and I went off with his state guests to talk airplanes, flying, and space until one in the morning.

One of my issues during my tenure in the Senate was nuclear nonproliferation. Five nations—the United States, the Soviet Union, Britain, France, and China—had acknowledged having nuclear weapons. I had been through World War II and Korea and knew what happens in conventional war. The horror of a nuclear holocaust was beyond imagining. I had resolved that I would join forces with whoever in the Senate was seeking ways to reach international agreement on limiting nuclear weapons and preventing their spread. Such an agreement had never been possible over conventional arms. But the issue had no champion when I arrived, so I took it on.

My approach was two-pronged: to seek agreement among nuclear nations to reduce stockpiles and bring them under control, while using a commonsense carrot-and-stick approach to persuade nonnuclear nations not to pursue the nuclear option. I thought we could promise cooperation in developing peaceful uses of nuclear energy, while using trade and economic sanctions to prevent the spread of weapons.

At the same time I recognized the importance of maintaining superiority in arms over the Soviet Union. Carter agreed. But the way to achieve this brought us into frequent conflict.

I supported the president's program for the most part, but I thought it was a mistake for him to halt production of the B-1 bomber. My test-pilot and Bureau of Aeronautics experience had taught me that the B-2, the Stealth bomber that was scheduled to replace it, with aerodynamic controls and radar evasion technology that had yet to be proven, would need extensive testing. I believed it would cost more and take longer to come on line than even the most pessimistic projections, while the B-1 was proven technology. And I worked hard, and successfully, to block his plan to withdraw American troops from South Korea. But the area where we clashed most bitterly was over verification safeguards of the Strategic Arms Limitation Treaty—SALT II—that the Carter administration was negotiating with the Soviet Union.

As a proponent of nuclear nonproliferation, I supported the terms of SALT II. But after the Iranian revolution, when the shah fled Iran in January 1979 and the caretaker government of Shahpur Bakhtiar was overthrown a month later, I knew that two key radar listening posts in northern Iran, required for verification of Soviet compliance with the treaty, were no longer available to us.

I was determined that the Senate not approve a treaty that could not be verified.

In April 1979 Annie and I were scheduled to travel to Groton, Connecticut, where Annie was to christen the newest Trident nuclear submarine, the USS *Ohio*, and I was to speak about the treaty. Rosalynn Carter was to be in Groton also. Her role in the program was to dedicate the keel being laid for the next submarine in the Trident line, the USS *Georgia*, and also to make remarks.

I sent my speech to the White House in advance, standard courtesy in politics. I had written that the treaty was unverifiable and that we should demand both advance warning of Soviet missile test launchings and American flyovers during the tests. That led to a phone call from the president that developed into a harsh exchange. No president before or since has ever talked to me that way, and I've never spoken that way to any other president. Carter said I could kill arms limitation while the treaty was still being negotiated.

"I support this treaty as much as you do," I told the president. "But I don't trust the Soviets, and I'm not going to vote for it without a way of verifying it." But I did agree to remove from my speech the demand for flyovers and advance warning of Soviet tests.

On the plane to Groton with Mrs. Carter, the coolness was palpable. I made my remarks, saying verification had to be better defined prior to signing or it would risk disapproval in the Senate. Mrs. Carter's demeanor was icy as she began her speech, saying, "Senator Glenn understands . . . premature public debate on issues such as this can be very damaging." It was as if the Senate had no role in the process. I knew about secret talks to replace the monitoring capability, but I didn't know if they would ever prove successful. My bottom line was, I would support the treaty when I knew it could be verified.

The president and Leonid Brezhnev signed SALT II in

June, but the Senate never approved it. Carter asked the Senate to delay ratification after the Soviet invasion of Afghanistan on January 3, 1980. And by then Carter had the takeover of the American embassy in Tehran to cope with, a nightly spectacle played out on television of mobs chanting anti-American slogans and holding dozens of our citizens hostage. The hostage crisis was beyond the president's control, but together with economic problems at home—inflation and high interest rates—it helped end his presidency.

The voters of Ohio reelected me overwhelmingly in 1980. The results reversed those of the national election, in which Ronald Reagan won a landslide victory over Carter and carried Ohio by nearly half a million votes. One Washington outsider had replaced another, but Washington embraced Reagan, with his cheerful bonhomie and simply stated, firmly held convictions, in a way it had never embraced Carter.

A side benefit of my substantial victory was a $100,000 campaign surplus. This allowed me to attend the kinds of state Democratic party events around the country that traditionally sent invitations to presidential hopefuls.

The presidency was not a dream of mine, as the Senate had been. Back in 1976, early in my first term as senator, Steve Kovacik, who had run my successful campaign against Metzenbaum, and Mike DiSalle, Ohio's former governor, wanted me to explore a run for president, but I thought it was far too soon to entertain such ambitions, if ever. Now, however, what I saw of Reagan's administration made me think about it anew. While I had felt at first that Reaganomics deserved time to work, I quickly realized that it was, as I put it in a speech, "economic laetrile," a reference to the much-ballyhooed cancer drug for which great claims were made but no results ever produced. I was not of the thinking, as Reagan was, that government had no place in people's lives. With a million workers out of jobs a year into his administration, and drastic cuts in key areas like education and basic scientific research, I thought the pain Reagan was imposing on the citizens least able to defend themselves, and on programs that would pay future dividends, was cruel medicine.

I also thought our political system was in trouble. Since Vietnam it had deteriorated into a name-calling contest between left and right. A candidate's stands on the Equal Rights Amendment, abortion, and school prayer had become more important than a comprehensive program that appealed to voters in the middle. The media were partly to blame for categorizing people, placing them in camps defined by words like conservative and liberal. I thought the center had been neglected in American politics, while the hot-button issues that people were shouting about should be left to individual choice.

The people of Ohio had looked at my views for a full term. If I had satisfied them, maybe I could try to do the same for the entire country. Ohio, with its industry and agriculture, its mix of ethnic groups, and its urban-rural balance, was a microcosm of the United States. I had had more attention and applause than any man needed in a lifetime. That wasn't what encouraged me to think about the White House. But I believed that the strident pull between left and right was potentially harmful to the nation. If I could bring the debate back to what I called the "sensible center," then I thought it would be worthwhile to try.

Three things had to be considered. Would my views benefit our country? Could we raise the money and put together a campaign? How would it affect our personal lives and what would it do to our family? Annie and I had talked all of this out. She had continued to improve her speaking, returning to Hollins for a second round of training and, most recently, working regularly with a speech therapist at the Walter Reed Army Medical Center in Washington. She also had spoken for the first time to a large audience, three hundred women in Canton, Ohio. Annie, who once would have been petrified at speaking to a crowd a tenth that size, told the story of her difficulties, fears, and humiliations as a stutterer. It was another first for our family, the first full-length speech she had ever given, and it reduced her audience to tears. I knew she was brave, dedicated, and determined. But we had both been around politics long enough to know how brutal a campaign could be. I needed to know if she was up to it. She told me she was game.

"I've put you through a lot," I pressed. "If this is something you don't want me to think about, or you don't want to get involved in, let me know."

"We made a life together. What you do is what I'll do."

In the fall of 1982 a group of staff members and advisors gathered with us over dinner at our house in Potomac, Maryland. It included Paul Tipps, the Ohio Democratic party chairman; Bill White, my administrative assistant; Kathy Prosser, my secretary; Mary Jane Veno— M.J.—my scheduler, troubleshooter, and political advisor; and Lyn, who had been close to all of my campaigns.

The thrust of the evening's discussion was whether I should seek the 1984 Democratic presidential nomination.

There were other meetings, too, with the result that I decided to put my toe into the water.

I approached the campaign preliminaries as a centrist. It was the position I believed in. Government should not run lives—it exists to set the principles and policies that help families and individuals achieve better lives. It is supposed to do the greatest amount of good for the greatest number, not cater to people on the fringes just because they are the loudest. It was also the position most likely to beat Reagan. But a pair of intangibles stood in my way.

With the loss of the Carter-Mondale ticket to Reagan and his vice president, George Bush, one of those intangibles was Teddy Kennedy, now a senator from Massachusetts, who had emerged as the unannounced front-runner for the Democratic nomination. A substantial core of voters saw Teddy taking his place after Jack and Bobby in his family's presidential lineup. And of all the potential candidates, a list that included Mondale and Senator Gary Hart of Colorado, Teddy was thought to be the one to beat, despite high negatives that had to do with the residue of Chappaquiddick fifteen years earlier.

The other factor was my reputation as a plodding speaker. I didn't see myself that way, but I had to admit that others did. The eloquence that I had been credited with following my address to the joint session of Congress after my flight in 1962 had been supplanted by the more recent memory of the 1976 Democratic National

Convention. Teddy, on the other hand, had the cadences and rhetoric of political speechmaking bred in his bones.

My relationship with Teddy was cordial. He was a colleague in the Senate, and I voted with him more often than not. As long as Teddy was in the race, or at least not out of it, he was a factor to be reckoned with.

But in December 1982 Teddy decided to spare himself and the extended Kennedy clan the intense and increasingly vicious media scrutiny that had become a part of presidential politics, and announced that he was not going to run.

This left the field wide open. It also forced me to accelerate my explorations. Mondale was probably the new front-runner. Hart had strength as a potential nominee. The other candidates included former Florida governor Reuben Askew, Senator Alan Cranston, Senator Ernest Hollings, and former senator George McGovern. I thought I had a broader base, wider name recognition, and a greater residue of support than any of them. I began to work on capitalizing on my strengths and trying to reduce my weaknesses.

Throughout my life I had always managed through hard work to neutralize my liabilities. I had sweated long nights to learn calculus when I was at test-pilot school, accumulated the credits for the degree equivalency that allowed me to become an astronaut, dieted and ran when I had to to make the weight requirement—in short, when I knew what had to be done, I was ready to do it. So that December I began working with a media consultant to punch up my speaking style and to prepare a speech that would articulate my theme: that many things remained right with America, that most people knew it, and that the answer to the problems we did have was not to splinter into factions of the right and left, hurling accusations at each other, but to buckle down and put our native ingenuity and generosity to work at solving them.

That month, on a visit to the border-straddling twin towns of Texarkana, Texas and Arkansas, I told business leaders that the United States remained "a land of unparalleled blessing, of unparalleled hope, of unparalleled opportunity. We remain that beacon of freedom that stands before the rest of the world." I invoked the scenes of fleeing Haitians who risked their lives, and lost, to reach the shores of America. And I returned to one of the quotes from Emerson I had seen in Bobby Kennedy's study on that bleak night in 1968: "This time, like all times, is a very good one if we but know what to do with it."

The speech was a success, and so through the end of 1982 and into the winter of 1983 I continued my exploration of the presidential possibilities. A poll in January showed Mondale ahead with 59 percent of potential Democratic voters. Twenty-eight percent picked me. That was a gap I thought I could close.

The more time went by in my unannounced campaign, the more I felt I was the one Democrat who could beat Reagan. My military background would counter Reagan's strident anti-Communism, which had led him to treat the Marxist Sandinista regime in Nicaragua as a major threat to democracy in Central America. I opposed funding the anti-Sandinista contras, on the grounds that they were little more than surrogates for the right-wing landowners who had held sway in the country for years. Although the Sandinistas were encouraging an insurgency in El Salvador, I didn't consider them much of a threat.

By March I had hit a good stride. Of the growing power of the Christian right in politics, I said, "Let us stop confusing the epistles of the New Testament with the apostles of the New Right." And I found an effective tag line in my call for a march toward "new horizons, guided by the light of old values." At the end of the month I marched into the lions' den by accepting an invitation to be the Democratic spokesman at the annual Gridiron Club dinner—an off-the-record gathering of the nation's media and political elite for a night of music, jokes, and laughter. Mondale had turned the invitation down, but I accepted. The Gridiron could make you or break you. James Reston of *The New York Times* wrote that I was making a mistake. But I felt that speaking there was a chance to reverse my image as a dull speaker before the opinion makers of the country, the very people who were keeping the old view of me alive.

I enlisted my friend Art Buchwald, among others, to help write the speech. I rehearsed it over and over. It was full of self-deprecating references, like the advice I supposedly had received from Democratic National Committee chairman Bob Strauss after telling him I was worried about making the Gridiron audience laugh: "Just give the speech you gave at the 1976 Democratic convention. The whole damn country laughed at that one."

The speech was a success. By the end of the evening I think I had erased the image among the most powerful journalists in the country that I couldn't make a speech or get a laugh.

On April 21, 1983, I announced my candidacy for the Democratic nomination at home in New Concord, in the gymnasium of the high school that had been renamed for me in the aftermath of my flight in 1962. The slogan I chose was "Believe in the future again."

Mondale and I were now the main candidates for the Democratic nomination. Polls in July placed us neck and neck; I had raised $1.6 million in campaign funds to his $1.2 million. And I was doing better in polls that pitted the Democratic hopefuls against Reagan.

That, in retrospect, created a false sense of security and pushed me toward organizational mistakes.

Rather than concentrate on the traditional early battlegrounds of a contested nomination—the Iowa caucuses and the first-in-the-nation New Hampshire primary—I started to focus on building a nationwide organization aimed at the "sensible center." But because I articulated positions that would appeal to Republicans and Democrats alike in a general election, I failed to excite passion among the core constituencies most likely to vote in Democratic caucuses and primaries.

On the advice of my campaign staff I moved ahead with plans to open offices in most of the states around the country.

Meanwhile, personal animosities that had been present in the campaign from the beginning became harder to control. Mutual suspicions marked the relationship between my Senate and campaign staffs and my campaign consultants. I was on the road too much to bring the warring factions at headquarters under control. I thought the problem would straighten itself out. It never did, but became a wound that drained life from the campaign.

Annie was by my side for all of the major campaign decisions and traveled with me most of the time, as did M. J. Veno. Lyn came in and helped when she could. One day during this time she was riding with me to my Senate office when I stopped for an event at the National Air and Space Museum. I went in alone while Lyn waited. Inside, surrounded by the artifacts of America's great leap beyond the bounds of gravity, I had a sudden feeling of displacement. The goals of Project Mercury and its successors had overwhelmed all the personal rivalries. We astronauts had had our differences, but none so great that we could not come together when it counted. Not only the country's prestige and glory were at stake, not just the amazing scientific strides we were making—lives were on the line. It had been the same in combat. Never had I trusted others so much, and I suddenly yearned for that ability to trust again, rather than having to find a footing in the shifting sands of politics, where alliances changed as quickly as the wind erases footsteps in a dune.

Back in the car with my daughter, I wondered out loud if I had made a mistake.

"No, Dad," Lyn said. "You got into it in the first place to try to make it better—to convince people that politics isn't bad and that the system can be improved. That's what you're trying to do now."

During the fall Mondale began to creep up in the polls again. The pundits placed much importance on the fact that the movie version of *The Right Stuff* was opening in the fall. Tom Wolfe's best-selling book was good, but what Hollywood did to it could have been titled *Laurel and Hardy Go to Space*. Somehow the movie's lukewarm reception had a chilling effect on the campaign. And with the Reagan recession weakening and the economy improving, the polls that showed me likely to beat the president also reversed, although I still fared better against him than any other Democrat.

In January 1984 I had offices open in forty-two states and announced on *Meet the Press* that I had opened number forty-three. It was an approach that was spreading my campaign resources—money, energy, and ideas—thin. I was looking ahead to the so-called Super Tuesday primaries in Alabama, Georgia, and Florida, so I didn't attend an Iowa Farm Forum all-candidate debate on farm issues. That was a serious mistake. I spent heavily on television in Iowa, but learned too late that caucus voters prefer personal contact, backed by an organization that impels them to the polls. I finished a disastrous fifth in the caucus voting on February 20, and I could have saved myself a lot of money by withdrawing then and there.

The Iowa caucus results sent a signal to New Hampshire and to the media. The headlines and commentary emphasized defeat. It was no longer my race and Mondale's. If I was to have some momentum going into Super Tuesday, I had needed a respectable finish in New Hampshire. But Iowa had painted me a loser, and Gary Hart vaulted over all the other candidates to win New Hampshire and supplant me as the Mondale alternative.

Once again I should have quit. But quitting has always been hard for me, and I gritted my teeth, borrowed money with the support of my strongest backers, and headed into Super Tuesday thinking that I might salvage my candidacy there. It was potentially my strongest area. But the defeats in Iowa and New Hampshire were too much to overcome, and my results in the three vital southern primaries were no better.

Three days later, on March 16, 1984, I played Kenny Rogers's "The Gambler" for my Senate staff. "You've got to know when to fold 'em," the country legend·sang. I announced my withdrawal from the race at a news conference in a Senate caucus room. "Although my campaign for the presidency will end," I said, "my campaign for a better America will continue."

CHAPTER 24

My campaign was almost $3 million in debt when I abandoned the race. It seemed to me that an electoral system so in need of money was ripe for abuse. As I turned back to the Senate, I decided to devote at least some of my attention to seeing that the voices of ordinary citizens with ordinary budgets weren't drowned out by the special interests and the rich.

The prospect of legislating once again reenergized me. It was just the tonic I needed to forget the mistakes of the campaign. Government Operations, now renamed Governmental Affairs, was not the kind of headline-grabbing vehicle that appealed to many senators, but I had seen in my nine years in the Senate that it could be an effective engine for improving the workings of government.

The committee had an investigations staff, but its resources were far outweighed by those of the General Accounting Office. The GAO is the investigating arm of Congress. All members of Congress can ask the GAO to look into areas of government that concern them, but the GAO shares with Governmental Affairs the specific mission of ensuring that the government is operating effectively and without waste. I thought it made sense to supplement the committee's resources by utilizing the GAO when we needed probes directed toward such goals.

Soon after Reagan was inaugurated for his second term—he defeated Walter Mondale in another landslide—a group of citizens from Fernald, Ohio, northwest of Cincinnati, came to my office to spell out their concerns about high cancer rates around the nuclear weapons plant there. Fernald is one of seventeen plants in eleven states that make up the chain of nuclear weapons production in the United States. Lisa Crawford, the energetic housewife who was the head of FRESH—Fernald Residents for Environmental Safety and Health—claimed that plant managers bypassed safety procedures, putting workers at risk, and that the plant emitted uranium dust over the small town. She also said that radioactive toxins were leaking from storage tanks into the water table.

I suspected the claims were overblown, but when I went to Fernald with some committee staff members, we quickly found the group's concerns were justified. Two hearings produced a litany of evidence that plant managers were grossly cavalier about safety and emissions. I requested a GAO investigation, and this revealed so many problems that I asked the Centers for Disease Control to begin an epidemiological study.

From those relatively small beginnings emerged a shocking picture of the entire network of nuclear weapons production within the Department of Energy. The abuses at Fernald, a plant early in the production chain, were relatively mild. With each step up, as the nuclear materials grew more refined and problems of radioactivity and toxicity more challenging, the safety and environmental problems worsened. The plants were far-flung and ranged from sites like Fernald to major installations at Hanford in Washington State, Rocky Flats near Denver, and the Savannah River plant in South Carolina.

The initial estimates to clean up the process and make it environmentally safe were $8 to $12 billion and a time span of several years. But as the GAO and the Department of Energy probed deeper, a frighteningly large problem emerged. As of mid-1998, it was estimated that it could take up to $300 billion and as long as twenty-five years to clean up past problems and make the plants safe for their workers and for the communities in which they are located—assuming the technology could be found to do it. The ongoing process of investigation and correction that began at Fernald and involves federal, state, and local governments is an example of the way I always envisioned government working in the service of its citizens.

As a final step, I introduced legislation creating the Defense Nuclear Facility Safety Board to provide ongoing oversight.

I was elected to a third term in 1986. Despite Reagan's continued popularity, the Democrats regained control of the Senate in this midterm election, and I became chair of Governmental Affairs. This allowed me to work toward laws that drew little attention, but that I'm enormously proud of.

The GAO, so effective in the nuclear weapons plant investigations, became a key factor in this new initiative.

Say "government efficiency," and the press is out the door and down the hall. Talking about it is as exciting as watching mud dry. But I saw no reason that efficiency in government had to be an oxymoron, and I was convinced that with the proper organization and accountability, a federal department could be as efficient as any business. Amazingly, I discovered that there was no requirement that departments be audited for the ways in which they spent their funds, or even that they lay out their plans and goals for the coming year. Every major business has a chief financial officer, but most departments of government had no such position. I sponsored legislation mandating CFOs for the largest agencies and requiring them to report not only to their own departments, but to the appropriate committees of jurisdiction in the Congress.

A companion measure expanded a corps of inspectors general assigned to root out fat, fraud, waste, and abuse in government. They, too, follow an independent reporting avenue, and they identify billions in potential savings every year.

I also pushed for streamlined procurement procedures and investigation of bloated inventories, like a ten-year supply of bar lights for military police vehicles, which led to new inventory controls in the Pentagon and other agencies. I'm convinced that measures like these have had a greater effect on the process of productive government than any amount of headline-making posturing.

My name cropped up as a vice presidential possibility again in 1988. This time the nominee was Michael Dukakis, the governor of Massachusetts. In the years since

Jimmy Carter had considered me, the political climate had changed and the scrutiny of potential candidates was intense. Attorney Zoë Baird was vetting me for the Dukakis campaign. She in turn hired the Fairfax Group, an investigative company. Their interrogators went into details and phases of my personal, medical, financial, and political life that even I didn't know about, and they were remorseless. Annie's ninety-three-year-old mother, the last survivor among our parents, had died that May, and they knocked on the door of an elderly woman who had cared for her at our home in Columbus to ask if she had ever observed me misbehaving when I came to visit. Looking for even a whiff of scandal, they didn't miss a trick. They found none, but Dukakis eventually chose Texas senator Lloyd Bentsen as his running mate and went on to be roundly defeated by George Bush in November.

With the initiatives I was pursuing through Governmental Affairs, I was more effective in the Senate than I had ever been. I was also active on the Armed Services Committee, to which I had moved in 1985, leaving Foreign Relations. I had done some productive work there, but I thought the committee had become a debating society that spent too much time in confirmation hearings and was having too little influence on the course of foreign policy. On Armed Services, despite starting over at the bottom of the seniority ladder, I thought I could apply my military experience to weapons procurement, readiness, and manpower needs, and also implement the same kind of efficiencies I had been promoting in Governmental Affairs.

But in 1989 the satisfying tenor of my Senate work was interrupted by singularly unpleasant and unwarranted press reports that for the first time in my life challenged my integrity. Long-simmering problems in the savings and loan industry, which had begun when it was deregulated earlier in the eighties, erupted into crisis early that year when those institutions started failing under the weight of bad debt. The behavior of the entire industry was suspect, and I found myself involved because of two meetings I and some other senators had attended back in 1987.

Sometime that March I had received a phone call from James Grogan, a former intern and now a lawyer representing and lobbying for Charles Keating, who had been a constituent and campaign contributor. Keating headed a California savings and loan called Lincoln Savings, which was being audited by the Federal Home Loan Bank Board. Grogan said the audit had gone on for almost a year and amounted to harassment. My office routinely opened twenty or thirty cases a week stemming from constituent complaints about the way they were being treated by the government. I asked Grogan to justify his charge, and he produced a senior partner at Arthur Young, the accounting firm employed by Lincoln Savings, who said such audits usually last no more than ninety days. So on April 2 I responded to an invitation from Senator Dennis DeConcini of Arizona, where Keating now lived, to meet in his office with Ed Gray, the head of the Home Loan Bank Board. John McCain, Arizona's other senator, and Alan Cranston of California were also at the meeting. The subject was the audit. Was it near a conclusion? What could Lincoln Savings expect? Gray said he didn't know the details and urged that we meet with the regulators overseeing the audit. They flew in from California, and joined by Don Riegle of Michigan, where Keating had business interests, we senators met with them on April 9. But when one of them said their investigation would result in criminal referrals, I closed my notepad and dropped all involvement with the matter.

In January 1988 Keating invited me to lunch with then House Speaker Jim Wright. At that time no criminal referrals had emerged from the Lincoln Savings audit. In fact, Grogan had told me the regulators and Lincoln Savings were working on a memorandum of understanding that would settle the investment practices and bookkeeping issues that the audit had raised. Jim Wright and I were friends. There was no reason not to have lunch, and rather than leave Capitol Hill, I arranged for us to eat at my Capitol office. The occasion was sociable and enjoyable. That May the memorandum of understanding was finalized and signed.

A year later the savings and loan crisis was full blown. Lincoln Savings became the biggest failure of all when its parent company filed for bankruptcy protection that April, and a renewed examination was indeed pointing toward criminal behavior. A firestorm broke in the press over the 1987 meetings with Gray and the regulators. Gray, who had left the bank board, had continually implied in interviews that Keating had brought undue influence to bear. Media reports focused on Keating's pattern of political contributions and whether he was trying to buy such influence. After a lifetime of adhering to scrupulous ethical standards, I found myself one of the "Keating Five." The Republican party in Ohio asked for an Ethics Committee investigation, and soon afterward Common Cause joined in the request.

Late in 1989 the committee sent a letter asking me to

respond to the complaints filed by the Ohio GOP and Common Cause.

My reaction to these slurs on my reputation was a combination of fury and resolve. I was going to fight the complaints with every fiber of my being. I had done nothing wrong, and I knew it. I provided the committee with my files for the entire period. They ranged from detailed personal and campaign financial records to scraps of paper from my desk.

By the summer of 1990 the Ethics Committee's tough-minded outside counsel, Robert Bennett, recommended that John McCain and I be eliminated from the investigation. But the partisanship that has increasingly infected the Congress on both sides of the aisle overrode common sense. The committee's Democrats, led by Howell Heflin of Alabama, would not excuse McCain, because he was the only Republican among the five—and if they didn't excuse him, they couldn't excuse me. The result was Ethics Committee hearings that started after the midterm election in 1990. Outside of people close to me dying, these hearings were the low point of my life.

Ultimately, on February 27, 1991, the committee issued a report saying I had "exercised poor judgment" in hosting the lunch with Jim Wright and Keating three years earlier. The issues about which the press, the Ohio Republicans, and Common Cause were up in arms failed to produce a single finding. I thought even this mild language was unjustified, but I accepted it because I wanted to put it all behind me.

McCain received a similarly mild rebuke. The other senators, who had been more involved with Keating, were judged more harshly. Keating was later found guilty of securities fraud and sentenced to prison.

The episode wound up confirming Bob Bennett's original recommendation, but it cost me $520,000 in legal fees and great personal anguish. It saddened and angered me, my staff, my friends, and every member of my family. But it fired my determination to return to the Senate for a fourth term. There was no way I was going to let an unwarranted inquiry end my commitment to public service.

The 1992 reelection battle was tough and rough. No senator from Ohio had ever served four consecutive terms. I still had campaign debts from 1984, and Ohio lieutenant governor Mike DeWine, a Republican, focused on those debts, and on Keating and the Ethics Committee hearings and report, in a viciously negative campaign.

It was also a presidential election year, and I campaigned hard for Bill Clinton in Ohio. I had met Hillary and Bill in Arkansas when I was running for the Democratic presidential nomination, and liked them both. I responded to his vigor and his ideas and the fact that he was always thinking of ways the government could be put to work on behalf of its citizens. In 1992 the so-called character issues didn't bother me. I've always believed that the press and the public have no business poking around in people's bedrooms. What mattered was whether the program put forward for the country was a good one. And at the Democratic National Convention that summer, Ohio put Clinton over the top as I announced the state's votes on his behalf.

The state's election campaign was a unified campaign. That is, a lot of the advertising and posters proclaimed the entire Democratic ticket—Clinton-Gore, Glenn for reelection to the Senate, and the state ticket. With less than three weeks to go before the election, Clinton came to Ohio for a rally in Springfield. I was to introduce him.

The Clinton–Bush numbers were close in Ohio; in fact, one of Clinton's national staff told me they had decided to concede the state, pull out of the unified campaign, and spend their money elsewhere. My polling said Clinton could win Ohio if he worked at it. So when I introduced him at Springfield, I told his managers I wanted some time with him after the rally. They put us into the limo together on his way to the airport. I gave him our polling figures and told him he could win in Ohio, but he had to do more, not less. I talked to him like a Dutch uncle, and kept on talking as we climbed on his plane, holding up his departure.

It was for a good cause. He took my advice, and within a day or so we got word they were staying in the unified campaign and maintaining their resources in the state. And on election night Clinton's victory in Ohio tipped the balance and commentators started declaring him the winner. He thanked me later, in an inscribed photo, for refusing "to let me give up on Ohio. Now we must deliver for Ohio and the nation."

I won my race, too, by a healthy margin over DeWine's spiteful campaign, which a news magazine called one of the dirtiest that year.

I had always been a staunch defender of the NASA budget in the Senate. After the Apollo program's spectacular series of moon flights and landings, NASA had entered a workmanlike phase with the space shuttle program. Public attention flagged when the shuttle flights became routine, but the *Challenger* explosion in 1986 reminded people that space exploration still required amazing courage, precision, and technical skill. When the flights resumed after *Challenger,* however, attention waned again and it often took forceful advocacy to persuade the Senate's budget overseers that neglecting NASA's scientific mission amounted to neglecting the future.

Its newest venture was in the planning stages: the international space station that would bring sixteen nations, including former Cold War enemies, together in an orbiting laboratory doing immensely valuable work. So in early 1995 I prepared for debate on the agency's budget by reviewing the latest NASA materials, including *Space Physiology and Medicine,* a book written by three NASA doctors. As I read, a chart jumped out at me.

The chart listed fifty-two different types of physical changes that happen to astronauts in orbit. Among them were muscle system changes, osteoporosis (bone loss), disturbed sleep patterns, balance disorders, a less responsive immune system, cardiovascular changes, loss of coordination, declines in drug and nutrient absorption, and a change in the body's blood distribution patterns. Back on Earth, subject again to the force of gravity, the astronauts recovered.

Years on the Special Committee on Aging—and the fact that I was then seventy-three—made me realize that these effects were familiar to the earthbound elderly. That weekend I pored through the *Merck Manual of Geriatrics* and confirmed the similarities.

I thought, "Here's something that ought to be looked into."

The more I considered these connections, the more fascinating they became. Questions started running through my mind. What would happen if somebody older went up? Would an older person be immune to the phenomenon of "space aging," since this had already happened to him? If he was affected, would he recover as younger astronauts did, or would there be lingering effects? Could NASA, by sending an older person into space, learn something that would help younger astronauts avoid some of these effects in the longer-term space flights that lay ahead? Could we figure out what within the human body turns these systems on and off? Would this allow us to erase some of the frailties of old age on Earth?

The answers to these questions seemed potentially quite valuable.

So I called the authorities who had written this book, Arnauld Nicogossian, Sam Pool, and Carolyn Huntoon. They told me about a pamphlet written by Dr. Joan Vernikos, the director of NASA's Life Sciences Division, which pointed out the same physical changes and said NASA hoped to research them. I talked to two doctors, Richard Hodes and Richard Sprott, at the National Institute on Aging, and learned that they also anticipated researching them in due course. I talked to Dr. Robert Butler, the founding director of the National Institute on Aging, with whom the Senate Special Committee on Aging and I had organized hearings during the 1970s on "the graying of nations" that had attracted gerontologists from all over the world. He, too, thought these subjects were worth investigating at some time in the future.

I wondered why the research had to wait. Shuttle flights were going up regularly. Why couldn't room be made on one of them for some experiments?

And then I began to think, "Why not me?"

I had always wanted to go back up. The fact that Bob Gilruth had never been able to put me on another flight had been a source of great frustration. Then, of course, I resigned and went on to other things. Now I realized I could do some very good science and have another flight at the same time.

The science would be the key. NASA wasn't going to give me a free ticket back into space. I'd have to convince NASA director Dan Goldin that the research was justified.

I was in regular contact with Goldin on budgetary and other NASA matters in the Senate, and I started using these meetings to point to the exciting possibilities of the research I had in mind. "Somebody should be looking into this," I'd tell him. "Maybe if you sent an older person into space, we could start to learn what turns these systems on and off."

Dan would just smile and nod and say it sounded interesting and maybe worth investigating.

"Well, if you decide to do it," I would press, "I'd like to volunteer."

At that his grin would widen, as if surely I was kidding.

I kept mentioning my interest over the next several months. Then, on a cold, bright December day in 1995, Annie and I attended a reception and luncheon in the Kalorama neighborhood of Washington. The guest of honor was Valentina Tereshkova, who as a Soviet cosmonaut in 1963 had been the first woman in space. The usual suspects with regard to space were there. Dan and I were talking at the upstairs reception, and I regaled him with my latest contacts at the NIA, how eager the institute was to develop more information, and the importance of my idea.

On our way downstairs to lunch, Dan turned to me and said, "You're serious about this, aren't you?"

"Serious as I can be," I told him.

The 1996 elections provided some of the best fun I've ever had in politics. President Clinton made half a dozen trips to Ohio during his reelection campaign, and I traveled and campaigned with him. His vigor on the campaign trail, and the way he engages people, is a wonder to behold. His grasp of issues is phenomenal. I took a lot of pleasure in introducing him at the State Fairgrounds in Columbus, where I said, "You don't move the country ahead by taking Big Bird away from five-year-olds, taking school lunches away from ten-year-olds, taking summer jobs from fifteen-year-olds, and taking student loans from twenty-year-olds." The president liked that line and threatened to steal it.

Before the event, on the way from the airport, I had told him about my conversations with Dan Goldin, my research, and the fact that I was seriously pushing to go up again. "You know, it will have to come to you for approval," I pointed out.

We were returning to the airport when he brought the subject up again. He grinned, leaned over, and slapped me on the knee. "I hope that flight works out for you," he said.

Clinton won reelection handily. His race against Bob Dole was the most expensive presidential campaign to date. Both sides carried accepted methods of raising money to extremes. Allegations of campaign financing irregularities took center stage after the 105th Congress convened in January 1997. Trent Lott, the Mississippi Republican who was now Senate majority leader, announced that the Governmental Affairs Committee would open hearings on the issue. Fred Thompson of Tennessee had become the chair of Governmental Affairs when the Republicans gained the majority in the 1994 election. I was the ranking minority member.

I went to Thompson's office to talk about the process. I saw the hearings as a golden opportunity to move toward real campaign finance reform. The corrupting influence of money in politics fed cynicism among ordinary citizens, who felt they had no voice in government and stayed away from the polls, undermining the democratic process. I advocated a wide-ranging, bipartisan probe that would look at irregularities across the board and perhaps lead to legislation to correct abuses. Thompson, who had been a counsel to the thoroughly bipartisan Watergate committee, agreed. I suggested that we combine majority and minority staffs and budgets into a unified committee, and review subpoenas before they were issued. Again he sounded supportive.

Several weeks later, without warning, I got a list of fifty-one subpoenas, all aimed at Democrats.

I called Fred, alarmed that an opportunity to move toward real reform was being lost. I knew he personally favored some version of campaign finance reform. He had cosponsored a reform bill proposed by John McCain. But he said "his people"—recently assigned to the committee by Trent Lott—"wouldn't go along with it."

When it became clear that Thompson and the Republicans were aiming only at Democrats and principally at the White House, I fought hard to expand the committee's focus. But it did no good. By the time the hearings concluded, near the end of 1997, that golden opportunity had again deteriorated into partisan squabbling, and campaign finance reform remained as elusive as ever.

Well before the battle over the campaign financing inquiry was fully joined, Annie and I had discussed whether I should seek another term.

In my twenty-two years in the Senate, I had watched the legislative process change. There was always partisanship—that was the nature of the system. Although it produced disagreement and debate, it ultimately forged budgets and laws on which reasonable people could differ but that worked for most. In general, lawmakers performed their duties in an atmosphere of mutual respect.

This was no longer the case. By the 1994 election, we had single-issue candidates, the demonization of government, the sneering dismissal of opposing points of view, a willingness to indulge the few at the expense of the many, and the smug rejection of the claims of entire segments of society to any portion of the government's resources. Respectful disagreement had vanished. Poisonous distrust, accusation, and attack had replaced it.

The partisanship that was so evident in the campaign

financing inquiry was a symptom of these changes, but this was not new. Back in August 1993, when the Democrats were still a majority, the Congress passed Clinton's proposals for overhauling government and budget priorities without a single Republican vote. Republicans decried them, yet they set the stage for what by 1999 had become the longest sustained favorable economic period in the history of the country—lowest interest rates, lowest inflation, and lowest unemployment in modern times—while reducing federal employment to its lowest level since Kennedy's administration.

The fact that no members of the opposition could see this was disheartening. Good proposals were vilified on the basis of personalities and party. I had never forgotten Bobby Kennedy's fervently stated belief that politics is an honorable profession, but a lot of politicians were acting as if they alone were the honorable ones.

What was worst of all was the prospect that young people who were entering the years in which they could make the greatest contributions to their country would be infected by the self-interest and hopelessness that had invaded political discourse. I thought with anguish of the degrading of public participation and public service as activities that somehow had become unworthy. I believed this cynicism could infect the body politic to such an extent that it could rob the country of its lifeblood—the constant reinvention that our democracy requires if it is to remain strong. The answer to our problems, it seemed to me, was not to savage the nation for its shortcomings, but to create inspiration through its strengths.

Annie and I discussed these concerns often. My belief in public service remained as strong as ever. But at the end of a fifth term I would be eighty-three, and I was content to leave Senate longevity records to Strom Thurmond. There were still some things I wanted to do, and I thought I could do them better as a private citizen.

So on February 20, 1997—the thirty-fifth anniversary of my orbital flight—Annie and I returned to Brown Chapel on the Muskingum campus, and I announced my intention to retire from the Senate at the end of my fourth term. "For the next two years," I said, "I intend to continue being one of the hardest-working senators in Washington. After that, I believe it will be time to serve my country in other ways. When I leave the Senate, it will not be an end, but a new beginning."

I took pride in what I had accomplished. The record even suggested a legacy in certain areas. Charles Bowsher, the U.S. comptroller general, had called the Chief Financial Officer Act the federal government's best step toward improved financial management in forty years. Three pieces of legislation I had worked on to curb the spread of nuclear weapons—the Glenn-Symington amendments to the Foreign Assistance Act, the Nuclear Non-Proliferation Act of 1978, and the Nuclear Proliferation Control Act of 1992—remain cornerstones of our nuclear foreign policy. Perhaps as important, I had encouraged U.S. and United Nations support for the International Atomic Energy Agency in Vienna, and made numerous trips to promote nonproliferation with government leaders and nations around the world. I had worked on the Taiwan Relations Act, which provides for a continuing relationship with Taiwan while recognizing the People's Republic of China. I authored the Congressional Accountability Act, which ended what I called "the last plantation" by bringing the

U.S. Congress under the same civil-rights, employment, labor, and health, safety, and environmental laws that apply to the rest of the country.

I foresaw that, by the time my fourth term ended, I would have cast almost ten thousand votes. These covered education, concerns for the elderly, environment and health, civil rights, regulatory reform, the future of science, military readiness, and many other areas. Each had contributed in small or large measure to the painstaking march of our democracy. I could not have asked for anything more rewarding.

My campaign for a shuttle flight had taken me to Dr. John Eisold, the U.S. Navy admiral who was the attending physician for Congress and had an office in the Capitol. Geriatrics was one of his medical specialties. I had sought his advice about the importance of the links between aging and space, told him I thought this research should go through whether I was involved or not, and confided to him my hopes of going up. I requested a thorough physical before I pursued it further. Beyond the normal workup at the Bethesda Naval Hospital, I took every heart exam known. I went through liver, kidney, and pancreatic scans, a whole-body MRI, and one for my head alone. When the results were in, Dr. Eisold told me he saw no physical reason why I shouldn't go up again.

That's when I called Dan Goldin and asked for an appointment. When I met with him at his office in NASA's headquarters in southwest Washington, I reiterated that the research warranted sending an older astronaut into space, and that I wanted to be the one.

Dan said, "If I'm going to consider this, there are two conditions. It has to be good science. And before I even think about your going, you'll have to pass the same physical requirements as all the other astronauts."

I already knew I could do that. So when I went to the Johnson Space Center for the annual physical that all former astronauts receive, I asked the doctors to give me every test current astronauts receive before being cleared for flight. Meanwhile, NASA undertook a stringent peer review of the scientific knowledge to be gained.

Annie was well aware of my hope to return to space. I wouldn't have wanted to conceal it from her even if I could. She knew my interest in the space program had not wavered and that I had even inquired about a place on one of the Apollo moon missions. She was used to my flipping on the television at all hours to watch a shuttle launch or a landing. But as I talked about the information that could be gained from sending an older astronaut on a mission, she reacted with what I would politely call a distinct coolness.

What she said, the first time I mentioned wanting to go, was "Over my dead body."

Dave and Lyn had similar reactions. I think they were primarily concerned for Annie.

But as time went on, she became as interested as I was in the scientific possibilities. And as she softened, Dave and Lyn, too, sounded reconciled to the idea. The entire prospect became more real when I endured the rigorous physical examinations at the Johnson Space Center that were reminiscent of those at the Lovelace Clinic in 1962.

We went to Vail for the 1997 Christmas holidays. We

had had a place there since 1980, a retreat where the family gathered to ski or hike, depending on the season, and to relish being together. Like many, perhaps most, families today, we were scattered across the country. Annie and I lived outside of Washington and kept an apartment in Columbus, Ohio. Dave and Karen and their two sons, Daniel and Zachary, were in Berkeley, California, where he had his medical practice and Karen hers as a psychologist. Lyn lived in St. Paul, Minnesota, where, after years as an alcohol and drug counselor, she had returned to her first love, art and painting. We got together eagerly whenever we could.

We spent a few days skiing and snowshoeing. Christmas came, and with it the exchange of presents. When we had finished our Christmas dinner, we turned to the tradition that began at my parents' table, talk of our activities and plans. It was then that Annie said, "John, you might want to tell the kids what may happen."

"Well," I said, turning to Zach and Daniel, who were then thirteen and fifteen, "your grandfather might be going into space again."

They were very quiet and looked to their parents as if for confirmation that this was a good thing. Then they listened as I explained—as I had explained to Dave and Lyn all those years before—the details of the mission. I told them about the medical similarities I hoped to explore as an elderly citizen in space. As Lyn had done, they asked hesitantly what could happen—and as before, I explained that NASA's safety record was a good one and that life was not risk-free if you ever wanted to accomplish anything.

Back in Washington, on January 15, 1998, an aide interrupted a meeting with the word that I had a phone call.

"You might want to take this one," she said when I started to object.

Dan Goldin was on the phone. "You're the most persistent man I've ever met," he said. "You've passed all your physicals, the science is good, and we've called a news conference tomorrow to announce that John Glenn's going back into space."

PART SIX

BACK TO SPACE

CHAPTER 25

I couldn't believe that was happening. "What in the world is going on?" I kept asking Annie.

NASA's announcement on January 16 that I would join the crew of the space shuttle *Discovery* on a mission officially dubbed STS-95 set off a hubbub that simply wouldn't die down. It was reminiscent of the beginnings of the Mercury program. We'd known there would be some attention, but we expected it to ease off in a week or two. We were wrong.

Some commentators accused NASA of staging a publicity stunt. Others said the White House was rewarding my stance toward the Republicans in the campaign finance hearings. Still others hadn't looked into the science, assuming this was just one more senatorial boondoggle.

Not the public, though. The vast majority of people seemed genuinely excited and intrigued by the possibility of beginning to find answers about the aging process.

I didn't know until two weeks after the announcement

that I'd misread my family's feelings about the venture. Dave was in Washington the last weekend in January, and I asked him to join me at the annual banquet of the Alfalfa Club. Alfalfa is a social club of two hundred members, including leading members of government, business and industry, entertainment, and the military. Dan Goldin was at the banquet, and when I introduced him to Dave, Dan said innocently, "You must be excited about your father going up again."

"Actually, no," Dave said. "I'm totally against it."

Dave, as a physician, certainly understood the medical benefits the flight was seeking. He was reacting on Annie's behalf. Even after she was convinced that the flight would be worthwhile, she had acknowledged that after sharing me with the rest of the country for fifty-five years, she'd hoped we would have more time to ourselves.

"I was a little jealous," she admitted after the announcement.

Rightly or not, I also sensed in Dave's response an attitude many adult children have about activities their elderly parents want to undertake. They wish the parents would sit home, play it safe, and not wander too far afield, because that way the children don't have to worry about them. It's a kind of role reversal, as if we were the children now. But we're not. All I can say to those children is, "Don't worry, we'll be fine. And if we're not, well, we were doing what we wanted to do."

I met my crewmates for the first time in early February. Curt Brown, the Air Force lieutenant colonel who had been designated the commander of STS-95, was at Hobby Airport south of Houston to help me get acclimated.

When he showed me the crew office at the Johnson Space Center, I had a sense of déjà vu. The office contained seven desks in one small room, and I thought I was back at Project Mercury.

The big difference was, these all held computers.

The next morning I was introduced at the regular Monday meeting of all the NASA astronauts as the newest—and oldest—member of the corps. It was a far more diverse group than the one that had taken the first steps into space. Not all, nor even most, astronauts today are test or fighter pilots. They're people with engineering, medical, and scientific expertise, and they come from varied walks of life. Twenty-five percent of them are women, and a number come from the space agencies in Europe, Japan, and Russia that are participating in the international space station.

The quality of the astronaut corps is reflected in the fact that every time NASA says it's seeking twenty-five new astronauts some three thousand candidates apply. Those selected for training are the best, and those selected for flight crews are the best of the best. The astronauts are a national resource, and it was absolutely terrific to be back in that kind of atmosphere again, free of pettiness and filled with purpose.

The Monday meeting included a conference call to the American astronauts assigned to the Russian space center at Star City outside Moscow. Curt presented me with the crew polo shirt that everyone wore during training. Like the rest of the world, NASA has gotten more casual in the years since Al, Scott, Wally, Gus, Deke, Gordo, and I traveled from place to place in suits.

I quickly got to know my STS-95 crewmates. Curt was low-key but very much in control. Both he and Steve

Lindsey—the pilot on STS-95 and, like Curt, a lieutenant colonel in the Air Force—had done weapons system testing at Eglin Air Force base in Florida. Talking with them was reminiscent of my Patuxent days. Steve Robinson was a Ph.D. in aerodynamics and astronautics who had been in charge of wind tunnel testing at the NASA laboratory at Langley, and had worked at NASA's Ames Laboratory in California. He was also the artist of the group. One of the traditions that has grown up around the shuttle flights is the creation of a special crew patch for each flight, and Steve's design for STS-95 combined a silhouette of the shuttle with a tiny orbiting capsule signifying *Friendship 7*. Scott Parazynski, a medical doctor, was known as "Too Tall." That came from when he was training for a Mir mission at Star City and tried to get into one of the Russian pressure suits. They were all too short. He was also one of the top U.S. lugers in the 1988 Olympic trials; when he didn't make the team, he coached the Filipino lugers. Chiaki Mukai, another bundle of energy, was also a medical doctor—a cardiovascular surgeon—and a member of the Japanese space agency. Pedro Duque, like Scott, had trained at Star City, and was the first Spanish astronaut; he had been assigned to NASA by the European space agency.

Our secretary, Polly Shroeder, who took care of administrative details, brought me up to speed in the new world of computers, E-mail, and the other changes time had brought to NASA since the early sixties. I needed survival training in computer use. A lot of the communication among the astronauts was done via E-mail. The Senate staff I had relied upon for computer expertise was no longer around to compensate for my shortcomings. Polly got me in shape.

Almost all of my crewmates were more experienced in space than I was. Curt was a real veteran; STS-95 would be his fifth flight and his second as commander. Scott had been up twice before, and Chiaki, Steve Lindsey, and Steve Robinson once each. The only one who hadn't been to space was Pedro, but he had trained extensively for a Mir mission that had been postponed.

I liked them all immediately. As the newest astronaut, all I wanted was to fit in and settle into the routine. At first they had a tendency to want to call me Senator. I told them to get used to calling me John.

I was the lowest-ranking member of the crew, payload specialist number two. That meant an astronaut trained to do specific experiments, as opposed to the commander and pilot, who fly the space shuttle, and the mission specialists, who are specifically trained to operate shuttle systems and perform experiments. Chiaki was my fellow payload specialist, but she outranked me by 349 hours in orbit. Her 354-hour mission on the space shuttle *Columbia* had come in 1994, my five-hour one more than thirty-six years earlier.

Curt stressed that he wanted us to adhere to a strict schedule for meetings and training sessions. Trying to get astronauts to meetings on time, he said, "is like trying to herd kitty cats." But training was different. The normal workday was eight to five, but as I got into the routine I saw that quitting time was rarely adhered to. Astronauts don't punch a clock.

At first I commuted back and forth between Washington and Houston, flying commercial most of the time, but my own plane a time or two. The early training was largely individual. I was going to do blood analysis, and centrifuge and freeze other blood samples, and I could learn how to do this on my own. Every astronaut takes pictures, and I could also learn the sophisticated range of cameras during one-on-one sessions with the experts. The same was true of extensive briefings on the orange pressure suit that astronauts wore now, and of water survival training, and of sessions during which I learned emergency egress from the shuttle by rappeling out the orbiter's roof hatch to the ground. All of this meant that early on, even when spending a week or two at a time in Houston, I could work around my Senate schedule, and even move ahead on matters that I wouldn't have time for later on. Between faxes and the telephone, I was able to keep up with business.

Later, we would be working on all-crew coordination and I wouldn't be able to juggle my schedule. But I had told Senate minority leader Tom Daschle that if a close vote on an important matter came before the Senate, he should let me know and I would be there. Most votes occur on Tuesdays through Thursdays, since members travel on the weekends. I was confident that even if I had to dash back to Washington, it wouldn't interfere with training, but as it happened, no such votes occurred.

Our sessions at the Johnson Space Center focused on Building 9N. It was a cavernous, white-painted space that housed a number of life-size mock-ups, one of the space shuttle in its entirety but without its wings, others of its fight deck and mid-deck, and still others of its various components. These included SpaceHab, a self-contained research laboratory carried in the shuttle's cargo bay, where many of STS-95's experiments would be performed, and the robot arm used to deploy and retrieve satellites, also in the cargo bay but controlled from the flight deck.

The space shuttle might seem cramped to most people. For somebody who had flown in the nine-by-seven-foot *Friendship 7*, it looked as accommodating as a Hilton. I was anticipating having the room to move around, to actually float, in space. My orbits in *Friendship 7* had allowed me to feel weightlessness, but not to experience it in the way that shuttle astronauts have since.

I used the mock-ups to learn my way around the shuttle. As you entered the ten-foot-by-ten-foot mid-deck, which would be my home for nine days during the mission, the galley was to the immediate left. It contained an oven, with spring-loaded shelves to hold food down, and above it was a device that would inject hot or cold water into the packets of rehydratable food we would eat. Also on the left wall—the wall toward the shuttle's nose—were thirty-three lockers for various tools and equipment. The far wall, the right one as you faced the nose, contained a four-tier sleep station consisting of long, narrow, boxlike compartments stacked on top of each other. To one side, there were more lockers. Along the back wall, a hatch opened to the tunnel leading back to SpaceHab and a ladder rose to the flight deck. Just inside the entry hatch was the waste management system, a great change from the low-residue diet and the condom, rubber tubing, and urine bag that had served in Project Mercury. In orbit, everything floats, so the shuttle's system included the use of an airflow device to draw everything into a collection tank, and foot straps and thigh bars to hold you in place while you performed. A canvas curtain provided privacy.

The flight deck above was more confined. The commander's seat on the left, and the pilot's on the right, with a large console of instruments, switches, and controls between them, dominated the nose. More instruments, dials, and switches crowded the panels in front of the two seats and above them over the windshield—at least twelve hundred in all. Four feet to the rear were the windows, hundreds more switches and dials, and controls for the payload bay doors and the robot arm.

It was impossible not to notice, especially during egress and emergency training but also just getting in and out of the cramped spaces, that I was a little older than the others. My flexibility was limited, and I didn't bend quite as well going through the shuttle hatch or into SpaceHab.

We were left on our own to exercise, as in Project Mercury, but now NASA had a fully equipped gym and a physical trainer. Beth Shepherd put me to work on a series of stretching exercises that would help me improve my ability to twist and bend as I got in and out of the shuttle mock-ups during training. As I strained against my body's resistance, I wished that I had paid more attention to stretching over the past twenty-five years.

Building 9N wasn't our only work site. NASA maintains other training facilities near Ellington Field, a few miles north of the Johnson Space Center. The Sonny Carter Training Facility includes the neutral-buoyancy laboratory, where astronauts training for space walks work underwater after being weighted to neutral buoyancy, using tools and equipment without the stability of gravity to assemble components. This wasn't a problem for us because we would be inside the shuttle, where there was always something to hold on to; we were using one end of the enormous pool to practice water landings. Wearing our bulky pressure suits, we were hoisted by a crane as if we were hanging from our parachute straps and swung out over the water. Then we were dropped into the water at descent speed with our life vests inflated. Once we bobbed up, we deployed our life rafts and other survival gear.

We also did balance testing in a phone-booth-like structure in which sections of the floor moved up, down, and sideways while we tried to stay upright; and we underwent tilt-table testing for orthostatic tolerance.

In cognition training, we reacted to numbers flashing on a computer screen for measurements of our reaction time. I was surprised, pleasantly so, to find that I wasn't far behind—and sometimes even ahead of—the younger crew members.

These exercises were demanding but fun. I had always stayed in good shape. Running to keep the weight off and to be in top shape during Project Mercury had become a habit, and as I grew older I switched from running to daily fast walks—power walks, they're sometimes called—plus the use of free weights and some stretching. I knew that regular, vigorous exercise is one of the keys to maintaining good health into old age, and it paid off—I was able to perform all the necessary tasks.

At Brooks Air Force Base in San Antonio, for the first time in thirty-six years, I was strapped into a centrifuge. But this ride was gentle compared with the ones I had taken earlier. The space shuttle, designed for research, had to enter orbit with payloads of scientific and other research materials, and rather than design all this equipment to high G tolerances, it was better to build a vehicle that took longer to reach orbit but subjected its cargo—and passengers—to fewer Gs. G forces at launch and reentry

are less than half of what they had been when Project
Mercury astronauts were riding Atlas missiles.

Many other things had changed since Project
Mercury, too. Versions of the silver pressure suits we had
worn, fitted to the quarter inch and worn over double-
layer long johns with air cooling in between the layers, had
been dropped from the launch wardrobe in the early days
of the space shuttle. Shuttle crews were launched in cover-
all-type flight suits. But after *Challenger* and NASA's re-
assessment of every potential hazard and possible response
to them, pressure suits were brought back to protect astronauts if the shuttle cabin should suddenly lose pressure
during launch or reentry. The purpose of a pressure suit is
to keep pressure on the body so that it retains blood gases
in case an emergency creates a vacuum.

Today's orange "pumpkin suits" are heavy and distinc-
tive. They're no longer fitted personally, but the five
generic sizes can be modified to fit almost everybody.
They're now cooled by tubes that carry cold water between the layers of double-layer underwear.

Along with the suit, the kit you wear during launch
and reentry includes a survival package consisting of a life
preserver, oxygen bottle, drinking water, flares, and various communications and signaling gear. This adds another
twenty-five pounds. (Back in Project Mercury, the survival gear we carried was part of the capsule, not the suit.)
The launch and reentry equipment also includes a back-
pack parachute with a life preserver attached, adding another twenty-seven pounds. The whole thing—suit,
survival gear, and parachute—weighs over eighty pounds.

Training under the weight of this array, I could see
why shuttle astronauts were happy to get into orbit so

they could get out of their suits and into comfortable clothes.

I studied stack upon stack of thick loose-leaf manuals that spelled out shuttle operations and launch procedures. There was also intensive classroom training focused mainly on the experiments. The science-rich flight of STS-95 would perform eighty-three different experiments and payload operations. They were extraordinarily diverse. One of the projects from NASA's program of high-school competitions was a look at the adaptation of cockroaches to space. Others, on behalf of commercial laboratories, would research microscopic capsules that offer the promise of improving the delivery of cancer drugs to the tumor site, and the growth of protein crystals. Still another would track electrode readings from a pair of toadfish to observe their vestibular systems in microgravity. And the government's Spartan satellite would be deployed to photograph the sun's corona before being recaptured by the orbiter. Each of us was responsible for some of the work that this involved.

Another research subject was a seventy-seven-year-old, 190-pound specimen who was also known as payload specialist two. In many ways, I would be as much of a guinea pig as I had been in 1962. Then, the other astronauts and I had been human subjects on the feasibility of space flight. In 1998 I would be a human subject seeking answers on the aging process.

That meant that in addition to the mission training I needed to get under my belt, I had to undergo medical tests that would establish an on-Earth baseline for the experiments to be conducted on me in orbit.

Sessions on both magnetic resonance imaging (MRI) and an advanced X-ray technique called DEXA scanned my bone and muscle structure for studies in osteoporosis and muscle loss. For these studies, I took regular doses of two amino acids—alanine, in pill form, and histidine via injection—and NASA dieticians kept track of every gram of food I ate, after which I gave countless blood and urine samples that would be used to measure the rate at which my body metabolized protein. The process of muscle loss and replacement involves protein turnover. Muscle loss in the elderly is thought to be caused primarily by lack of exercise, but in space astronauts lose muscle from decreased protein production. My baseline measurements would be compared with samples taken later on in space to see how my rate of protein production compared with that of younger astronauts.

The same blood and urine samples would also be used for immunology analysis. Both aging and space flight suppress the human immune response. Again, comparisons between what happens in older and younger bodies would be a start toward understanding immune suppression in both cases.

Older people often sleep poorly, and so do astronauts in space, perhaps because some change in their body clocks disturbs the circadian rhythms that govern patterns of waking and sleeping. So for sleep and other studies, I swallowed a pill that contained a thermometer and transmitter that sent readings of core body temperature to a receiver I wore on my belt. I wore a head net and an instrumented spandex vest with twenty-one different monitoring devices that recorded what my brain and body were doing as I slept. These instruments measured the volume and rate of my respiration, my eye movements during sleep, my brain waves and heartbeat patterns, my heart rate and pulse, and my blood oxygen content. After sleep sessions, I sat at a computer keyboard to perform reaction and cognition tests. All these tests, save the MRI and X rays, would be repeated in space, and the entire array at intervals after I returned. This would create a comprehensive portrait of my body before, during, and for several months after the flight, to look for information on "space aging" and aging on Earth.

Annie came with me almost every time I went to Houston. Rather than leave her waiting in the hotel, Curt invited her to classroom and training sessions. She sat in on my classes, observed my training, and learned along with me. She also talked to the scientists who were conducting the aging studies, and all this enhanced her feelings about the flight. She became genuinely enthusiastic.

As time went on, the two of us tried to analyze what kept the media and the public so interested in my return to space. The Cold War competition that existed in 1962 was history, and STS-95 was the ninety-second shuttle flight. To my thinking, its excitement was in the science it hoped to accomplish, not in any one member of the crew.

But the interest kept on building, and it wasn't the science that provoked it. Annie told me it was because people in America needed somebody to look up to—and it was true that personal scandal in high places was dominating the headlines and the national agenda in the first half of 1998. It also seemed as though people of all ages were being caught up once again in the excitement of space flight itself. The space program was a place to which they could turn to bring back the sense of purpose that had driven human exploration since before recorded history. That was good for the program, although it re-created the problem Annie and I had had long ago of trying to maintain a private place for family.

There was another factor, too. People my age weren't really expected to do much. Even when older individuals had higher expectations than that, society tended to take a dim view. The old folks should be slowing down, not trying to act as if they were young. The general idea was for them to stand aside and get out of the way. And that's what too many older people did, just sat around like couch potatoes and waited out their years. The idea of an ancient guy like me going into space was exhilarating.

The truth is, the old stereotypes no longer fit, if ever they did. Older people are increasingly active. While the processes I was going to study in space do tend to slow people down as they age, increased longevity and better health mean more older people are doing more things than ever before. My scheduled return to space helped bring this trend into the open. Older people were gratified at the evidence that they remained important. Young people, like our grandkids Zach and Daniel, just thought it was neat. A lot of people have grandparents. Apparently it wasn't a great leap to project them onto the space shuttle.

Wherever we went after the announcement, Annie and I were approached by people of all ages. They stopped me on the street to wish me good luck. Kids asked for autographs. And time and time again, I received the confirmation that my flight meant something special to those of a certain age.

When I said during a television appearance that older people have just as many hopes and dreams and ambitions as anybody else and that they should continue to pursue them, I got an outpouring of supportive letters and E-mails. The elderly were agreeing with me wholeheartedly and eloquently. Later, I read that an Elderhostel offering a space camp was receiving record numbers of applications.

One day at Houston Intercontinental Airport, a couple about my age stopped me. The man said, "I just wanted to tell you that you've changed my life."

"How's that?" I asked.

"All my life, ever since I was a boy, I was fascinated with Mount Kilimanjaro. I read about it, and I wanted to climb it. Then I got married, and we had kids, and I just kept putting it off. Then my wife told me I was too old." He glanced at his wife, who was laughing and rolling her eyes. "But now I've been telling her that if John Glenn can go back into space at seventy-seven, I can climb Mount Kilimanjaro, and we've got tickets to Africa next month."

"That's great," I said, and it was.

More than anything, I think the excitement surrounding my return to space was due to that redefinition of what people could expect of the elderly, and what the elderly could expect of themselves.

In April, President Clinton came to Houston for a meeting. Back in Washington, I had invited the president to come by and look at our training facilities if he was in the area, and he sent word that he was now accepting the offer.

Before he arrived, his staff and the Secret Service let us know that he was recovering from a knee operation and climbing around the shuttle mock-up might aggravate it.

"Don't ask him if he wants to get into the simulator," I was told, "because he will."

When the president arrived, he was all enthusiasm. I knew he would be watching some of the mission on television and despite the warnings, I thought he might want to know what the orbiter's interior was like. So when we were inside Building 9N and he was looking at the mock-ups, I told him I didn't want him to hurt his knee but wondered if he was up to a look around. Predictably, he was. He crawled through the hatch into the mid-deck, and after we talked for a few minutes he looked at the ascending ladder.

"What's up there?" the president asked.

"The flight deck. Would you like to go up?" I asked.

"Let's go," he said, and we climbed up one after the other while the presidential aides and bodyguards looked on nervously. But he really enjoyed it, and asked a lot of questions that showed a serious interest in what the mission would be doing.

The president also got a taste of space meals, which along with everything else had become much more sophisticated since 1962 and my "meal" of applesauce squeezed from a tube. I knew the food had improved, but I wasn't entirely prepared for the menu of forty-two different kinds of foods, including scrambled eggs, oatmeal with raisins, beef with barbecue sauce, turkey tetrazzini, beef Stroganoff with noodles, various vegetable combinations, shrimp cocktail, all sorts of fruit, cookies, candy, brownies, and chocolate.

About the only thing we didn't have was bread. Bread makes crumbs, and in the orbiting shuttle the crumbs would float around and get into people's hair and eyes as well as into instruments and the air recirculating system. On the shuttle, tortillas are substituted for bread.

Most of this food, with the exception of apples and carrot and celery sticks, is rehydratable, thermostabilized like canned peaches or applesauce, or prepackaged in snack packs like nuts or dried apricots. It's not gourmet fare by comparison with Le Cirque 2000 in New York, but it's a cut above what I remembered from the days when aerospace medicine wasn't even sure people would be able to eat and swallow in space.

The president sampled green beans with mushrooms, rehydrated with hot water, and one of the cold-water-rehydrated shrimp cocktails that had become a favorite on shuttle flights. I advised him to stir it up unless he wanted a full blast of horseradish.

During the Senate's month-long August break, Annie and I virtually moved to Houston. We had a car driven down, and lived in a hotel suite.

This allowed us to catch up on family time. Lyn, and Dave and his family, had already been down in May to go through the facilities, meet the crew, and get up to speed on the mission. I took them into the shuttle mock-ups, explained what I would be doing in the mid-deck and SpaceHab, showed them the flight deck, and explained how the robot arm was controlled. There was more of that in August, because in the coming months, my time would get scarcer.

As the calendar moved closer to the flight, the crew began working on timeline training. A shuttle flight with multiple crew members is an extraordinarily complicated job of coordination. And one like this, even though it wasn't one of the longer flights, was nevertheless more complicated than most because of the number of experiments that were to be done. Each crew member had a series of tasks that had to be accomplished during the nine days of the mission, along a timeline that incorporated what the other crew members were doing as well.

On September 28, with the flight barely a month away, I cast my final vote in the Senate. It was a relatively innocuous one, a procedural motion on a bill that would place a moratorium on states taxing Internet sales, and I voted in favor of it. It brought my total number of votes in the Senate to 9,414.

After almost twenty-four years, I regretted not being able to deliver a farewell speech, but there just wasn't time to write one. Two other members who had entered with me—Wendell Ford of Kentucky and Dale Bumpers of Arkansas—were also retiring. Tom Daschle gave us a farewell luncheon, but I was in Houston and had to attend by a satellite video hookup. Still, the emotions of our long and collegial association were palpable. Both Wendell and Dale had earned the respect and affection of their colleagues and of their constituents, and it was hard to say goodbye after so many years.

There were a lot of things I would miss about the Senate, and others I was happy to be rid of. But some of the people were the finest I have ever known, and seeing and working with them on a daily basis was something I would truly miss.

Three weeks before the launch, the crew flew to Cape Canaveral for a series of exercises that would end with a run-through of what would occur on launch day—a terminal countdown test. Our transportation was five T-38s, fast jet trainers that NASA's pilot-astronauts use to maintain their flying skills. They're short-range planes that require a stop for fuel on the way.

Once we reached the Cape, we all practiced driving

the "tank," an M-113 armored personnel carrier that waits near the base of the launch pad in case the crew needs to make an escape from the top of the orbiter—an exercise we would practice in part later. In the real thing, we would slide down in canvas buckets—called escape baskets—along a 1,152-foot cable to a bunker, but if whatever emergency occurred didn't require that we stay in the bunker, we'd drive away in the M-113. Since it could fall to any one of us to drive it, we all took turns. I'd driven it before; the same diesel-powered machine—with its huge accelerator pedal, two-handle steering, and a high seat so the driver could see out—had stood by for bunker departures during Project Mercury. Maybe it was even the same "tank."

This was a press and photo opportunity. It was already clear that the press presence at the Cape was going to rival that of the most-attended launches of the past—Al Shepard's and my Mercury flights, the Apollo moon flights, and the launch of *Discovery* to restart the shuttle program more than two and a half years after the *Challenger* explosion.

I was irritated at the press session at the bunker and I let it show. There were seventy-five to a hundred reporters and camerapeople there, more than had ever appeared for tests like these before, and most of the questions were directed at me. History was repeating itself in that the questions were personal and had little to do with the scientific value of the mission. I said I wished the media would focus on the science and pay some attention to my fellow astronauts and their accomplishments, but it did no good. I ended up doing most of the talking anyway.

Even though there were three weeks to go, the beaches north and south of the Cape and the campsites along the Indian River were already filling up with watchers, just as they had in January and February of 1962. Matters came to a head during a training exercise at the Cape, when launch crew members and people working on the gantry started asking for autographs from the whole crew. Curt finally laid down the law, and said we were there for training, not signing autographs.

I rode with Curt when he flew some shuttle-style approaches in a Gulfstream jet specially modified to mimic the way the orbiter acts when it's returning from orbit to a landing. The orbiter, which with payload averages 230,000 pounds, is unpowered when it returns from orbit. It's a big, heavy glider. After it descends into the atmosphere, with its belly of heat-resistant tiles taking the friction and heat of reentry, it drops on a twenty-degree glide slope, at ten thousand feet per minute, until it flares—eases its descent—for landing. By contrast, a commercial airliner descends at about three degrees.

Coming down with Curt on a shuttlelike approach, I told him, "The last time I saw a runway from this angle I was dropping bombs on it."

The Gulfstream's modifications allow reverse thrusters to be used in flight, in contrast with a commercial jet, which can reverse engine thrust only on the ground. Reversing thrust in flight is the only way to slow it when it's coming down so steeply. The shuttle's bulky profile has the same effect. When it flares for landing, it shallows out on the glide path and slows to about two hundred knots, or 230 miles per hour, when it hits the runway. NASA gives it plenty of room; the shuttle landing facility at the Cape is fifteen thousand feet long and three hundred feet wide.

The capstone of all this activity was the terminal countdown test itself. We got up just as we would on launch day and worked our way through everything that would take us through launch. The kitchen already had our orders for launch morning. I told them, "I want steak and eggs. It worked the last time, and I don't want to break a winning streak."

Coming out of crew quarters in our pumpkin suits, we saw that all the Cape staff we worked with were there, as they would be on launch morning, watching as we got into the van that would take us to the launch pad. A large media contingent was also there to see us off.

The van took us straight out to the pad, where another flock of reporters and photographers waited. They would be replaced by NASA photographers on launch day.

We went up the gantry 195 feet to the white room, the dust-free transition zone through which astronauts have always entered their spacecraft. Scott Parazynski and I were the last two on the gantry, and he pointed out a phone on the superstructure. I had the bright idea that if it connected to the outside, I would call Annie. It did, and nobody was more surprised than I was when she picked up the phone at our hotel room back in Texas. Scott talked to her, too. The reporters who were watching via closed-circuit television demanded to know whom we had spoken to. The whole thing sent ripples all the way to Houston and back again. It's the last time I'll pick up a gantry telephone.

Climbing into the shuttle and getting strapped down requires contortions. It's like taking a seat on an airplane, only the airplane is standing on its tail. I was on the mid-deck, between Chiaki and Steve Robinson, in seats that would be removed and stored after launch. The practice countdown proceeded as it would on launch day, except that there was no propellant in the tank. The control center in Houston was manned the way it would be, the launch director was at the Cape, the built-in holds during the countdown took place. Most of the work at that point fell to Curt and Steve Lindsey. Lying back in position, I noticed that one of the Cape's notorious mosquitos had made it into the orbiter with us.

Curt had told us that he wanted a silent cockpit when the countdown reached the automatic nine-minute hold. That means that unless there's an emergency, nobody talks but the commander and pilot. Then the countdown continued to zero.

Of course, we didn't lift off, but I think all of us would have been happy if we had.

At the end we did a simulated mode-one egress, the exercise that would have tied in with the "tank" if we had done it in its entirety. If something happened to the booster, we would have to get out by ourselves. Fire was likely, so water would be drenching the top of the gantry. We would activate our emergency ten-minute supplies of oxygen and follow a yellow line—the "yellow brick road"—to the escape baskets waiting on the gantry. These baskets would slide down the cable to the bunker near the pad, but for the exercise they were chained in place.

After a debriefing with the vehicle integration training team, the test was over. We had little doubt that we would be ready to go come launch day.

Returning to Houston was the most fun of all. We stopped in Arcadia, Louisiana, to refuel. I was flying with Curt. We took off in formation, and then I took the controls and flew wing on Steve Lindsey, tucked up in formation, for the forty minutes back to Ellington Field. Most of my flying in recent years had been restricted to my twin-prop Beechcraft Baron. Joining the shuttle crew had given me a chance to fly jets again, backseat at least, and I was really beginning to feel at home in the cockpit.

I told Annie, "I haven't had this much fun in a long time. I don't know what I'm going to do when I grow up."

Back in Houston, I confronted the latest well-meaning but overkill response to the upcoming flight. NASA 1, the divided roadway that runs from Interstate 45 past the space center entrance and east toward Galveston Bay, had sprouted a forest of red-white-and-blue banners with my picture on them, proclaiming the road "John Glenn Parkway." The change was temporary, but even so, I thought things had definitely gotten out of hand.

I still had a few matters to brush up on. One of them was the use of the vastly more sophisticated cameras NASA now employs, in conjunction with the equally sophisticated means of reproducing the images they took. The agency had come full circle from the days when I had trouble talking management into letting me take a handheld camera on my flight. Now it wanted all the astronauts to take pictures when they had free moments at the windows. The spectacular images of Earth taken from space, and of the reaches of space taken from outside the atmosphere, have become a staple of NASA's repertoire. Photographs from space give scientists vital information on weather and geographic changes such as shifts in river deltas, snow patterns, and evidence of rain forest destruction. And nothing is more effective at reminding people what a fragile planet we inhabit than an image of Earth, with its thin film of atmosphere, poised against the darkness. Now the agency was also using digital cameras, with images fed from diskettes into computers.

We continued to practice experiments in our SpaceHab mock-up over and over, to be certain that we could maintain the timelines necessary to complete them once in orbit. Most of the SpaceHab practice had come at the Cape, where the real thing was being loaded with experiments, and we kept it up in Houston.

I also kept up almost daily sessions providing baseline medical measurements, and with the rest of the crew, I continued to work at mission simulations. The flight-training specialists still delighted in thinking up emergencies for the commander and pilot to respond to. That was one thing that hadn't changed since Project Mercury. The simulations that challenged the rest of us meant following our timeline of experiments and duties starting at any point during the flight.

At a crew news conference on October 15, the questions were once again aimed largely at me. But a Japanese reporter asked Chiaki how she felt about flying with a "hero of America" like John Glenn. Chiaki was one of the liveliest and brightest of the crew members, and she displayed a deft touch with the media. She said it was as good as flying with a hero of Spain like Pedro Duque and all the other heroes on the flight.

I was tickled.

A week later, after dinner with Annie, Lyn, Dave, Karen, and the grandkids, I moved to crew quarters at the Johnson Space Center. Preflight quarantine is designed to ensure that astronauts don't carry randomly acquired ailments into a shuttle mission. Once the quarantine goes into effect, only family members and other people pre-screened by NASA doctors are allowed to have contact.

I was glad for the quarantine. For one thing, it meant

the mission was just a week away. After the training and all the attention, it would be a relief to get into orbit. For another, it freed us from all the hullabaloo outside.

But I had a special reason to be pleased we were entering the quarantine. The news had broken that I had been removed from an aspect of one of the experiments that would focus on me. This involved the hormone melatonin, produced by the pineal gland. Many travelers take melatonin, which is available over the counter in any drug or nutrition store, to reduce the effects of jet lag and to counter insomnia. Chiaki and I were both going to take tablets containing small amounts—about one-tenth of the amount over-the-counter forms contain—to gauge its effects in space. It was one small part of the sleep experiments, and during some of the sleep periods, the pills would be placebos. But Charles Czeisler, the Harvard Medical School researcher on the project, had found that among a long list of exclusionary criteria in this FDA-approved study was one that precluded my participation. NASA treats astronauts' medical records with doctor-patient confidentiality. Beyond desiring a level of medical privacy myself, I felt that to release the information on my own would set a bad precedent and create similar demands on future astronauts. I continue to think so. But some of the press treated my removal from the melatonin aspect of the study like a coverup akin to Watergate, and I was glad to get away from the hoopla and concentrate on preflight preparations.

On Monday, October 26, with three days to go before the flight, we flew to Cape Canaveral at last to begin the final countdown.

CHAPTER 26

The Cape hadn't seen reporters and photographers in such numbers since the Apollo moon launch days. Some thirty-seven hundred had applied for credentials. And the crowds on the beaches and along the waterways were huge, too.

In the back of my mind I was preoccupied by the fear that my history of delayed flights might come back to haunt the mission. Early in the week, the weather forecasts weren't that good. But they improved day by day, until on Wednesday NASA's meteorologist said there was a 100 percent chance of good weather for the launch the next day, October 29. He said it looked as if Mother Nature wanted me to go back into space as much as everybody else did.

This time Annie was at the Cape, along with the rest of the family. NASA has begun taking astronauts and their families to the launch pad for a viewing the night before. Afterward we went to the beach house conference center for a quiet dinner. That was where Annie and I said our goodbyes—the pack of gum again, and her reply: "Don't be too long."

The launch time was civilized by Project Mercury standards—1400 hours. We awoke in crew quarters, suites that were an improvement over the bunk beds I remembered. Their walls had no windows, since shuttle schedules sometimes require crews to shift their normal wake-sleep routines in advance by way of artificial light, but outside we found the bright, clear morning that the meteorologist had predicted.

We put on our crew shirts for the traditional breakfast photo opportunity. I reprised my meal of steak and eggs with orange juice and toast. Looking around at what my fellow crew members had ordered, it seemed that steak and eggs had also become a launch day tradition.

The atmosphere was businesslike as the launch approached. We were eager to get going.

After breakfast, we went back to our rooms to tidy up. We also packed two small bags with basic clothing and personal effects, shoes, and a flight suit and toilet kit. One of them would be shipped across the Atlantic if we didn't achieve full orbital speed or something else went wrong and we had to land at one of the TALS—transatlantic landing sites. There's one in Spain, and another in Morocco. NASA would send the second bag out to Edwards Air Force Base in California, our alternative landing site, in case conditions weren't right for landing at the Cape when we came down.

Suiting up, each of us worked with the same small crew of suit technicians who had helped us during training. My crew was Jean Alexander, Carlos Gillis, and George Brittingham. We each sat in a big leather chair, and the suit techs hovered around us as if we were actors being made up for our stage appearance.

Getting into the suit took forty-five minutes. I had to

be something of a contortionist as I pulled on the special underwear rigged with cold-water tubes for cooling. It wasn't easy at my age to get into the suit itself, either— feet first into the legs, then maneuvering to get my head and torso into it before the suit techs zipped it up the back. They fixed the gloves so they were pressure-tight, and fastened the helmet to the neck ring. When the visor was sealed, the entire contraption was pressure-tested to make sure there were no leaks. Around the suit room, the crew looked like Poppin' Fresh doughboys in bright orange.

Then I loaded my pockets, one on each thigh and each shin, one on each shoulder. You have to know where your emergency radio and signaling equipment are—left-leg pocket. And your knife and other survival equipment— right-leg pocket. The rest of them held various tools and gear.

Suited up, we headed toward the elevators, past the technicians and cooks and workers who had helped us throughout training. The suit techs followed, holding our helmets. This, too, was a trip I was familiar with. But the expressions were different this time. When I took my walk from crew quarters on the day *Friendship 7* was finally launched, I was going solo and it was a first flight. There was more uncertainty on the faces then.

Still, there was the same silent acknowledgment that we were going to be riding a rocket that could kill us if anything went wrong. The ground team was there to say goodbye and wish us luck. Their expressions said they were pulling for us. They wanted us to have a safe, trouble-free, and successful mission. The spirit of teamwork and camaraderie was written on each face. It was as if their thoughts and wishes were going to be riding on that rocket, too, and none of us could have thanked them enough for everything they'd done.

A pool of reporters and photographers watched behind the ropes as we walked from the elevators to the transfer van. I don't think there was room for a single person more in the crowd.

The atmosphere in the van was casual and jocular during the six- or seven-mile ride to the launch pad, though as I looked around at my crewmates I could see that we were getting ready to be serious. Then we reached the gate to the pad. The guard stepped into the van, and Curt said, "Launch passes, everybody." The crew reached into the pockets on the left shoulders of their suits and pulled out small blue cards. I felt in my pocket, thinking somebody must have put mine there, but there was no card. Pedro was doing the same thing. Amid our fumbling, I was about to ask when the cards had been issued when I noticed that the rest of the crew—shuttle veterans—were looking at us rookies, trying to hide their grins. We had bitten, hook, line, and sinker. They all had a laugh, Pedro and I had our initiation rite, and the van proceeded toward the pad.

At the pad, we walked back out along the ramp and looked up at the shuttle. That's another launch day tradition, and it's quite a sight.

The space shuttle is the most complex machine ever made. It has two million parts, and a million of them move. Its wiring laid end to end would stretch 230 miles, and it has six hundred circuit breakers. The orbiter itself has three eighty-thousand-horsepower engines that each develop 393,800 pounds of thrust. They are fed by the huge rust-orange tank to which the orbiter and the boosters cling during launch, and the two solid-fuel rocket boosters each develop 3.3 million pounds of thrust. The weight at liftoff is about 4.5 million pounds, and total thrust at liftoff is over 7 million pounds.

It was up there ready to go, and the liquid oxygen that oxidizes the liquid hydrogen fuel venting out the top in wisps of vapor adds to the sense of drama. It's a huge machine containing an almost unfathomable amount of power. That's the point when it hits you. It's for real—you're going up.

The elevator took us up. It was a beautiful day, and I paused to glance around at the Cape and the space complex that had changed so much since the time of Project Mercury. As I looked south to the Canaveral lighthouse, the Atlas and Titan launch gantries that are the remaining occupants of Heavy Row were reminders of the early days. Pad 14, where *Friendship* 7 and the rest of the Project Mercury Atlas flights had launched, was still there, but its gantry had been dismantled long ago. The blockhouse is a museum. It was hard to imagine that virtually the entire history of space travel had occurred between my first ride and my second. Somebody had pointed out that more time had passed between *Friendship* 7 and this *Discovery* mission than had passed between Lindbergh's solo transatlantic flight and *Friendship* 7. It didn't seem that long to me, but that is the way lives pass when you look back on them: in the blink of an eye.

I don't think anyone was scared. Apprehensive? Yes. I felt the same constructive apprehension I'd felt as a forty-year-old, keyed up and ready to go. Everybody knows something could go wrong, but you just put that behind you and go do what you've been trained to do.

Chiaki had said that I ought to remember that in

Japan, seven is a lucky number, and my age, double seven, was doubly lucky. That was a good way to look at it, too.

I couldn't have been happier that morning. This was about to be the culmination of a very long effort, both a chance to go up again after I thought that chance had been lost forever, and the beginning of a precious opportunity. I was a data point of one, but it was a start, and I saw the flight as the first step in a process that I hoped would lead to a new area of research that could eventually benefit tens of millions of people.

Curt was the first into the spacecraft, and he climbed up to the flight deck, followed by Steve Lindsey and Pedro. I was next to last. No phone call from the gantry this time. Steve Robinson and Chiaki were already in their seats there on the mid-deck. They were being strapped in as I got there, and Scott came in after me and went on to the flight deck.

I hoisted myself into the seat by way of a strap hanging from the lockers overhead. Seated for launch between Chiaki and Steve, I was on my back with the wall of lockers less than three feet from my face.

Launch was two and a half hours away as the strapping-in proceeded. The best thing to do is just lie there and let the technicians do the work. The seats aren't the body-conforming contour couches of the early flights; they're flat bench-type seats that are padded but not all that comfortable. The only way to adjust them is by pumping a bladder that provides lumbar support to your lower back. The early seats were designed to help us endure eight times the force of gravity, but a shuttle launch produces only three Gs.

Carlos and Jean did the finishing touches, making sure my straps were tight, the emergency oxygen was plugged in and tested, and everything was good to go.

After that, we all ran through a checkout of the communications system. Curt was talking back and forth with the launch control center at the Cape and mission control in Houston, which would assume control at liftoff. We went through intercom and radio checks. Everybody answered in order: the commander, the pilot, the three mission specialists, Chiaki as payload specialist one, and then me, "PS two, loud and clear."

At twenty minutes, the countdown stopped for the first of the two built-in holds, designed for last-minute catch-ups and adjustments. Then it resumed and ticked down to the second built-in hold at nine minutes. This one was supposed to last ten minutes, but it went on longer than anticipated because an alarm had gone off when the cabin pressure was brought up. When the countdown resumed, we breathed a collective sigh of relief. After that, Curt came on the intercom to say, "Okay, everybody, we're going on silent cockpit." At that point, you stayed off the loop unless you really had something to communicate. The next comments we'd make would be in orbit.

But we all could hear Curt's and Steve's communications with the launch center and with Houston.

At five minutes the countdown stopped again because two airplanes had entered the restricted area. We heard the irritation in Curt's and Steve's voices. How on earth could you get to this point and have airplanes in the area? Nobody knew how long the hold was going to be. The FAA should yank flight licenses over something like that because there's no excuse for it.

After a few minutes, the count resumed. As it went down, all I wanted was to get going.

About six seconds from zero, the orbiter's three main engines lit. I felt the shuddering and the resonance as they built toward full thrust. The shuttle bent as if it was starting to bow, then straightened. The push of the orbiter's engines is straight up, but the center of gravity of the whole launch assembly, including the solid rocket booster engines and the external tank, is a point a few feet into the tank, so the assembly, held down by eight massive bolts, flexes in that direction.

As it came back to vertical, the solids lit. We were going someplace. The shaking and the shuddering and the roar told us that. In rapid sequence the solids built up power, the explosive hold-down bolts were fired, and over seven million pounds of thrust pushed us up at 1.6 Gs.

I hit the timer on my knee and the one on my wristwatch. The wristwatch gave the mission elapsed time starting from launch, and would also count days. The timeline for all our activities, including research experiments, required us to know the day as well as the hour and minute from launch.

The vehicle was moving at a hundred miles an hour by the time it cleared the launch tower. It was accelerating far more rapidly than the *Atlas,* and its shaking and vibration were much more pronounced.

Max Q, and the worst shaking and shuddering, came about sixty seconds after launch. The main engines throttled back automatically to keep the vehicle within its structural limits. Then came the voice from the ground, "Go at throttle up," which meant we were through the area of maximum aerodynamic pressure and the main engines had returned to full throttle.

The solid-fuel boosters run for two minutes and six

seconds. Everyone looks forward to the moment they burn out and detach. They're the one thing in the launch vehicle you have absolutely no control over. You can't throttle them back, you can't shut them off, and you can't detach them. There are no emergency procedures if anything goes wrong. You just hope everything keeps working right. I had told Annie and Dave and Lyn, who still worried, that when the solids were gone we were home free.

They burned out. I felt a sudden loss of thrust, then heard a bang like a rifle shot as the explosive bolts holding them to the external tank fired and detached them. They would cartwheel down until their parachutes deployed to bring them down for retrieval and reuse.

With the solids gone, the ride eased out. The orbiter's main engines run smoothly, and you ride into orbit accelerating as the fuel in the external tank is burned, making the vehicle lighter. You hit three Gs just before you reach orbit.

Then another bang, more muffled than the first, signaled that the spent external tank was jettisoned. It would burn up reentering the atmosphere over the Indian Ocean. After that, we were operating on the fuel that was stored within the orbiter itself for the final sprint to orbital velocity.

Once we hit orbit and had main engine cutoff, we got busy right away. Chiaki and I were responsible for getting people out of their suits and stowing the suits and all the equipment on them into net bags, color-coded for each crew member. That was more complicated than it sounds. Each item had to wind up in the bags in the order in which it would be removed as we resuited for reentry at the end of the flight.

I took my helmet off and put it down, and it came

floating right up past my face. It moved much more than I anticipated. I had to stick its communications cord under my legs to hold it down until I could get a bag to put it in. Stray gloves and equipment were floating around. Even releasing my seat harness, I found I had to be careful because I had a tendency to take off. Foot loops kept my feet on the floor, and bungee cords against the front of the lockers helped me corral stuff floating by. I kept my suit on while Chiaki and I helped the others out of theirs, wrapping my legs around the seats for leverage. By the time I finally got out of my suit, I had worked up a pretty good sweat.

We stowed the bagged suits and equipment temporarily in the sleep stations until we could transfer them later to the airlock that led to the SpaceHab. Then we folded and detached the seats, including the two rear seats from the flight deck, and got them out of the way. It was a lot easier than on the ground, where they weighed seventy pounds. Now, the flick of a fingertip would move them where they had to go.

Because everything floated, Velcro, duct tape, and bungee cords were invaluable. Things had to be held down, and those were about the only devices to do it.

Floating around took a little getting used to. When I moved across the mid-deck or through the twenty-five-foot tunnel leading to SpaceHab back in the payload bay, just a tiny amount of pressure was enough to start the process. Pushing off without the right alignment could send me spinning. The tunnel to SpaceHab was only three feet wide, and I learned to adjust my course as I floated through it. Reaching for items that were hovering nearby, sometimes I bumped them and then had to chase them down. I learned right away not to push too hard off the wall or to reach for things too fast. And all the switch plates had guards that prevented us from turning something on or off inadvertently when we bounced off the walls.

One of my main concerns was whether I was going to be sick. Space sickness affects about a fifth of astronauts initially, and while I had felt fine during my Mercury flight, I didn't know how I would react in the shuttle. I had taken Phenergan, which many astronauts use before going up, and I adapted rapidly. I couldn't have felt better, and three hours into the flight I reprised an old line in my first transmission from orbit: "Zero G and I feel fine."

For the first hour of the flight Chiaki and I worked hard down on the mid-deck, so we weren't able to see out a window. Everyone except Curt had come down from the flight deck. He had to perform the orbital maneuvering system (OMS) burn that put us from an elliptical orbit into a circular one. He established the shuttle in a tail-down attitude, with the radiator surfaces of the payload bay doors open to dissipate heat, and by then he was ready to take his suit off and get into other clothes. When he went back up, I followed to look out. By that time, we had made a full circuit and were coming back into daylight again over the Pacific.

Discovery was at an orbital height of 300 nautical miles, or about 348 statute miles, the highest continuous orbit for a shuttle mission. It gave us a rare view for a shuttle flight. We were more than twice as high as I'd been in *Friendship 7*, and I could see entire weather patterns beneath me even better. Once again I looked out at the curve of the horizon and the bright blue band that is our atmosphere—the thin film of air that makes life on Earth possible—and I realized how much I'd missed being in space all those years.

Curt described it when he radioed to Houston, "Let the record show that John has a smile on his face and it goes from one ear to the other and we haven't been able to remove it yet."

I wanted to do a good job. We were at the start of a nine-day mission and had come through the first phase with things well organized, but there wasn't any time to waste. The timeline called for starting a number of experiments immediately after we entered orbit.

Scott and I floated back through the tube to SpaceHab to activate several experiments that held the potential to improve medical treatments on a wide range of fronts. The BioDyn payload was a commercial bioreactor that contained work in several areas: protein research that could aid in ending transplant rejection; an investigation into cell aging, seeking tools to fight various geriatric diseases that cause immune-system breakdown; improved ways of making microscopic capsules to deliver drugs directly to the site of a disease; tissue engineering aimed at making synthetic bone to improve dental implants, hip replacements, and bone grafts; and heart patches to replace damaged heart muscles.

Then I moved on to ADSEP, part of a series of experiments in separating and purifying biological materials in microgravity, with aims such as producing genetically engineered hemoglobin that may eventually replace human blood. Starting ADSEP meant moving its various modules from storage into active bays and setting switches and turning dials according to detailed instructions in our flight-data files. These experiments were only a fraction of the science we would do during our nine days on board.

By the time we returned to the mid-deck, I was hungry. It was then five and a half hours into the flight, longer than the total flight of *Friendship* 7. I hadn't eaten since breakfast, and hadn't had time to grab a snack from the pantry, a shallow drawer near the mid-deck ceiling that was loaded from the bottom, like a kitchen drawer at home but upside-down, with the contents secured with netting.

Eating involved first injecting hot or cold water into the rehydratable packets, then waiting three to five minutes. As it absorbs the water, the food thickens and won't float out of the packet. We all carried scissors for cutting the packets open as part of our regular equipment. The packets had small Velcro patches on their surfaces, so you could eat anywhere and stick your meal onto one of the orbiter's hundreds of Velcro strips if you wanted to put it down.

I ate a full meal, starting with a shrimp cocktail and moving on through macaroni and cheese, peanut butter and jelly in a tortilla, dried apricots, banana pudding, and apple cider.

After eating, it was time to prepare for sleep. We had been up since six that morning, and working in space since midafternoon. The schedule called for a two-hour presleep period that gave us time to wash up, send E-mails, review the next day's work, or gaze back at Earth from one of the windows. A few of the crew put on headphones and listened to music. We all had the opportunity to bring a selection of compact discs along. My choices included music by Henry Mancini, Peter Nero, and Andy Williams. Peter and Andy are good friends, and Annie and I had been especially close to Hank and Ginny Mancini, visiting and vacationing with them on many occasions before Hank died in 1994. I also took along a disc of barbershop chorus harmonies by the champion Alexandria Harmonizers, a taste I inherited from my dad. After that, the entire crew slept. Space days and nights lasted the same forty-five minutes I had experienced in *Friendship 7*, and since the shuttle orbited through five of these days and nights during an eight-hour sleep period, its windows and portholes were shaded while we slept. Chiaki and I bedded down in our sleeping bags in two of the sleep stations. Steve Robinson took another, and we reserved the fourth in the tier for storage. It was like being tucked into a long pine box with a sliding panel for a door.

The rest of the crew hooked their sleeping bags to walls or ceilings wherever they pleased. Curt slept on the flight deck, Steve Lindsey in the mid-deck, and Scott and Pedro found space back in SpaceHab or the tunnel.

I used a block of foam for a pillow, even though my head—and the rest of me, for that matter—needed no support in weightlessness. It was just a way of making sleep in space familiar, even though it meant bringing the pillow to my head instead of putting my head down on the pillow.

When we awoke, in the so-called postsleep period during which we washed with foamless soap and brushed our teeth with foamless toothpaste, I noticed that we all had fat faces. This resulted from the fluid shift that weightlessness causes. The body senses it no longer needs the same fluid volume it has in a gravity environment, and you eliminate the excess through urination. The fluid that's left moves from the abdomen and legs into the upper body and face. We all looked comical, Steve Robinson even more so because his hair was standing up like Dagwood Bumstead's. But the facial effect isn't permanent; it would recede in another day or two. Steve's hair, however, would keep floating.

At breakfast, I put into my mouth the largest, fattest, longest jelly bean anybody ever tried to eat—and I wasn't allowed to chew it up. It was the thermometer pill that transmitted core body temperature readings to an external monitor. The readings would constantly chart fluctuations in my body temperature.

After another day of work, meals, and a sleep period, day three began with the first of my orbital bloodlettings. Scott, as the flight doctor, took the almost daily blood draws used for the protein turnover, immunology, and blood chemistry studies for which Pedro and I were subjects. Each draw produced two samples, one that I would analyze with an in-flight blood analyzer, another that I would separate by running through a centrifuge and freeze for later analysis. I had attached the centrifuge to the ceiling with duct tape. The centrifuge spun at 3,000 rpm, and once when I tried to move it off its axis of rotation I found this was impossible. Its torque was enough to send me spinning.

I'd discovered on the ground that a semipermanent intravenous catheter to supply the blood had proven too uncomfortable after a full day's activities, so I decided I'd rather take the needle sticks. Scott became my Count Dracula after he floated in my direction for a blood draw wearing a set of plastic Halloween fangs. By a few days into the mission, he started grinning whenever he came my way with the syringe—or maybe it was just my imagination that he got to look more maniacal than ever.

The protein turnover study, the mission's experiment in muscle loss and rebuilding for which I was a prime subject, required me to take alanine pills and histidine injections several times during the flight, just as I had in preflight testing. The researchers would compare the findings with the baseline studies done back then, and also with on-Earth readings taken after the flight.

Night four of the mission saw me and Chiaki rigged up in our head nets and instrumented vests. The twenty-one leads from the apparatus fed into boxes we wore on our waists, where the information was recorded for later analysis. We repeated everything the next night. These procedures, too, were bracketed by blood draws and urine samples, and were followed by cognition testing.

Sleeping with the elaborate head net and vest turned out to be easier in orbit than on the ground, where the electrode leads were uncomfortable. Imagine sleeping with a dozen buttons over half an inch thick stuck on your head that you feel every time you roll over. Weightlessness removed the irritating pressure.

On night six I donned a Holter heart monitor that I wore for twenty-four hours to provide a constant electrocardiogram. Anomalies in heart function in some of the other astronauts during space flight made NASA doctors decide to look at the action in a seventy-seven-year-old's heart.

All the while, I kept track of other experiments back in SpaceHab and on the mid-deck. The one that fascinated me most was Aerogel, a superthin, light, translucent substance with marvelous insulating qualities—a microscopic layer insulates as well as thirty thermal windows. It was my job to activate it simply by turning several switches. It's thought that manufacturing Aerogel in microgravity might solve the problem that keeps it from being in common use on Earth. So far, it's been impossible to make it as clear as glass.

On nights seven and eight Chiaki and I put the sleep nets and vests on again for two more sets of readings.

·   ·   ·

The Spartan satellite we were to deploy was our biggest payload, and the reason for our high orbit. It weighed a ton and a half, and was designed to photograph the sun's corona and the effects of solar winds from outside Earth's atmosphere. Solar winds produce interference that affects communications, electrical grids, and electronics on Earth, an effect that is heightened during times of high solar activity.

On the third day of the flight, Steve Robinson took the controls of the fifty-foot robot arm and maneuvered it to connect with the Spartan, lifting it out of the payload bay and away from the orbiter. This was a delicate operation, requiring great care.

Once the Spartan was on its own, Curt used the orbital maneuvering system to move away from the satellite. The satellite would orbit independently for two days, taking pictures, until Steve retrieved it again on day six. To accomplish this retrieval, Curt maneuvered the orbiter to within a few feet of the Spartan, a flawless rendezvous that put Steve in a perfect position to bring the Spartan back on board. I was in the SpaceHab with the best view in the house as he nestled Spartan gently back into its cradle.

On November 3 I briefly donned my political hat. It was the first time in years I didn't go to the polls on Election Day. I and the rest of the American crew had filed absentee ballots—but I broadcast my normal Election Day get-out-and-vote message to the voters back home.

The next night, Curt, Steve Lindsey, and I did a live shot with Jay Leno on *The Tonight Show*. Curt was a big Jay Leno fan—we all were, but he really shone. He spoofed me and California drivers, and even brought the comedian up short after Leno asked him what we could see from orbit. "Well, Jay," Curt said, "sometimes, if the lighting is good, we can see the Great Wall of China, but we just flew over the Hawaiian Islands and we saw that. And Baja California. You can see the pyramids from space, and sometimes rivers and big airports. And actually, Jay, every time we fly by California we can see your chin."

Mission control radioed that we had futures as comics if we got tired of space.

We communicated with Earth by radio, television, and E-mail. We did a televised news conference from space, and a hookup with schoolkids from all over the country who asked better questions than the reporters did. John Glenn High School in New Concord was one of the schools. Another was the Center of Science and Industry, a learning center in Columbus headed by Kathy Sullivan, a former astronaut and deep-sea explorer.

I found E-mail, which was still new to me, a fast and effective means of communicating. I E-mailed Annie and the rest of the family, who were staying in Houston during the flight, and then I decided to try for a different first. Steve Robinson was my tutor, and once while I was slowly pecking out a message he asked if I was sending another E-mail to Annie.

"Nope. To the president," I said.

"What?"

"An E-mail's probably never been sent to the president of the United States from space," I said. "And he'd appreciate it, too."

He did. He replied the next day, and described an eighty-three-year-old woman who had told him space flight was okay for a young fellow like me.

. . .

The importance of the cameras that waited at the ready on Velcro patches beside most of the shuttle's windows came to the fore with Hurricane Mitch. It had made landfall in Honduras on the day before our launch, and hung over Honduras and Nicaragua for several days, dumping twenty-five inches of rain, causing mudslides that swept away entire villages, and killing over seven thousand people. A few days into our flight, mission control called for photographs of the devastated area.

One of the laptops on the flight deck was set up to track *Discovery* on its orbits around the world. By following the track on the screen, you could anticipate when you were approaching an area that needed to be photographed. You couldn't wait until you recognized Honduras, for instance, because at 17,500 miles an hour—five miles per second—the photo angles you wanted would have slid by already. We got the shots we wanted.

In some cases, the higher orbit of *Discovery* meant more spectacular views than I had seen from *Friendship 7*. Coming over the Florida Keys at one point in the mission, for example, I looked out toward the north and was startled that I could see Lake Erie. In fact, I could look beyond it right into Canada. The entire East Coast was visible— the hook of Cape Cod, Long Island, Cape Hatteras, down to the clear coral sands of the Bahamas and the Caribbean, south to Cuba, and beyond.

A night of thunderstorms over South Africa produced a view of a field of lightning flashes that must have stretched over eight hundred or a thousand miles, the flashes looking like bubbles of light breaking by the hundreds on the surface of a boiling pot.

All the while, our views of Earth were stolen from the

time we gave the eighty-three experiments on board. Each crew member kept on his or her timeline, and as we neared the end of the mission all of the experiments were working and successful. This remained our primary mission, and we were confident that we were making real contributions to science.

As *Discovery* approached the end of the mission, the crew wrapped up the various experiments and began preparations for reentry. It was like spring cleaning in a house in which every wall and ceiling were just more floors onto which things had been tossed. Although we had done a fairly good job of keeping the shuttle's interior tidy as we went along, notes, copies of our timeline tasks, and flight-data files detailing our work on the experiments were stuck to Velcro and duct tape and behind bungee cords all around the mid- and flight decks, SpaceHab, and the tunnel leading back to it.

Once the cabins had been policed, Chiaki and I set up one of the seats for resuiting. We retrieved the helmets and suits, started with Curt, and then helped the rest of the crew get ready. Then we got the rest of the seats in place and suited up ourselves, while Curt and Steve Lindsey closed the payload bay doors and oriented *Discovery* for the de-orbit burn that would begin its descent into the atmosphere. We were all suited and strapped in before the burn.

Down at the Cape, chief astronaut Charlie Precourt was aloft in a Gulfstream testing the crosswinds at the shuttle's three-mile landing strip. Crosswinds at the Cape put off the decision about starting the burn until the last minute. The big glider gets only one chance to land and conditions must be right; crosswind limits are set relatively low. The clock ticked down, and I worried that we might have to go around again and land at Edwards. But with only twenty seconds left, a voice from mission control came through the headphones: "*Discovery,* you have go for burn."

The OMS engines fired over the Indian Ocean a little over an hour before landing. It wasn't the dramatic kick I had felt in *Friendship* 7. It was smoother, though still definite. The slight dip in speed, from 24,950 feet per second to 24,479, was enough to take *Discovery* out of its orbital equilibrium and start it toward Earth. We flew over Baja California at Mach 24 and an altitude of forty miles. The Gs never reached more than two.

As we descended, we gulped various high-salt concoctions that were supposed to help us adjust to gravity again. Reentry and return to gravity would reverse the fluid shift we had experienced. At the moment we didn't need the fluid, but the high salt content was meant to fool our bodies into retaining it until we were on the ground, when gravity would take over and increased fluid would be necessary. For reentry, under our pressure suits we each wore G suits, the leggings and lower-torso wrappings that we would inflate to keep fluid from rushing to the lower body from the brain. All of this was supposed to keep us from getting light-headed and dizzy when we were first back on Earth. The stuff I was drinking was lemon-lime flavored, and by the time I'd downed three of the five eight-ounce bags, it tasted awful.

Falling through the atmosphere in *Discovery* wasn't the dire experience it had been in *Friendship* 7. This time there was no possibility I might burn up. The tiles on the underside fended off the heat, and they didn't boil away like the

Mercury capsule's heat shield. A glow but no fireball enveloped us as we descended. Even if it had, it wouldn't have been visible from the windowless mid-deck.

Curt took the orbiter through a series of banking maneuvers to reduce speed and altitude and bring *Discovery* onto its final glide path. He told mission control he had the runway in sight. Two minutes later, I felt the orbiter flare and then touch down on the long Cape Canaveral runway. The main gear hit first, and the nose wheel a few seconds later with a bang right under our feet on the mid-deck floor. The mission elapsed time was eight days, twenty-one hours, and forty minutes, and it was 12:04 P.M. Eastern standard time on Earth. We had made 134 orbits and traveled 3.6 million miles before we rolled to a stop.

Curt thought I should give a homecoming statement. "Houston, this is PS two, otherwise known as John," I said. "One G and I feel fine."

That wasn't strictly true, however. My stomach was revolting against all that salt-loaded lemon-lime gunk. A fair number of astronauts get sick on landing whether they fluid-load or not; I might have been stricken anyway. The flight surgeon asked if I wanted to come out on a stretcher. Astronauts had done that before. It was perfectly legitimate. I said, "Absolutely not." I made it from the orbiter to the crew transport vehicle with the rest of the crew, got unsuited, and then the stuff all came up. I had absorbed none of it, and my body was now demanding fluid in order to feed oxygen to my brain for equilibrium and balance. I was dizzy and shaky.

But I knew one thing. I was going to walk out of there onto the runway if it killed me. Annie, Lyn, and Dave and his family were waiting with the other families and the welcoming delegations, the ground staff, and the television cameras—and through those cameras an audience around the country and the world. Going back to space, I had defied the expectations for my age. I was going to defy them again by getting out of the transport vehicle onto the ground under my own power and joining my crewmates for the traditional walk-around under the orbiter. I drank some water and began to feel better.

Out on the runway, under a bright midday sun, Dan Goldin was saying nice things that I heard about only later: that my flight had inspired the elderly, changed the way grandchildren look at their grandparents, and made future flights safer for future astronauts.

Almost two hours after landing, I gripped the handrails of the vehicle stairs and climbed down to the sun-flooded runway. I needed to keep my feet wide apart for balance. The crew stayed close, Curt especially. It was that same mutual concern and camaraderie that make NASA and the space program so special.

Curt said a few words. He thanked the launch and ground crews at the Cape, mission control in Houston, the payload teams who organized the experiments, and the rest of the supporting players. We did the walk-around, but kept it short. Dan and Charlie Precourt walked next to me as I made my duck steps. I noticed vaguely that Curt had put *Discovery*'s nose wheel right on the runway's center line. Then I encountered a six-inch hose carrying air into the shuttle. I wanted to jump over it—jump for joy. I had gone back into space again; I had completed my checklist. Now I was home. Annie was waiting. I stepped over it instead. I was being forced to act my age, but only for a moment.

EPILOGUE

The story doesn't end there, though perhaps the drama does. Back on Earth, three weeks of medical checks mirrored those made immediately before the flight, and I returned to Houston several times over the next six months for more MRIs, DEXA X rays, blood draws, urine samples, and other follow-up work to assess my body's longer recovery processes in the wake of the flight.

The medical results were still being analyzed and tabulated as this book was being put to bed the following September. An early reading suggests there is no reason that older astronauts cannot continue to go into space as active mission participants and research subjects.

Meanwhile, I left the Senate without fanfare. The postelection recess had already started when I returned from space, and continued until my term ended on January 3, 1999.

The crew of STS-95 was feted at a big parade in New York City. We toured Europe and Japan in January.

Whether we met royalty and heads of state, or saw and heard the enthusiasm of the people who filled auditoriums for the presentations we made, the attention the flight had received was everywhere evident.

The two flights I was privileged to make stand as bookends of the history of space flight thus far. Now new volumes are being written. Last December, barely a month after the flight of STS-95, the first two components of the international space station were joined in orbit. The station, a testament to international cooperation, will make possible quantum advances in research. I hope that the studies on aging to which I opened the door on *Discovery* will be expanded on the space station.

At home, I turned to organizing my life out of the Senate, to writing, and to a venture that promises future generations a view of public service that fits my own. The Ohio State University has established the John Glenn Institute of Public Service and Public Policy in Columbus. Muskingum College in New Concord, and COSI, the Center of Science and Industry in Columbus, will have special cooperative arrangements with the institute.

The John Glenn Institute will give our archives and artifacts a permanent home that will be available to scholars and the general public. But more important, the institute will convey to young people the importance and the rewards of public service. It will bring together officials at all levels of government, diplomats, heads of state, and scholars for seminars on vital areas of international and domestic policy.

I'm looking forward to teaching and leading some of those seminars, and especially to working with the young people who will in future years fulfill the roles so vital to our bold and unique concept of democracy.

While Annie and I will continue to work together in public service, we will also spend more time with Lyn, and Dave and Karen, Zach and Daniel. Family remains the center of everything.

I expect the future to bring new rewards and challenges. And as always, going forward requires touchstones in the past. My parents' legacy was honest hard work, sweat and dirt, effort and the grasp of opportunity. They believed in themselves and in their country, and they had faith in God. That legacy has been my guide through all the places I've been and all the things I've been fortunate enough to do, and I've tried to pass it along.

For years now, I have kept one of Dad's plumbing wrenches on my desk. It's an eight-inch wrench, its steel heavy and darkened with oil and age. I keep it there as a reminder of just where and what I came from. That wrench tells me all I need to know about myself and about the country that I love.

AFTERWORD

Now, approaching two years since I returned to space as a somewhat grizzled guinea pig for investigating similarities between natural aging processes on Earth and what happens to the bodies of young astronauts during space flight, the medical data are providing some interesting insights. All of the electrodes pasted to my head and chest, the weird helmet I had to wear when I was sleeping, the horse pills I took that transmitted my internal temperature to a receiver on my waist, all the blood draws and urine samples, produced a lot of fine print and statistics. Scientists and doctors can read about them to their hearts' content in the peer-reviewed journals that keep track of developments in the various fields and disciplines.

For the rest of us, the big news in the medical details of my flight was the absence in most cases of significant anomalies between me and the rest of the crew. I was seventy-seven when I returned to orbit in the space-shuttle *Discovery,* in October and November 1998, and I turned seventy-nine in

July 2000. Yet the data taken during and in the months after the flight showed that, for the most part, my body reacted as well as most younger astronauts to the stresses of launch and reentry and nine days in the weightless environment that temporarily causes effects similar to aging.

Dr. John Charles, who was in charge of life sciences for STS-95 and coordinated the medical experiments in which I was a subject, joked that I did pretty well for a fit forty-year-old.

One secret, of course, is to come from a good gene pool. But on a broad scale, the message is one that health professionals stress routinely—a balanced diet and regular exercise are the keys to good health and longevity, and to the ability to perform at a high level of physical and mental competence well into what (at least until recently) has been thought of as old age.

One of the markers of old age, for example, is a lessened sense of balance. People tend to fall more as they get older, which with the weakened bones caused by osteoporosis can result in debilitating hip fractures. Causes of these balance problems vary, including deterioration in eyesight, inner-ear function, and other sensory mechanisms, and the loss of brain cells (neurons), that accompany aging. Older people are also susceptible to a sudden drop in blood pressure if they change position abruptly, especially when moving from sitting to standing. This drop in blood pressure can cause dizziness, or fainting.

Mission scientists gave me balance tests before and after the flight. They expected "age-degraded balance" at both ends, as well as a slower recovery from the effects of the flight. I was happily surprised that my EQ—Equilibrium Quotient—scores before the flight were comparable to those of astronauts roughly half my age. And although my balance on landing day was on the seriously shaky side, it wasn't out of range for astronauts of all ages returning to earth's gravitational pull, and in the days following the flight I recovered my balance and orientation just as quickly. The results suggested that age alone does not disrupt "human adaptive postural control"—balance—induced by lost gravitational inputs during space flight.

If that's the case, age may not be the sole factor behind balance problems on Earth, either. The scientists called for further study of the factors—genetic, psychosocial, lifestyle, and others—that contributed to my quick recovery, with the aim of improving the quality of life and reducing the physical risks of aging.

The series of MRIs—magnetic resonance imaging—I was given before and after the flight to look for changes in my muscles likewise revealed few differences between me and seven younger astronauts who were tested on STS-95 and on an earlier spacelab mission. My results showed no significant muscle atrophy during the flight, while the others tested showed small but significant muscle losses. A caveat—another experiment I was involved in made a post-flight measurement impossible until five days after landing, which might have masked any atrophy I may have suffered.

A few weeks after landing, all others had recovered their muscle volume, while I showed a small increase in muscle volume two to four weeks after the flight.

One cautionary note was a bone marrow spectroscopy test that revealed, in my case, bone marrow changes similar to those experienced by younger crew members on much longer flights. The scientists cautioned that this was "interesting, but not understood at this time," and went on to say that in general, the MRI results revealed no unusual changes in me from the flight.

My sleep patterns did change during the flight. My own estimates, and the sensors I wore during four sleep periods during the flight, indicated persistent insomnia. The instruments showed that my total time of sleep, and my sleep efficiency, were reduced. This, too, corresponds to the experience of other astronauts on STS-95 and other flights, probably because of weightlessness, the absence of natural daylight and darkness, and the orbital pattern that produces a sunrise and a sunset every hour and a half. The sleep sensors showed that I frequently woke up and went back to sleep during my sleep periods, but I didn't remember it.

I wasn't aware of this during the flight, either, but measures of caloric intake showed that I ate more than the other members of the crew. I've always had a tendency to gain weight, and so it seems I'm dieting about half the time. The flight scientists said eating might help combat nausea in space, so maybe I was following orders. Or maybe I was just using the flight as an excuse to take a vacation from my diet.

One significant area in which I differed from the other crew members most was in the rate at which my body produced stress hormones, which affect the immune system's ability to respond to infections. My blood levels of plasma cortisol and ACTH, as well as urinary cortisol, were extremely elevated compared with the rest of the crew. What was apparently happening was that my immune cells weren't circulating through my vascular system as efficiently as in the younger astronauts, perhaps due to alterations in cell-surface adhesion molecules. When these data are analyzed furthur, they may provide clues as to why the elderly are more susceptible to infections.

The protein turnover study in which I participated with one of the other astronauts assumed reduced protein synthesis during the flight, so that it would be outstripped by protein breakdown and muscle loss would result. Hormones are again a factor—cortisol, which promotes muscle breakdown, increases during orbital flight, while testosterone, which increases muscle protein building, decreases. But pre- and post-flight measurements to determine lean body mass—muscle—revealed no change in either case, and the results were indistinguishable based on age.

This was a data base of only two, however, and a look at broader trends requires a cautionary note. Testosterone is normally lower and cortisol normally higher in the elderly. My short flight held the effects to a minimum, but the combined effects of age and prolonged weightlessness may mean that older astronauts who fly in the future should be limited to shorter missions as opposed to long stints in orbit in the International Space Station.

My medical data from STS-95 will continue to be analyzed for some time to come. It's my hope that other older astronauts will be scheduled for flights in the future, as the earth's growing aging population creates the need for new solutions to the problems of aging and a larger data base to help us attack them. My flight, and the data it produced, are only a toe in the door. With more research, and more data, NASA and its scientific partners may grasp the keys to sending younger astronauts on flights to Mars and beyond, routinely employing older astronauts in space, and easing or erasing many of the frailties of old age here on earth.

Until then, the best lesson to be learned from my return to space is not a new one. It's simply this. Take care of yourself.

ACKNOWLEDGMENTS

Many people deserve thanks in connection with this book. Attempting to name them in no way diminishes the contributions of those I meant to name but in the press of time and the failure of memory somehow overlooked.

My children, Dave and Lyn, shared their recollections of our extraordinary times.

Mary Jane Veno, my former Senate staff director, has been my aide-de-camp and alter ego for more than twenty-five years. Her political instincts have been unfailing, her devotion unstinting, her friendship unsurpassed, and her persuasiveness in urging me to write a book undeniable. She and Lyn organized the personal archives that went into it.

Tonya McKirgan and Nicole Colette Dauray are valued associates and friends who helped ease my transition from the Senate into private life.

Capitol attending physician Admiral John Eisold was

a trusted confidant and advisor through all the preparations for my return to space.

Special thanks to Dan Goldin, whose decision made possible a PS-2 Glenn, and to Mike Mott and Jack Daley at NASA headquarters, whose personal interest and influence played a vital role.

Thanks to the great crew of STS-95—Curt Brown, Steve Lindsey, Steve Robinson, Scott Parazynski, Chiaki Mukai, and Pedro Duque—for their very special friendship and help before, during, and after the mission. They made a lot of dreams come true.

Friends Tom and Ida Mai Miller were, as always, cordial and helpful in responding to inquiries about my life, and Tom unearthed from his files some of the photographs that appear in this book.

Scott Carpenter provided interviews, as did Bob Voas and the late Pete Conrad. Pete died too soon, and the world's a less cheerful place without him.

Brien Williams conducted oral history interviews that helped start the process of my remembering.

Mike Gillette, director of legislative archiving at the National Archives, produced order from voluminous files.

B. J. Andrews provided invaluable assistance in the late stages.

Ethel Kennedy, Jim Whittaker, George Plimpton, and Walter Cronkite helped recall the friendships and events we shared in the glorious and tragic sixties.

Henri Landwirth and Hallie and John Quinn retold stories of other long and constant friendships.

Ron Grimes and Jack Sparks, of my Senate office, annotated my Senate career.

Many people helped fill blanks in the research. They

include my sister-in-law, Annie's sister Jane Hosey, and her husband, Jim; Tom Gregory at the Lone Star Flight Museum in Galveston, Texas; Judith Whipple, ship's historian and curator at the USS *Lexington* Museum on the Bay at Corpus Christi; Udit Gandhi, who aided in research about Muskingum College and New Concord; Janice Tucker, Muskingum College director of public relations; Lorle Porter; the Public Information Office of the Ohio State Patrol; the Zanesville Times Recorder; Wallace Spotts, my first flying instructor; New Concord's informal historian, Dr. Kenneth V. Kettlewell; the Western Reserve Historical Society; Ted Williams and the Ted Williams Museum; Espen Ronneberg of the Republic of the Marshall Islands Permanent Mission to the United Nations; Dr. Roger Launius of the NASA history office, and archivist Mark Kahn; Carmen Melendez and Marilyn Mode of the New York Police Department public information office, Sgt. Tom Gambino and the NYPD Museum, and former officers Kevin Hallinan and Henry Buck Jr.; Ann Ferrante of the Marine Corps History Museum; Senate historians Dick Baker and Don Rickey; John Collins and Doug Wickland, who explained the workings of the Cutts compensator; Dr. Ruthan Lewis of NASA; David Schwartz of the Game Show Network; Colleen Mason; Peter Mersky; the American Medical Association.

The more recent details of my return to space were enlivened by, among others, Polly Shroeder of the astronaut office at the Johnson Space Center, Houston; Barbara Zelon and Doug Ward of NASA's public affairs office at the JSC; Dr. Robert Butler of the International Longevity Center.

The process that became this book began with my

friend David McCullough. He led me to Morton Janklow, our literary agent, who placed this book in the best possible hands at Bantam Books. There the editorial work of Ann Harris and the late Beverly Lewis gave it shape and clarity under deadline pressure and trying conditions.

Warm thanks also go to Barbara Nevins Taylor, who gave up her husband, Nick, for long stretches of time during the writing of this book; to Nick's assistant Raina Moore; to Susan Raines and Jim Duffy, whose home in Washington was a home away from home while Nick and I worked together; and to Mary Margaret Walker and Alan Aiches, Washingtonians who also were gracious in their hospitality.

Finally, words cannot adequately express my great appreciation to Nick Taylor. A distinguished author, his friendship, wisdom, judgment, and literary skills helped make it all come to life.

INDEX